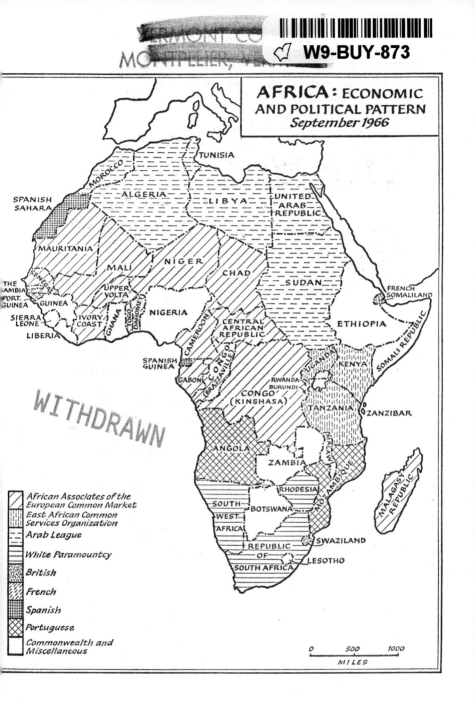

AFRICA: ECONOMIC AND POLITICAL PATTERN
September 1966

TUNISIA
SPANISH SAHARA
MOROCCO
ALGERIA
LIBYA
UNITED ARAB REPUBLIC
MAURITANIA
MALI
NIGER
CHAD
SUDAN
FRENCH SOMALILAND
THE GAMBIA
SENEGAL
PORT. GUINEA
GUINEA
UPPER VOLTA
SIERRA LEONE
IVORY COAST
GHANA
TOGO
DAHOMEY
NIGERIA
CAMEROON
CENTRAL AFRICAN REPUBLIC
ETHIOPIA
SOMALI REPUBLIC
LIBERIA
SPANISH GUINEA
GABON
CONGO (BRAZZAVILLE)
CONGO (KINSHASA)
UGANDA
KENYA
RWANDA
BURUNDI
TANZANIA
ZANZIBAR
ANGOLA
ZAMBIA
MALAWI
MOZAMBIQUE
MALAGASY REPUBLIC
RHODESIA
SOUTH WEST AFRICA
BOTSWANA
REPUBLIC OF SOUTH AFRICA
SWAZILAND
LESOTHO

WITHDRAWN

Legend:
- African Associates of the European Common Market
- East African Common Services Organization
- Arab League
- White Paramountcy
- British
- French
- Spanish
- Portuguese
- Commonwealth and Miscellaneous

0   500   1000
MILES

A SYSTEMATIC
REGIONAL GEOGRAPHY

VOLUME 9
AFRICA

A

A SYSTEMATIC REGIONAL GEOGRAPHY

General Editors
J. F. UNSTEAD M.A. D.Sc. formerly Professor of Geography in the University of London, vols. 1–5
E. W. GILBERT B.Litt. M.A. Professor of Geography in the University of Oxford, vol. 6 and subsequent volumes

Volume 1 (SBN 340 09132 0)
THE BRITISH ISLES

Volume 2 (SBN 340 09133 9)
EUROPE

Volume 3 (SBN 340 09134 7)
A WORLD SURVEY—From the Human Aspect

Volume 4 (SBN 340 09135 5)
AUSTRALIA, NEW ZEALAND AND THE SOUTH-WEST PACIFIC

Volume 5 (SBN 340 09136 3)
MONSOON ASIA

Volume 6 (SBN 340 09137 1)
ANGLO-AMERICA

Volume 7 (SBN 340 09138 x)
THE SOVIET UNION

Volume 8 (SBN 340 07997 5)
SOUTH-WEST ASIA

Volume 9 (SBN 340 07998 3)
AFRICA

Further volumes are in active preparation

A SYSTEMATIC
REGIONAL GEOGRAPHY

VOLUME 9
AFRICA

N. C. POLLOCK
B.A. (Cape Town) B.Litt. M.A. (Oxon.)
Lecturer in Geography, University of Oxford

UNIVERSITY OF LONDON PRESS LTD

SBN 340 07998 3

Copyright © 1968 N. C. Pollock
Maps and diagrams copyright © 1968 University of London Press Ltd

University of London Press Ltd
St Paul's House, Warwick Lane, London, EC4

Printed and bound in Great Britain by
C. Tinling and Co. Ltd, Liverpool, London and Prescot

Preface

THE task of writing a regional geography of Africa is almost herculean, not least being the problem of keeping up to date with the swift pace of change. African scenes are usually multi-toned and vivid and the subject matter is of great variety. In an attempt to capture the colour and vigour of the scenes more clearly more emphasis has been laid on the regional and descriptive approach than on the systematic, and a certain amount of detail omitted in order to clarify the regional picture. Historical and political factors have also been stressed so that the reader may appreciate more fully the importance of Africa's past and its deep, underlying links with the present. Despite the increasing flow of material on Africa it is often difficult to obtain accurate information, particularly of a statistical nature, and some of the detail on demography, crop production or other social and economic matters is of doubtful value.

I have been on a number of field trips to various parts of Africa, lectured in a South African university for six years and have taught the geography of Africa for many years.

I am most grateful to the Institute of Commonwealth Studies and the School of Geography in Oxford, and the Board of Management of the Vaughan Cornish Bequest for research grants which have enabled me to visit East and West Africa. I appreciate greatly the generous hospitality extended to me by many friends, the help received on numerous occasions from officials, and the general friendliness experienced on journeys through Africa. I am deeply indebted to Professor Gilbert for his encouragement, advice and editorial skill and to those of my colleagues who have kindly read various sections of the book in manuscript form and offered helpful criticism.

N. C. POLLOCK

Contents

Maps and Diagrams

Plates

PART I: GENERAL

CHAPTER ONE

Discovery, Exploration and Colonization

EUROPE had plotted Africa's coastline reasonably accurately by the sixteenth century, but only discovered the interior during the nineteenth century. Prior to these dates her knowledge of Africa was confined to the Mediterranean coastland of North Africa, although garbled reports of Prester John and his Christian kingdom, the Mountains of the Moon, great rivers flowing east and west and weird tribes and headless people, had penetrated the Arab cordon. But it was only with the probing voyages of the Portuguese that knowledge of the coastline was gained from the fifteenth century on and through the exploring, humanitarian and commercial zeal of English and other explorers of the nineteenth century that the interior was opened up. Why was it that Africa had remained the unknown continent, unknown at least to Europe, for so long? Does the answer lie in Africa's physical hazards, the defiance of her peoples or lack of economic attraction?

PHYSICAL FACTORS

Africa's interior has never invited penetration. There are few embayments or natural harbours and to the seaborne invader a landing on the surfbound shore is hazardous. There are long stretches of desert coastline from Morocco to Senegal and from Angola to the Cape, or a swampy coastline backed by mangrove-forest as in West Africa. There is little coastal plain and the ascent to the interior plateau in east and south Africa is steep and abrupt. No great rivers afford a long and uninterrupted passage into the interior as in North America, Europe and Asia. The great plateau close to the coast causes numerous falls, rapids and cataracts, for example the Kebrabasa rapids on the river Zambezi and the Crystal Falls on the Congo. Sand bars and distributaries at the mouths of rivers such as the Niger and Zambezi and, further upstream, seasonal shallow water as on the Benue also impede navigation up the rivers.

Great heat and humidity in the tropical lowlands and diseases such as malaria, yellow fever and sleeping sickness have long been a barrier to European penetration, witness the failure of early nineteenth-century attempts to explore West Africa and the high mortality rate among adminis-

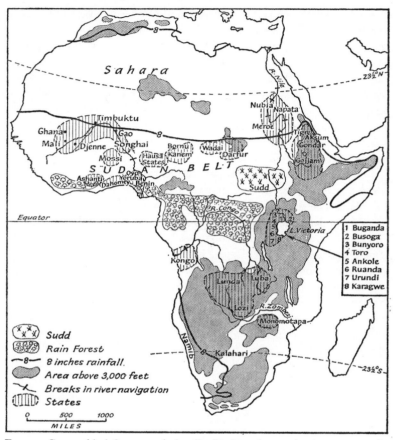

Fig. 1—Geographical factors and the distribution of some important states in tropical Africa before 1850

trators and missionaries in Sierra Leone. Movement on a large scale in the interior was also limited greatly by the widespread incidence of tsetse fly and ignorance of the use of the wheel. Animal transport by horse, ox and camel was only practised in the northern and southern savanna zones, elsewhere movement was on foot along forest paths.

The Mediterranean fringe of Africa has been affected by the world's great civilizations and religions, some knowledge of which has filtered through the Sahara along the caravan routes, or up the Nile Valley to the tropical lands of the Sudan zone. To the south of the Sudan belt the great barriers of the Congo forest stretching east–west, and the Nile sudd and desert south of Ethiopia have further sifted the cultural influences from

15

the north. Even such cultural influences as came south were stifled by the forbidding physical background of Africa which is the most tropical of all the continents (Fig. 1). The bulk of it has seasonal rainfall which varies greatly in amount and incidence. Huge tracts, such as the Sahara which covers one-third of Africa, are arid or semi-arid and continuously affected by drought. In those areas where rainfall is adequate the soils are highly leached while its seasonal incidence has probably encouraged lateritic development. The combination of erratic rainfall, prolonged drought and easily exhausted soils has caused man and animals to move constantly in search of fresh ground for crops and livestock.

Some eight to ten thousand years ago parts of the *Sahara* may have had a savanna vegetation with nomadic pastoralists and a wide variety of fauna and flora. The southward spread of the Quaternary Ice Sheet in Europe caused pluvial periods and a higher rainfall than there is now, with an integrated river system where there are now disconnected dry wadis and a desert landscape. The herdsmen, hunters and animals of this period are portrayed in prehistoric paintings and graffiti found recently in many Saharan wadis, caves and rock shelters in the high plateaux of the Central Sahara, such as the Ahaggar and Tibesti. With the final retreat of the ice sheet in Europe and the establishment of a subtropical high pressure zone the Sahara became increasingly arid, forcing men and animals to retreat to the central highland masses, the desert edge, the oases and southern forests and plateaux and the Nile Valley. The Sahara became a considerable barrier until the introduction of the camel some 2,000 years ago and the Arab invasion of North Africa in the seventh century, although ox-drawn carts crossed the desert before this.

POPULATION AND SETTLEMENT

Migration has occurred over thousands of years, groups of people hiving off periodically from a parent group and moving away over Africa's wide spaces in search of fresh land and water. Over the last 2,000 years movement into Africa has probably funnelled out from the Nile Valley and Horn of Africa, skirting the Nile sudd and Congo forest in a triangular sweep, west across the Sudan belt and south down the plateau savannas of East Africa, as the map (Fig. 1) attempts to indicate. There has also been much migration into East Africa of Negroid peoples from West Africa. East Africa was probably an early cradle of man,[1] so that there has also been much inter-African movement. Movement has not been

[1] Much evidence has been discovered by Dr Leakey at the Olduvai Gorge site in Tanzania to support this view.

confined to westward and southerly surges, however, and has gone back and forth, so that the palimpsest of man's occupation of Africa records many layers of settlement in places. Strong, centralized states have only tended to develop in certain areas such as the Sudan belt, forest fringes and Lake Victoria region which are favoured by accessibility to cultural influences, availability of water and more fertile soils. These states have harnessed natural resources, achieved a diversity of economic activity and complex social organizations and supported an increasing density of population. An environment that before could only support a few hunters and collectors or comparatively small numbers of pastoralists now supported a population ten times as great. It was the spread of agriculture and the use of iron that above all enabled these regions to support far greater populations.

A knowledge of the working of *iron and copper* and the change from hunting and gathering to the growing of food crops developed almost simultaneously about 2,000 years ago. This knowledge took a long time to filter into Africa, but probably came via Meroe and Kush in what is now the Sudan. Kush had a considerable iron smelting industry and has been called the Birmingham of ancient Africa. The introduction of iron meant a much greater control over the tough African vegetation for the stone axe could not cope with tall bushes and trees; iron implements were essential for the slash and burn technique of shifting cultivation. The rapid spread of iron-using village farming communities was in the nature of a revolution, understandable when it is realized how difficult hunting and food collecting could be in regions with a long dry season. It was in the savanna belt where the cultural revolution was most successful and wherever crops like millet could be grown during the wet season and stored for the dry season.

Africa possesses a rich proto-historical and historical legacy. Increasing evidence is being amassed in support of this from archaeological discoveries such as the 2,000-year-old Nok culture of Nigeria's Middle Belt with its fine terracotta figurines; the ruined buildings and terraces of stone found over a vast area in East and Southern Africa; smelting ovens and various types of pottery; anthropological sources such as folk lore and legends; tribal customs and languages and from more recent art forms such as the magnificent Ife and Benin bronzes and historical sources such as the medieval Arab writers and Portuguese archives.

European penetration of the interior of Africa occurred mainly during the nineteenth century, when vast areas were recovering from the ravages of the slave trade, depopulation and breakdown of tribal society and economy. Most European travellers saw ample evidence of this and not the strength of Africa's cultural past and, understanding little of the complexities of tribal society, stigmatized it as primitive and backward.

AFRICA PRIOR TO THE FIFTEENTH CENTURY

Long before the spread of iron working techniques across the Sudan belt, the North African coastal belt had been controlled by various groups. The *Phoenicians* were among the first, followed by Greeks, Romans, Arabs and many others; their hold was generally limited to the regions with a Mediterranean type of climate and winter rains of at least 10 inches which occur discontinuously along the coast. Besides occupation of the Mediterranean coast of North Africa, the North Atlantic and Red Sea coasts were also explored. Regular voyages were made by the Phoenicians to the west coast of Africa and commodities such as ivory, gold and slaves brought back to Carthage. Knowledge of Africa south of the Sahara was only gained by travelling up the Nile and first by the Greeks, and later by the Ptolemies, considerable knowledge was gained of the Upper Nile. Expeditions probably travelled as far south as the sudd and swamp regions of what is now the southern Sudan in latitude 10° N.

Roman hegemony was established over the whole of the North African coast extending as far south as Wadi Halfa and Mauritania. The name of the Roman province of Africa (approximately the modern Tunisia), a Berber word, has been extended to describe the whole continent. The coastal areas were important agriculturally and much corn and wine was produced for export to Rome. Wheat, olives and other Mediterranean crops were produced in what is now desert country. The wide dissemination of Roman farms and settlements can be traced from air photographs. Archaeological investigations have also revealed the vigour of urban life and the prosperity of the rural community in Cyrenaica and parts of the Barbary coast. A Roman expedition from Tripoli was able to cross the Sahara and reach Agisymba or modern Sudan. Nevertheless, knowledge of West Africa was scanty and the Phoenician exploitation of West Africa ceased with the fall of Carthage. The Romans were far more interested in trade with the Far East and spices, silks, and other commodities were imported from India in great quantities.

The Roman carrying trade with the East was undertaken largely by the *Greeks* who by A.D. 40 had learnt of the advantages of the monsoon wind system in the Indian Ocean, knowledge possessed by the Hindus hundreds of years before this. From December to February winds blow from the north-north-east, veering to south-south-west from April to September and alternating with the movement of Asiatic high and low pressures. For thousands of years traders have come down the Red Sea and Persian Gulf, from Arabia, India and elsewhere, and sailed into East African ports on the winter monsoon in December, returning to their home ports with the summer monsoon in April and May, their holds laden with cargoes of ivory, ambergris, gold and slaves.

The Greeks were thus able to strike across the Indian Ocean without being forced to hug the Arabian coastline and, avoiding the interference of the Sabaeans and other Arabian people, to become the main carriers for the trade with Rome. Greek knowledge of the East African coastline was considerable, as can be seen from the *Periplus* or merchants' guide book for the Erythrean Sea (the Indian Ocean), published in the first century A.D. This describes accurately the East African coast as far south as Zanzibar. Ptolemy's geography was far less accurate than that of the *Periplus*, and the geographical features of East and Central Africa were placed too far south, while a great tongue of land was supposed to extend eastwards from Africa and connect with Asia, turning the Indian Ocean into a great lake.

With the decay of the Roman Empire commercial interest in Africa and Asia diminished greatly and European knowledge of Africa was further reduced by the rapid spread of Islam along the North African coastline in the seventh century A.D. *The Arabs* swept along the coast destroying the Graeco-Roman-Byzantine culture and uniting the northern periphery of Africa with the rest of the Moslem world. Cut off from access to the Red Sea Greek trade disappeared along that channel also. Arab influence penetrated along the Saharan desert routes into the great Sudan belt. The first Arabic reference to West Africa is to the 'Land of Gold', and exploitation of the gold resources of the forest zone to the south of the savanna was one of the major stimuli in the establishment of the remarkable *Sudanic states*.

Some of these states covered a great area; Mali, for example, stretched for a thousand miles east–west from the Senegal river to the great bend of the Niger in the mid-fourteenth century. They had a complex organization, specialization of labour, a literate class, scholars, merchants, a standing army and cavalry, extensive trade with North Africa and they exacted tribute over a wide area. Ibn Batuta visited Mali in the mid-fourteenth century and describes the security and ease of travel in the country, showing that law and order were preserved over a great area.

The trade with North Africa in salt, copper, silks, swords, horses, gold, ivory, slaves, kola nuts and other commodities flourished in the Sudan zone. There were other states, too, besides Mali; Ghana arose in the eleventh century, and Songhai on the middle Niger, while there were smaller states in Hausaland, but none attained the power and extent of Mali. After the sixteenth century and the devastation of the Moroccan attack the Western Sudan declined in power with the rise of the slave trade along the coast and the widening of commercial links between Europe, the coast of west Africa and the Americas.

The Sudan states could not extend south into the forest zone where tsetse fly, disease and dense vegetation precluded the use of horses and cavalry. Freedom from their domination allowed other strong states such as Benin, Dahomey and Ashanti and the Yoruba states to arise in the forest zone; Benin reached the height of its power in the sixteenth century after the decline of the Western Sudan states.

East Africa

West Africa, situated between sand, river and ocean, was accessible to Europe with its market for the production of tropical crops and to America with its great demands for slave labour, but East Africa had no triangular run, as in the North Atlantic, to pile up the profits and encourage exploitation. The physical background of East Africa is also less favourable than in West Africa. Situated on the eastern side of Africa astride the equator only a limited area receives well distributed, adequate rainfall, the bulk of the plateau having a seasonal and highly variable rainfall, which limits economic activity to extensive stock rearing. Contact between the coast and the more favourable parts of the interior, such as the Lake Victoria region, the Rift Highlands and the volcanic mountains, such as Mount Kilimanjaro, was difficult because of the wide expanse of the semi-arid Nyika, and the lack of navigable waterways. There was also little to attract outside interest in the interior, no gold as in Manicaland or the Guinea coast, no spices or other tropical commodities in demand in Europe, and only the coastal strip with its many small harbours and Indian Ocean trading and cultural connections developed economically before the nineteenth century.

Following on the Greeks and Sabaeans, Indian, Persian and Arab influence was felt along the coast. By the twelfth century *Arabs and Persians* had established themselves in a string of fortified trading settlements, many on island sites, at Sofala, Sena, Quilimane, Mozambique, Kilwa, Zanzibar, Mombasa, Lamu, Pate, Barawa and Mogadishu. Kilwa, established in the tenth century, became the predominant centre on the coast, until it was replaced eventually by Mombasa and Zanzibar. Trade was the raison d'être of these ports and little attempt was made to settle in the interior. Southern Arabia was the centre of an important maritime trade and each year fleets sailed down the East African coastline to trade in ivory, slaves, ambergris, gold and other commodities. A mixed Arab-African population developed along the coast speaking a distinctive language, Swahili, which has become the lingua franca of much of East Africa.

VOYAGES OF DISCOVERY

The next stage in the opening up of Africa was the series of *Portuguese voyages* along the west coast in the fifteenth century. An important figure was Prince Henry the Navigator (1394–1460) who collected information and fitted out maritime expeditions, many of which were, however, not only the work of Prince Henry but a national effort by the Portuguese as well. Some of the reasons activating this maritime expansion were the desire to circumvent the Moslems' hold on the North African coast by probing and trading voyages to the south, to engage in a crusade against the Moors and to search for the legendary Christian Kingdom of Prester John, which might prove an ally against Islam. Other important factors were the improvements in navigation and the design and construction of ships. Ships became larger and could accommodate larger crews and hold more stores for a long voyage, while the multiplication of sails enabled ships to sail close to the wind and tack off a dangerous shore. Navigational instruments, the mariners' compass and astrolabe, and charts were made more accurate, thereby reducing the hazards of sailing imperfectly known waters.

Steady progress was made in the first half of the fifteenth century and, by 1460, the date of Prince Henry's death, the Portuguese had mapped the coast of West Africa from Cape Bojador to Sierra Leone. By 1462 the Portuguese reached the Gulf of Guinea, and in 1482 the fort of Saõ Jorge da Mina (Elmina) was built to safeguard supplies of Guinea gold. Two voyages by Diego Cam in 1482–4 and 1485–6 reached the mouth of the Congo River and the modern Walvis Bay. The doubling of the Cape of Good Hope was actually achieved by Bartholomew Diaz in 1487, helped considerably by the information that Covilhaõ had gained previously on his journey from Egypt to India. In half a century the Portuguese explorers had discovered and plotted some 5,000 miles of the African coastline, a notable feat. They had overcome the geographical barriers of the Saharan desert coast, the adverse currents of Cape Bojador, the hot fever-stricken mangrove coast of West Africa, the desert coast of South-West Africa, the hazards of the Benguela current and the stormy South African waters.[2]

The Portuguese mariners before Vasco da Gama had carefully hugged

[2] The Cape Rollers are still dangerous at times.

the African coast, but Vasco da Gama pushed boldly out into the Atlantic Ocean to the south of Cape Verde which he reached at the beginning of August. His next port of call was the South African coast near St Helena Bay.

Natal was reached on Christmas Day 1497 and Malindi only in April 1498. The Portuguese found flourishing trade centres, harbours thronged with shipping and a complex maritime trade in minerals such as gold, iron and copper and other goods such as beads, ivory, cloth and porcelain and, as on the west coast, slaves. The last part of the voyage with the help of an Indian pilot from Malindi and the summer monsoon took a short time and da Gama arrived in Calicut on 20 May 1498.

Just over twenty years later a Portuguese fleet under D'Almeida destroyed Arab power on the east coast of Africa, ensuring Portuguese supremacy for the next hundred years. Arab trading posts and towns along the coast were occupied and destroyed and their trading links broken.[3] Forts were built at Sofala in 1505, Mozambique in 1508, Kilwa Kisiwani 1505–12 and Mombasa 1593–1698. Fort Jesus still stands as an impressive stone structure on the eastern side of Mombasa island guarding the entrance to the old dhow harbour. Mozambique became the Portuguese headquarters in 1550, as it was in a more central position to tap the resources of the interior of south-east Africa.

Despite its hot and unhealthy climate, the *Zambezi Valley* became the gateway to Central Africa and its mineral resources of gold, iron, copper and tin. Gold is widespread in schist belts of the gold-bearing Archaean rocks; ancient mine workings are found all over Rhodesia and the Transvaal now. Minerals have been the main reason for the exploitation of south-east Africa's resources for over a thousand years. Tradition, indeed, associates the gold mines of Monomotapa with those of Ophir, and the ruins at Zimbabwe, Dhlo Dhlo, Kami and elsewhere were thought to belong to a very early civilization. However, recent archaeological investigations and radio-carbon dating suggest that these impressive ruins are of Bantu and not Indian or Arab workmanship, the largest possibly being constructed in the seventeenth century. Stone ruins of dry masonry are scattered all over Rhodesia. The Inyanga area, for example, has many hill forts, channels and terraces indicating a considerable knowledge of irrigation and cultivation and a marked degree of cultural development compared with most of southern Africa at this period.

The Portuguese knew this part of south-east Africa as *Monomotapa*, the

[3] Mathews, Davidson and other writers suggest that in fact the east African coastal culture was African and not Arabic, as African as Timbuktu and Goa.

name given to the paramount chief of the Makaranga people. Gold was obtained from shallow workings and panned and, with iron, copper, tin and ivory, exchanged with Indian and Arab traders for cotton, beads and luxuries. The incursion of the Portuguese and subsequent warfare destroyed the complex trading pattern that had been established between the coastal towns and Shona inhabitants of the plateau and led to a great decline in the export of minerals.

The first *Portuguese explorer* of the interior was Antonio Fernando who made at least two journeys before 1514 and explored much of Manica, recording distances between chiefs' kraals and the presence of gold and grain.

Although military expeditions such as Barreto's failed, Portuguese traders and missionaries added greatly to geographical knowledge, for example dos Santos who spent ten years in Zambezia at the end of the sixteenth century. He says of Portuguese traders, 'from Tete they goe with their wares thorow a great part of Manamotapa to three marts Massapa, Luanza, Manzovo, in which the inhabitants of Sena and Tete have houses and factories thence to store all the Countrey'. Sena was 160 miles upstream on the Zambezi and had a stone fort, factory and church and acted as the base of a Portuguese official known as the Captain of the Rivers. The Portuguese population of Sena was about 50, mostly traders. Tete was over 300 miles up the Zambezi and also had a fort and some 40 Portuguese who traded with Monomotapa.

Activities further upsteam and to the north were barred by rapids and hostile tribes, so that the main Portuguese sphere of influence lay south of the river over a great area of the plateau, the lower Zambezi valley and the delta.

Further north in the eastern horn of Africa is the great dissected plateau of Ethiopia. Various Coptic Christian kingdoms spreading southwards from Axum have been able to survive Moslem attacks largely because of the natural protection afforded by the inaccessibility and high altitude of the plateau. The Portuguese were of assistance during the sixteenth and seventeenth centuries.

The Congo River and Angola

The Portuguese were also active on the west coast near the Congo river mouth and the coast to the south. Posts were established in 1501 at Mpinda at the mouth of the Congo and São Salvador in the Kingdom of Kongo. Portuguese interests spread to the interior because of the slave trade and reports of rich silver mines. The slave trade soon became dominant and during the period 1575–91 over 100,000 slaves were exported. A number of fortified posts were set up along the Cuanza river which provided a suitable route for some distance into the interior.

The Portuguese sphere of influence was extended further south with the construction of a fort at Benguela in 1617, and by the seventeenth century the Portuguese were active over a wide area stretching for nearly 1,000 miles from Cabinda to south of Benguela and nearly 200 miles into the interior. As in Mozambique a search was made for reputedly rich silver and copper mines, but the great copper deposits in the regions now known as the Katanga and the copper belt of Zambia were too inaccessible then for exploitation by the Portuguese. Angola's main economic function until the late nineteenth century was to supply slaves to her great neighbour across the South Atlantic, Brazil.

South Africa

By the eighteenth century Portuguese influence in southern Africa had declined greatly, partly owing to Portugal's own decay as a maritime power and the rise of the Netherlands, England and France, and partly owing to renewed activity by Arabs on the east coast. European powers, Holland particularly, became interested in the great curve of territory lying to the south of the Portuguese territories. This region, half desert or semi-desert was neglected until the mid-seventeenth century when the *Dutch* established a refreshment station at the Cape of Good Hope. The Portuguese avoided the stormy waters of the Cape in their caravels, which by the seventeenth century were less seaworthy than the sturdier Dutch East Indiamen. The Portuguese swept round the Cape and up into the Mozambique Channel to make for Sofala, Mozambique or Mombasa, there to re-equip and wait if necessary for the summer monsoon to India. The Dutch and English, however, used St Helena and Table Bay as convenient half-way houses before sailing far to the south to pick up the monsoons in June and July, so avoiding the tricky currents and winds in the Mozambique Channel.

The Dutch settlement at *Table Bay* and under the lee of Table Mountain expanded slowly. By the end of the seventeenth century the white population was a few thousand in number, consisting of Dutch East India Company officials and Dutch and Huguenot settlers. Although its original function was solely to supply passing Dutch ships with wine, wheat, fresh vegetables and meat, it had become a colony of settlement by the early eighteenth century. Passing ships supplied only a limited market for wheat, wine and vegetables, but the demand for meat was not satisfied. Many of the settlers thus became stock farmers or *veeboere* and trekked to the east and away from De Kaap with its small area of Mediterranean type of climate to the summer rainfall areas of the eastern Cape.

The original inhabitants of the south-western Cape were *Bushmen* and *Hottentots*, the former, hunters, who disputed bitterly Dutch possession of their hunting grounds; the latter, nomadic pastoralists, who were soon vanquished by disease or later absorbed in the Cape Coloured population. In the eastern Cape the trekboers met the Xhosa, Pondo and other *Bantu tribes* who had moved south from the region of the Great Lakes hundreds of years before. A Dutch hunting party visited the Xhosa as early as 1736 and by the end of the eighteenth century much of southern Africa had been explored in the search for wild animals, fresh grazing and water.

West Africa

But Africa south of the equator was of minor economic importance until the mineral discoveries of the late nineteenth century. It was West Africa with its vast and accessible supplies of slave labour that became of paramount importance to the economy of Europe and the Americas. The *Portuguese* were the major European power on the west coast until they were supplanted by the French, Dutch and English in the seventeenth century. The Cape Verde Islands colonized by the Portuguese in 1460 were an ideal base for trading along the coast. Elmina, founded in 1482, became the Portuguese headquarters for the Gold Coast, while the island of São Thomé became the main base for the Niger delta region in the 1490s.

The two main areas of Portuguese activity on the west coast were in *Senegambia* and the *Gold Coast*. In Senegambia Portuguese influence was able to extend inland and part of their sphere of influence survives now as Portuguese Guinea. On the Gold Coast, however, their activities were limited to the coastal strip and did not penetrate beyond the first wooded ranges. This stretch of the coast to the east of Cape Three Points is the driest in coastal West Africa, Winneba receiving less than 30 inches of rainfall on an average per annum. It attracted a remarkable concentration of forts and trading activity in the seventeenth and eighteenth centuries, partly for physical reasons, as it had a dry climate and a rocky, sandy coast-line which afforded good sites for forts and harbours. The Grain[4] and Ivory coasts to the west of the Gold Coast tended to be neglected as they lacked natural harbours.

At first there was a considerable variety of trade on the west coast, with

[4] It derived this name from the grains of malagueta pepper obtained along this stretch of the West African coast.

gold, ivory, pepper, gum arabic and slaves. Gradually, though, with the economic development of plantations in the Americas and the Caribbean and the production of tropical crops such as sugar and tobacco which demanded great supplies of labour, the human rather than the natural resources of West Africa were tapped. Faced by an acute shortage of labour the Spanish communities in America began to import *slaves* in the first half of the sixteenth century. By the end of the century nearly a million slaves had been sent from West Africa to Central and South America and the Caribbean islands; the Portuguese were the main carriers.

Attacks on the Spanish and Portuguese monopoly were made by the *Dutch* at the close of the sixteenth century. They realized the value of a permanent base on the Gold Coast, and so captured Elmina in 1637 and gradually extended their hold on the coast. The Dutch became the main carriers in the slave trade and supplied the Spanish settlers in America and the Caribbean, and the French and English in some of the West Indian islands. After the middle of the seventeenth century the *French* and *English* engaged increasingly in the trade and the ports of Bristol and Liverpool throve. The lucrative and valuable triangular trade developed with the exchange of goods such as textiles, firearms, powder, shot, knives, spirits and iron and copper bars for slaves who were sold in the Caribbean. The last stage of the triangular run was the export of tropical plantation products—sugar, rum, molasses, tobacco, etc., to Europe.

For a period of a hundred years from the mid-seventeenth to the mid-eighteenth centuries a large number of European forts and trading posts were set up along the West African coast. At the beginning of the eighteenth century there were thirty European forts, mostly Dutch and English, on the Gold Coast. The Dutch headquarters was at Elmina, the English at Cape Coast Castle. The English share of the Gold Coast slave trade gradually increased during the eighteenth century at the expense of the Dutch, Britain carrying more than half the estimated 60,000–70,000 slaves a year. Slaves were bought from the coastal tribes who acted as middlemen. As the demand increased strong states, such as Dahomey and Ashanti, developed rapidly, raiding for slaves far into the forested belt.

The *great states of the Guinea forest zone*—Benin, Dahomey, Ashanti and to some extent Oyo—were closely associated with the slave trade which aided their rise to power and eventually helped to bring about their downfall.

During the seventeenth century *Benin* reached the height of its prosperity based on the export of blue cotton cloth, coral beads, pepper and slaves. As in the case of Oyo, the expansion of Benin was based on the use of firearms, while its commercial predominance lay in the fact that it was the main centre for trade between Yorubaland and the Europeans. But the destruction wrought by widespread slave raiding, the revolt of tributary Yoruba states anxious to obtain some of the spoils, the intervention of European powers and other factors eventually brought about the downfall of these states.

It is extremely difficult to estimate the total number of slaves taken from West Africa to the Americas. The trade went on almost uninterrupted until the mid-nineteenth century. Estimates of the number of slaves who were transported to countries such as Brazil, Mexico, the southern U.S.A. and the Caribbean islands vary widely from 15 to 50 million.

AFRICA IN THE NINETEENTH CENTURY

The history of the nineteenth century is marked by such important events as the abolition of slavery and the European exploration of the interior. Britain, the main carrier of slaves during the eighteenth century, was equally zealous in attempting to suppresss it during the nineteenth century. Anti-slavery agitation led by Buxton and Wilberforce, the wave of humanitarianism and interest in missionary enterprise, changing economic conditions and the growth of industrialization in Britain were some of the main reasons for the *abolition of slavery* throughout the British Empire in 1833. By the end of the nineteenth century most countries had followed Britain's example. Two of the main motives for British interest in Africa during the nineteenth century were the desire to suppress slavery and to replace it by a more legitimate trade in tropical products. The rapid growth of industrialization in Europe greatly increased the demand for products such as palm oil, cotton and rubber. As West Africa is only 3,000–4,000 miles from Europe it was in a strategic position to tap European markets.

Interest in the abolition of slavery took a practical form in the founding of colonies for freed slaves on the coast of West Africa. The Sierra Leone Company was formed in 1787 and after great initial difficulties Freetown was founded in 1792 and Sierra Leone became a Crown Colony in 1808. Similarly, freed slaves from the United States of America were settled at Monrovia, Libreville and Maryland which were amalgamated in 1857 to form the republic of Liberia.

Exploration

At much the same period as the anti-slavery agitation interest quickened in *exploring* Africa's unknown interior. In 1788 the African Association of

London was formed to encourage African exploration and in particular to solve the problem of Niger drainage. Another major drainage problem was that of the Nile. Until the nineteenth century maps of Africa often showed the *Niger* rising in the east, flowing west in three branches and entering the sea as the Senegal, Gambia and Rio Grande rivers. From 1788–93 three unsuccessful expeditions were sent out by the African Association. European attempts to penetrate the forest barrier were defeated by the hostility of the coastal tribes and the deadly nature of the climate. This barrier was only turned by determined efforts from the Sudan zone. The fourth attempt by *Mungo Park* in 1795 was successful, when he reached Sego on the Niger. On his second journey in 1805 and 1806 he travelled down the Niger from Koolikoro to Bussa where he was killed. His journeys were of great importance and he was able to prove that the Niger flowed to the east and not the west.

It was not until twenty-five years later that the Lander brothers setting out from Badagri sailed down the Lower Niger from Bussa to the delta long known to traders as the Oil Rivers. In 1823 the Clapperton Expedition crossed the Sahara from Tripoli to Lake Chad. Clapperton visited *Kano*, but was disappointed in the city, although fascinated by the market of which he says, 'the *soug*, or market, is well supplied with every necessary luxury in request among the people of the interior'. For centuries Kano had been one of the main markets of the Sudan zone.

Timbuktu, another great market of the Sudan, attracted European attention for a long period. A Portuguese embassy may have travelled there in 1565 and many attempts were made to reach it subsequently, Laing being the first to do so in 1825, followed by René Caillié in 1828. The city did not recover from the collapse of Mali and by the nineteenth century its population had shrunk greatly and its university had decayed. One of the greatest African travellers was Barth who travelled extensively in the mid-nineteenth century, visiting Timbuktu, Sokoto, Lake Chad, the Niger and many other places. He provided a wealth of information about the Sudan belt of West Africa.

East Africa

In East Africa Portuguese influence had been displaced by the Arabs at the end of the seventeenth century, although further south in Mozambique the Portuguese were able to maintain a somewhat tenuous hold on their possessions. The *Omani Arabs* became predominant especially when Seyid Said the Sultan of Oman transferred his court from Muscat to Zanzibar in 1840. Prior to this date Arab caravans had penetrated the interior of East Africa, and travelled from coastal ports such as Tanga, Bagamoyo and Kilwa along a number of routes, followed approximately now by the Central Railway, via Dodoma and Tabora to Lake Tanganyika or south to Lake Nyasa. *Slave raiding* was the major economic concern, Arabs and African tribes, such as the Yao, carrying out the slave raids and financed by the Indians, or Banians, from Zanzibar. As in West Africa slaves had been exported from East Africa for a very long period, particularly to Arabia and Persia, but the extent and area of slave raiding increased greatly towards the latter half of the nineteenth century as more of the interior was opened up. Zanzibar became the great entrepôt for East Africa and the headquarters of the slave trade, replacing Mombasa in importance.

The excesses of the East and Central African slave trade were revealed, in particular, by *Livingstone*, perhaps the greatest of African explorers, who working from the south discovered Lake Ngami in 1849. In 1851 he moved to Sesheke on the Zambezi river and later was kindly received at Linyanti, the Makololo capital.

Livingstone's first major African exploration was a journey from Linyanti up the Zambezi in a north-westerly direction to its headwaters, then west across the high Angolan plateau watershed to Loanda which he reached in 1854. Instead of returning to England he retraced his steps to Sesheke and decided to travel eastwards down the Zambezi to Quilimane, so crossing Africa from west to east. His most famous African discovery was made on this journey, the magnificent waterfall on the Zambezi, known to the Africans as 'the smoke that thunders', but named by Livingstone the Victoria Falls. It is possible that the Falls had been visited previously by a Portuguese trader or Boer hunter, but this is the first recorded discovery. On his return to England he exposed the horrors of slavery and advocated its destruction by a combination of Christianity and commerce. He realized that the only way to eradicate slavery was to replace it by trade and other economic activities. African chiefs were anxious to obtain guns, cloth, beads and other trade goods, and slaves and ivory were the only commodities they could offer in return. Slaves were also indispensable as human porters in a region so greatly lacking in communications.

The purpose of Livingstone's second journey, on behalf of the British government, was to explore the Zambezi Valley and ascertain the value of the river as a navigable waterway. The expedition was only a partial success, passage up-river being impeded by the Kebrabasa Rapids and delayed by the inefficiency of the launch. The Shire River was explored and Lake Nyasa (now renamed Lake Malawi) rediscovered.[5]

On his last journey in 1865 Livingstone attempted to find the source of the Nile. After three years' wandering, during which he explored the region north-west of Lake Nyasa and discovered lakes Bangweulu and Mweru, he arrived at Ujiji where he was found by Stanley. After they parted company Livingstone went back to the swamps of Northern Rhodesia which he was convinced contained the source of the Nile, four ancient fountains. This waste of swamp and lake is the indeterminate watershed between the Congo and Zambezi drainage, a region with few recognizable landmarks and typical of the slight swells which act as major watersheds in Africa. It was thus little wonder that Livingstone, hindered as well by incorrect longitudinal calculations, lost himself in the swamps. Overcome by disease, long continued exposure to fever and lack of adequate medicines, he died at Chitambo on 27 April 1873.

The problem of the *Nile drainage* was only solved by Speke, Baker and Stanley after 1860, a riddle that had puzzled the world for over 2,000 years. Lake Tana, the source of the Blue Nile, was discovered by Jesuits in the seventeenth century and revisited by Bruce in 1788. Krapf and Rebmann were the first Europeans to travel far into the interior of East Africa. In 1848 and 1849 they saw Mounts Kilimanjaro and Kenya and brought back reports of the great lakes. In 1856 the Royal Geographical Society sent *Burton and Speke* on an expedition to discover these great lakes. Their route, which followed a major slave caravan route, was from Bagamoyo via the Ruvu river valley on to the East African plateau, thence via Tabora to Ujiji on Lake Tanganyika. On the return journey Burton was left ill in Tabora while Speke went north to Mwanza and the southern end of Lake Victoria. Despite Burton's disbelief Speke was convinced that this was the source of the Nile.

On a second visit to this region in 1860 Speke and Grant went round the western side of the lake and found the outlet of the Nile at the Ripon Falls. He wrote, 'I saw that old Father Nile without any doubt rises in the Victoria Nyanza . . .'. He was the first European to enter Buganda, then a powerful state, and meet Kabaka Mtesa whose capital and palace he described: 'the palace or entrance quite surprised me by its extraordinary dimensions . . . the whole brow and sides of the hill on which we stood were covered with gigantic grass huts.'

The problem of the Nile drainage was further elucidated by Baker, who followed the course of the Nile upstream and discovered Lake Albert,

[5] Gaspar Bocarro, a Portuguese explorer, travelled from Chicova to Kilwa in the early years of the seventeenth century and antedated Livingstone's discovery by over 200 years.

and also by *Stanley* who circumnavigated Lake Victoria, sailed down the Lualaba and thence down the river Congo to the sea.

The Cape

Before the nineteenth century there were only a few thousand Boer pastoralists and wine and wheat farmers in the *Cape*. During the Napoleonic wars the British government, appreciating the Cape's strategic value, occupied it, first in 1795, then continuously in 1806. The small white population was augmented considerably in 1820 by the arrival of some 5,000 British settlers in the Eastern Cape. Increasing dissatisfaction with British administration and other factors caused nearly 10,000 Boers, known as the Voortrekkers, to leave the Cape and trek northwards on to the well watered grasslands of the high veld. Passage to the west and north-west was restricted by aridity and to the north-east by hostile Bantu tribes held in check along the Border by the British army. The high veld had been cleared of its Bantu population by the Zulu wars of the 1820s. After the defeat of Msilikazi and the Matabele the Boers were able to set up embryonic states, which later became the Transvaal and Orange Free State republics. Initially successful in Natal, where Dingaan and his Zulu impis were defeated decisively, the Boers were forced to submit to British rule when Natal became a British colony in 1843. The majority of the Boers retreated on to the high veld and by 1854 the British recognized the Boer communities on the plateau as independent.

By 1880 most of the interior of Africa was explored and no longer was it necessary for cartographers to people the hinterland with fabulous tribes, legendary animals, vague lines of rivers and molehill mountains erupting, according to artistic fancy, anywhere on the map. As something of Africa's economic potential was revealed economic rather than humanitarian considerations actuated further exploration. Great chartered companies were formed, not unlike the earlier trading companies, for example the Dutch East India Company with its multifarious interests. Companies such as the Royal Niger Company, Imperial British East Africa Company and the British South Africa Chartered Company traded widely and virtually ruled over large parts of West, East and Southern Africa.

Until the 1870s European commercial exploitation of Africa was largely confined to the coast and centred on the slave trade, particularly on the west coast where Africans acted as middlemen and prevented European entry into the interior. After the 1870s, however, as exploration opened up the interior and the exploitation of minerals and tropical crops replaced slavery as economic activities, European interest in Africa as a whole intensified. By the 1880s there was a scramble by Western European powers to obtain control of Africa's natural resources. The next period from 1884–1910 thus marks the Scramble for Africa and the revolutionary impact of European culture on African tribal society on a far wider scale.

THE SCRAMBLE FOR AFRICA

Stanley's discovery of the vast Congo basin helped to precipitate the rush, as it became known that the Congo river led into the heart of equatorial Africa and its supposedly vast resources of tropical products. Stanley was convinced of the great importance of the Congo basin, but failed to interest Britain. King Leopold of the Belgians, however, realized its significance and sent an expedition under Stanley in 1879 which led to the formation of the Congo Free State in 1884. The *Berlin Conference* was also held in 1884 to consider the question of the Congo and other matters of common interest. An agreement was signed in 1885 which stipulated, among other clauses, that navigation of the Congo and Lower Niger rivers should be unrestricted and, most important, that no new European annexations of territory were valid unless there was effective occupation in the form of direct administration.

West Africa

British and French spheres of influence were rapidly turned into colonies ruled by the metropolitan countries. The British sphere of influence in the Niger delta was hastily converted into the Oil Rivers Protectorate some 20 years after the British had formed the colony of Lagos in 1861. Sir George Goldie and the Royal Niger Company, which he formed in 1866, helped to extend British power northwards from Yorubaland to Ilorin and the Fulani emirates, a task which was furthered by Lugard. In 1900 the administrative and military functions of the Royal Niger Company were taken over by the British government. In 1914 the Northern and Southern Provinces of Nigeria were united to form the great country of Nigeria with an area of nearly 400,000 square miles.

On the Gold Coast, the centre of the British slave trade in West Africa for 150 years, the Colony was formed in 1874. To the north in the heart of the forest belt was the powerful Ashanti Confederacy which fought against the British in three campaigns in 1874, 1896 and 1900. By 1902 with the addition of Ashanti and the northern territories to the coastal strip the area of the Gold Coast Colony was enlarged to some 90,000 square miles.

The *French* pushed north from their coastal footholds and east from

Senegal through the Sudan savanna zone to overcome Arab, Fulani and negro resistance and form a series of new states, for example, the Ivory Coast and French Guinea in 1893 and Dahomey in 1900. By the early years of this century the French had acquired a wide area of West Africa consisting of nearly 2 million square miles. French West Africa was composed of a continuous block of territory, whereas British West Africa with an area of less than a ½ million square miles was in four separate units— Gambia, Sierra Leone, Gold Coast and Nigeria.

Germany acquired two regions, the Cameroons with an area of 200,000 square miles and Togoland with 33,000 square miles. Portuguese influence was confined to the 14,000 square miles of Portuguese Guinea.

Southern Africa

Farther to the south the *Portuguese* had long been interested in the Congo region, and thus pushed their claims against the Belgians and French. A compromise was reached and the left bank of the Congo plus Cabinda to the north was given to Angola. The Congo Free State obtained the north bank and a narrow strip between Angola and Cabinda which formed a vital outlet to the Congo Basin. Portugal also hoped to establish a Central African empire stretching across Africa linking up Angola and Mozambique. Although encouraged by Germany and France her ambitions were thwarted by the extension of British influence northwards from Bechuanaland onto the high plateau crossed by the Zambezi.

Before the discovery of diamonds along the terraces of the Vaal river and later in volcanic pipes at Kimberley in 1870, the great interior plateau of South Africa was occupied by the *Boer republics* of the Transvaal and Orange Free State. This great area was thinly populated by Afrikaans pastoralists practising a largely subsistence economy on very large farms. There were no railways and only rough tracks for ox waggon traffic and the few towns such as Pretoria, Potchefstroom and Bloemfontein were very small.

The discovery and exploitation of the Kimberley diamond mines in 1870 and the vast Witwatersrand goldfields in 1886 caused an economic revolution which helped to bring about a union of the separate colonies and republics of the Cape, Natal, Orange Free State and Transvaal in 1910 and the emergence of a modern state in less than thirty years. It was hoped that further rich mineral fields would be discovered in the Basement rocks of the plateau north of the Transvaal. The occupation and exploitation of this area was largely the work of *C. J. Rhodes*. He was a man of great financial acumen, imagination and energy who used a vast fortune acquired on the Kimberley diamond mines to further his dream of creating a British Africa. In order to further this aim he became Prime Minister of the Cape and persuaded the British government to take over control of the strategic Missionaries' Road through Bechuanaland to the north. Rhodes obtained a concession from Lobengula, paramount chief of the Matabele, to prospect for minerals in his domains. He formed the *British South Africa Company*, and in 1890 equipped and sent out a Pioneer Column which skirted Matabeleland and occupied the north-western part of the plateau inhabited by the Mashona. After wars against the Matabele and Mashona in 1893 and 1896 the whole of the plateau between the Zambezi and the Limpopo, as far east as the Great Escarpment, was formed into the territory of *Southern Rhodesia*. Below the Escarpment on the widening and unhealthy coastal plain lay the Portuguese territory of Mozambique.

The Chartered Company also operated in a great area to the north of the Zambezi. The northernmost limit to their activities was set by the southerly extension of the Congo Free State and the rich copper mining region of the Katanga. Sir Harry Johnston helped greatly to confirm the British sphere of influence round Lake Nyasa and in 1891 *Nyasaland* became a British protectorate.

East Africa

The *Germans* became interested in acquiring a colonial empire in the 1880s. They swiftly occupied a long stretch of the barren coast and more productive interior plateau of south-west Africa between the Cunene and Orange rivers and to the west of the Kalahari. Likewise, on the east coast they were able to seize an area as large as Nigeria that lay to the south of British East Africa and north of Mozambique and between the curve of the great lakes and the sea.

As in West Africa, one of the major factors influencing the British participation in *East African* affairs was their effort to control the slave trade centred on Zanzibar. The Omani Arabs had occupied the island of Zanzibar for a long period, but it was only in the early nineteenth century that the Sultan transferred his court from Oman to Zanzibar. The British alliance with Oman was of long standing, and so a British Consul was appointed to Zanzibar with the object of persuading the Sultan to curtail the slave trade. The influence of the Sultan on the mainland of East Africa did not stretch much beyond the coastal strip. The British sphere of influence was also vague, and the Germans claimed a considerable part of what is now Kenya, besides the area to the south of Mount Kilimanjaro which became German East Africa and later Tanganyika.

The whole of East Africa might have been drawn into the German colonial orbit, but for the work of individuals like Sir Harry Johnston and Lugard and the British East Africa Company which received its charter in 1888. The boundary between the German and British spheres of influence was drawn from the coast to the north of Mount Kilimanjaro, then in a straight line north-west to Lake Victoria at 1° S. latitude. As in West Africa the British government took over the British East Africa Company's administrative functions and British protectorates were declared over *Uganda and the future Kenya*.

Buganda and the densely populated region on the northern and western shores of Lake Victoria were soon connected with world markets by the construction of the Kenya–Uganda railway from Mombasa to Kisumu, which it reached in 1901. Sir Harry Johnston was sent out as Special Commissioner to organize the economy and make the territory self-supporting. He concluded an agreement with the Buganda chiefs whereby they were given possession of some 9,000 square miles of mailo land under a form of individual tenure that paved the way for the rapid development of cotton as a cash crop, to be followed by coffee. A vigorous economy based on peasant cultivation has been built up in Uganda, whereas in Kenya and Tanganyika *white settlers* have played a major role in the economic development of these territories.

The Germans were slow to build a railway across the plateau to Lake Tanganyika and the Central Tanganyika line was started from Dar es Salaam in 1904, only reaching Kigoma on Lake Tanganyika in 1914.

A few of the *Asians* brought to East Africa for the construction of the Kenya–Uganda railway stayed on after its completion and added to the original nucleus of Indian population along the coast. The Asian population has spread over East Africa, increased considerably and played an important role in the commercial development of the three territories.

Southern Africa

The great *Congo basin* region remained for long a private possession of King Leopold, but in 1908 was taken over by the Belgian government and became the Belgian Congo. Its mineral resources are considerable and were of primary importance in developing the economy and communications of the new state. Tin was first worked in the north-east in 1904, copper in the Katanga in 1911 and diamonds in the Kasai region in 1913.

In Southern Africa the exploitation of minerals, diamonds and gold in South Africa, and gold in Southern Rhodesia caused the rapid development of an exchange economy. Within ten years after the Rand auriferous deposits were discovered railways had been built to connect the coastal ports of Cape Town, Port Elizabeth, East London, Durban and Lourenço Marques with Kimberley and Johannesburg. The European population of *South Africa* quadrupled in little more than 30 years and was over 1 million in number in 1904. Political rivalry between the British and Boers in the Transvaal caused the Anglo-Boer War of 1899-1902, but less than ten years later the four colonies—Cape, Natal, Transvaal and Orange River Colony were joined together as the Union of South Africa.

The development of Southern Rhodesia was far slower as gold resources were much smaller and more scattered than in South Africa and in 1904 the white population was only 12,628. Railways were extended northwards from Mafeking to Bulawayo, and at great cost of life, because of the unhealthy climate, a railway was built from Beira on to the plateau via Umtali to Salisbury, which was linked to Bulawayo in 1902. A line was also built north-west from Bulawayo, which reached Livingstone in 1905, and was extended to the Congo border by 1909. The development of Northern Rhodesia was far slower and the copper deposits on the copper belt were only worked extensively in the 1930s.

North and north-east Africa

While the destiny of Africa south of the Sahara lay in the hands of the West European nations *North Africa* continued under the dominion of Islam, which had already lasted for a thousand years, but two events of major importance in Egypt caused a revival of European interest in this region during the nineteenth century. The first was the short-lived French occupation during the Napoleonic wars which caused a realization of Egypt's historic past, and the second was the opening of the Suez Canal in 1869 which caused the Mediterranean to become once more a sea route of major importance. The British considered it essential to safeguard this vital route to the Far East and intervened directly in Egyptian affairs in 1882 when disorders appeared to threaten communications. At the end of the nineteenth century British influence was extended into the Sudan, where a Condominium with Egypt was set up in 1899. The French intervened in Algeria in 1830 and in 1834 it was declared a French colony. Settlement by Europeans was encouraged and in 1960 there were over a million settlers, mainly French, in Algeria. Expansion of French influence occurred to the west into Morocco, east into Tunis and also south into the desert to link up with the Sudan belt. The French attempted to extend their control across Africa in the Sudan savanna region, but met with

British opposition in the Sudan and so were forced to withdraw from the Nile basin in 1899.

Italy entered the colonial field late and was left only the arid regions of north and north-east Africa. Tripolitania and Cyrenaica were acquired from Turkey in 1911, while in the 1880s the northern extension of the Ethiopian plateau was invaded and became the Eritrean Protectorate. A large stretch of desert coastline in the horn of Africa was acquired as Italian Somaliland and attempts to consolidate the Italian hold in north-east Africa by the invasion of Ethiopia were thwarted by the defeat of the Italians in 1896, only to be successful for a brief period some 40 years later. French and British strategic interests were furthered by the acquisition of French and British Somaliland. Until the mid-nineteenth century *Ethiopia* had been a collection of warring districts and provinces. The work of consolidation was undertaken by the emperors Theodore II, John IV and Menelek. The latter played an important part in the Scramble for Africa and united Abyssinia effectively, subdued the Galla tribes, defeated the Italians and doubled the size of his territory.

AFRICA IN THE TWENTIETH CENTURY

The penultimate phase of Africa's history was brought to a close by the outbreak of the First World War in 1914. The very brief period of ten years from 1884–94 marked the partition of Africa among the European powers. Before 1884 the interior of Africa was largely free from European political control, but by 1895 over 90% had been parcelled out among the European maritime powers. The older colonial powers of Great Britain, France and Portugal expanded into the interior from their coastal footholds, while the new states Belgium, Germany and Italy (the latter two countries only recently unified) filled in the gaps. The French controlled most of the great north-western bulge of Africa from Algeria to the Congo, and the British most of the eastern third of Africa from the Suez Canal to the Cape of Good Hope, guarding the vital approaches to the Indian Ocean. Portugal was left with an enlarged Angola and Mozambique in the southern hemisphere and the tiny Portuguese Guinea and the islands of São Thomé and Principe in the northern hemisphere; Spain had Spanish Sahara, Ifni and Rio Muni. The German territories were in German East and South-West Africa, the Cameroons and Togoland. Belgium eventually acquired the bulk of the Congo basin and Italy was restricted to the north and north-west.

The political map of Africa was drawn in all too short a space of time, artificial boundaries sliced across natural regions and tribal groupings and

European systems of administration were imposed on African tribal society. Contemporaneously with the establishment of law and order came the building of railways and the growth of economic exploitation. Economic progress was often slow if mineral resources were scanty, distances were great, an adequate labour force was lacking and capital difficult to obtain.

The last period is *1920–50* as the events of the last ten years belong more properly to the chapter on political development. Territorial changes in Africa after the First World War were concerned mainly with the ex-German colonies. It was decided at the Versailles Conference to adopt the principle of international supervision and Great Britain, South Africa, France and Belgium were entrusted with the administration of ex-German territories as Mandates under the League of Nations. Great Britain was given a mandate over German East Africa and parts of Togoland and the Cameroons; the Union of South Africa a mandate over South-West Africa; France the bulk of the Cameroons and Togoland, and Belgium the grass-lands of Ruanda Urundi, formerly part of German East Africa. With the revival of the principle of trusteeship after the Second World War the mandated territories became United Nations Trust territories. Since 1945 the administering authorities, with the exception of South Africa, have prepared these territories for self-government.

During this last period an increasing proportion of the African population was drawn into the exchange economy, although the majority of the population were still engaged in subsistence agriculture. There has been a great increase in the production of export crops such as palm oil, cocoa, groundnuts, cotton, coffee, sisal, etc. Minerals, particularly gold, diamonds, copper and tin, have played a major role in economic development. But the economy of African territories was, and still is, largely dependent on a restricted range of primary commodities, for example Gambia, ground-nuts; Uganda, cotton; Zambia, copper. During the depression of the early 1930s there was a sharp drop of 50% and more in the value of commodities exported from most territories. This had a serious effect on government revenues and social services and affected particularly European farmers who had to rely on governmental help. In Southern Africa the depression hastened the adoption of the 'civilized labour' policy designed to protect the economic interests of the white against the non-white inhabitants.

During the *Second World War* the Italian hold on East Africa was broken, and North Africa was the scene of prolonged desert campaigns until the battle of El Alamein saw the tide of war swing against Germany and Italy. The first invasion of Europe occurred from Tunisia in 1943 and North Africa ceased to be a battleground for European armies. The war had a profound effect on the colonial peoples of Africa and Asia. European prestige sank low and her hold on a wide range of colonial territories was irreparably weakened. The political freedom of first India and Pakistan, then other Asian countries, proved a great stimulus to the fires of nationalism or Africanism. Britain put in train a programme for self-determination and soon after Ghana became independent France suddenly granted independence to her African territories. The map of 1950 showed the bulk of Africa still under the political control of Europe; that of 1960 reveals that most of Africa north of the equator is governed by indigenous rulers. The influence of economic forces, the great rise in volume and value of agricultural exports, increased urbanization and industrialization and the rise of a middle class in some territories, and the political influence of returned soldiers, middle class and a small educated minority who have taken advantage of the winds of change blowing over Africa, are some of the reasons for this dramatic transformation in Africa's post-war political scene. The Scramble for Africa took some ten to fifteen years at the end of the nineteenth century and only half a century later the Retreat from Africa of the European colonial powers is in full swing. It took thousands of years to complete the exploration of Africa, but the Colonial Period has lasted barely a hundred years.

With the withdrawal of European powers Africa has become increasingly open to world-wide influences and a battleground between the rival faiths of Christianity and Islam, and the ideologies of capitalism and communism.

CHAPTER TWO

Physical Environment

RELIEF, STRUCTURE AND DRAINAGE

Africa is less complex structurally than the Americas, Asia and Europe; there are no new intricately folded mountain systems stretching across or down the length of the continent. Africa's new folded mountains are found only in the north-western and south-western extremities of the continent, forming in the south an abrupt termination to the vast spade-shaped plateau which stretches uninterruptedly from the Cape to Ethiopia. There is considerable compactness of form and regularity of shape, and this, combined with Africa's large area, is emphasized in Figure 2 which shows China, Brazil and India fitting comfortably into Africa. Two-thirds of Africa projects north of the equator, the great bulge of West Africa and the Sahara causing it to extend nearly 4,000 miles on an average from east to west, compared to a width of less than 2,000 miles in the south. In comparison with Europe's 19,000 miles of coastline, Africa has only 15,000 miles, despite the fact that its area is three times that of Europe. There are no inlets, inland seas or great navigable rivers penetrating the heart of the continent as in Europe or North America.

It is Africa's tropical massiveness that impresses: nearly 80% of Africa is within the tropics, and the part of the continent situated between 37° N. and nearly 35° S. is bisected by the equator. If Africa is compared with Australia, which is bisected by the Tropic of Capricorn, or with South America where the equator cuts across the northern part of the continent, it is obvious that Africa is the most tropical of the continents. However, the high plateau in the east and south, much of it above 4,000 feet in altitude, does mitigate considerably the effect of the heat. Africa consists of a vast *plateau* or shield, somewhat similar to other old shield areas of the world such as Brazil and Western Australia, which has been emergent since pre-Cambrian times, apart from brief marine invasions in the north. It is only at the southern and north-western ends of the continent that folded mountains have been tacked on to the rigid block forming the plateau.

The core of the continent is composed of Archaean schists and gneisses

FIG. 2—Africa's compact form

intruded by granites which, when stripped of their later sedimentary covering, form an irregular surface. The fundamental complex is overlain extensively by continental sedimentaries of various ages which are horizontal or slightly warped. Where erosion has been more active near the edge of the basin or plateau protective cappings of harder rock such as sandstone, dolerite or basalt, may give rise to a mesa or butte whose steep slopes rise abruptly from the gently sloping pediment or rolling plain. The characteristic African landscape is either one of wide spaces with endless horizons or an alternation of level plain and plateau, punctuated by a tumbled array of table-topped mountains and conical hills.

Africa has been affected predominantly by epeirogenesis rather than orogenesis, except for the Atlas and Cape ranges, and this updoming with its attendant fracturing and faulting has caused a number of extensive plains, basins and swells, volcanic peaks and highlands, block mountains and rift valleys (Fig. 3). Towards the south and east the plateaux are higher and more continuous than in the north and west. Africa may thus be divided into two major physical regions by a line from Eritrea to the mouth of the Congo river. The first region of *Low Africa* to the north and west of this line is largely north of the equator, and is aligned east and west. It is far lower in altitude and has a much greater area of basins of late Mesozoic and Cainozoic sedimentation, some well watered such as the Congo and Southern Sudan section of the Nile, others arid areas of inland drainage such as Chad and El Juf.

The second region, *High Africa*, is mainly south of the equator, and is aligned mainly north and south. It has a greater extent of high plateau with an average altitude of over 3,000 feet, the greatest escarpments, extensive areas of volcanic highlands, rift valleys and more deeply incised river valleys than Low Africa. The outward tilting of the plateau has caused deep downcutting and entrenchment of rivers and a well marked fall line.

The southern plateau section of High Africa is encompassed by an impressive escarpment, known as the *Great Escarpment*, which marks the tilted edges of the plateau. In the south-east the Drakensberg, or Dragon Mountains, reach a height of over 11,000 feet and form the highest point of the escarpment. Precipices of dark Stormberg basalt 6,000–7,000 feet high rest on and are intercalated with cream and light-coloured Karoo sedimentaries.

Below the escarpment wide terraces, bearing marks of former erosion cycles, step down to the sea, while the Tugela and its tributaries and other swift coastal rivers on the plateau edge cut their way deeply to nearby base level, the Indian Ocean. The escarpment is an erosional feature and is being pushed back steadily by the short but vigorous coastal streams imbued with erosive power by the abundant summer rains from the wide stretches of the Indian Ocean. Great embayments have been carved out of the plateau by the Limpopo, Sabi, Zambezi and other major rivers. The *Zambezi* rises far to the east on the plateau of Angola, and curves southwards through the sandy plains of Barotseland, to turn and twist westwards again through and down the restricting rim of highland, via the great Victoria Falls and faulted trench of the lower Zambezi.

The watershed of the Zambezi, Congo and other African rivers is an indeterminate, very gently undulating region of swamps and marshes marking the imperceptible summit of a great swell or dome. It was in regions

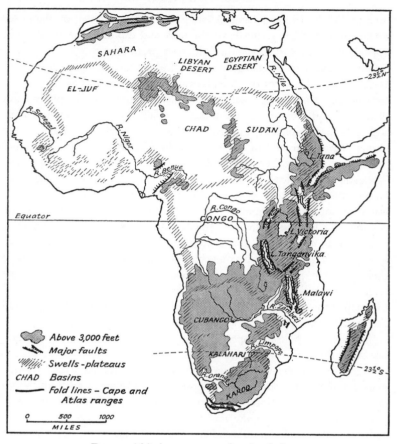

FIG. 3—Africa's structure and major drainage

like this that Livingstone travelled extensively in his first expedition across
Africa, and where later, while seeking the fountains of the Nile, that he
died. He describes these depressing expanses of swamp as follows: 'Our
progress is distressingly slow. Wet, wet, wet; sloppy weather truly, and no
observations, except that the land near the Lake being very level, the rivers
spread out into broad firths and sponges.'

These coastal rivers that are fed by the copious waters of the high plateau
edge, and that breach the plateau, have occasioned much river capture of
slower moving plateau-based rivers. Thus, for example, the upper reaches
of the Cunene which used to flow to the Etosha Pan now flow westwards
into the south Atlantic. The Zambezi has captured most of the water that

used to flow into Lake Ngami. Another example is afforded by the lower reaches of the *Niger* which have cut back and captured the middle and upper waters of a former eastward draining proto-Niger.

Other African rivers rising on the high outer edge of the plateau, such as the Orange, Niger, or the Nile, flowing from the tropical and equatorial zone of high rainfall, meander in wide mature valleys on the plateau and then plunge off it in a series of cataracts and falls, such as the Victoria Falls and Kebrabasa rapids on the Zambezi, and the Aughrabies Falls on the Orange river. The end of the road for an African river that debouches swiftly on to a narrow coastal plain is thus often a stage of seeming youth. Navigation of African rivers is greatly hindered by rapids and falls. Despite the fact that the *Congo* has many thousands of miles of navigable waterway spread out behind the Crystal Mountains, access to the Atlantic is impossible because of the plateau edge. African rivers are thus bunched behind their narrow exit channels, and on reaching the ocean tend to build deltaic mouths, or become blocked by sand bars and only burst their way through the barrier after floods. The pattern of some African rivers is not unlike that of a tree whose branches and leaves are represented by the plateau course, the trunk by the short plateau foreland course, and the roots by the splayed-out alluvial delta.

The *southern plateau* slopes gradually from the high eastern reaches, known as the high veld, to the wide sands of the Tertiary filled basin, to rise again to the ancient highlands of South-West Africa. Here the Archaean platform of granite and gneiss is exposed rising like an island from the sea of Kalahari sand and Karoo sedimentaries. Over part of the Kalahari there is no drainage, and over the rest of the area the drainage is interior, a reflection of its aridity and basinlike form. It is only at the northern edge that the Zambezi and tributaries skirt the basin to reach the sea eventually. The western edge of the plateau overlooks the desert Namib coast. Here are no swift perennial rivers to push back the plateau, so that its edge is much closer to the coast than in the more humid east.

To the south of the Kalahari basin is the smaller basin of the Upper Karoo crossed by the Orange river, while south of the Great Escarpment is the Great Karoo consisting of a number of small drainage basins. The western and southern borders of the Great Karoo are formed by the Cape folded mountains of Palaeozoic sandstones, conglomerates and shales. Much of the plateau is covered by thick Karoo sedimentaries of Carboniferous to Triassic age, which further north have been stripped off to reveal the Archaean basement rocks. The rich mineral-bearing belt of granites, gneisses and schists with vast stores of gold, copper and many other minerals forms a great curve to the east of the Kalahari.

The narrower *northern extension* of the plateau in East Africa ends at the downfaulted trench of the Red Sea. It is first bisected and then trisected by the great *Rift Valley* which stretches for 70° of latitude from the borders of Turkey to south of the Zambezi delta. Near Lake Rudolf the Rift Valley splits into two trenches which roughly encircle the recent crustal sag of Lake Victoria, to unite again near Lake Malawi. The western arm of the Rift Valley is occupied by a series of large and deep lakes, for example Lakes Tanganyika and Malawi are each over 300 miles long and thousands of feet deep. The eastern arm traversing drier regions has far smaller lakes. The uplifted blocks at the edge of the rift are tilted, causing an alteration of the drainage pattern. In the Lake Victoria region rivers that drained west during the Pleistocene age are now reversed and drain into the lake or the Victoria Nile. The peculiar leaflike shape of Lake Kyoga is also due to this eastward tilting.

It is not the actual downfaulted trench of the rift that is impressive, but the assemblage of accompanying highlands, horsts, volcanic masses and peaks. The Abyssinian plateau and Kenya highlands have been built up by a great mass of volcanic rocks resting on the Archaean platform and associated with rifting. From these plateaux rise still higher volcanic peaks, some of which like Mount Kenya reach the permanent snowline. The Ruwenzori massif is not of volcanic origin, but an upthrust horst of basement rocks with a number of peaks over 16,000 feet in height. Recent discoveries have shown that rifting extends back over a very long period to pre-Cambrian times, and is not only post-Tertiary.

Various theories have been put forward to explain the origin of the rift valleys. Wayland's compressional theory, which is no longer tenable presupposed that the downfaulted rift was overridden and kept down by the force of the upthrust sides. Gregory's tensional theory put forward the thesis that a long thin zone or wedge sank like the keystone of an arch as opposing forces tore the surface of the land apart. Gregory considered that the East African rift system was connected with the foundering of the Indian Ocean, while du Toit linked it with the disruption of Gondwanaland. Dixey suggests that the problem of the Rift forms part of the greater problem of the rising East African swell and the sinking Mozambique geosyncline, and these in turn of the whole basin and swell structure of Africa.

As in South Africa the plateau in East Africa is close to the sea, but unlike the south falls to the coastal plain in a number of step faults. The coastal plain widens out in northern Kenya and Somaliland. In Ethiopia the faulted edge of the plateau presents an impressive wall thousands of feet high when viewed from the valley of the Awash.

As Africa has been free from folding for such a long period a number of erosion cycles have run their course and left extensive traces of their former existence in the shape of vast *peneplanes* at various altitudes. There are well recognized changes of level at 2,000, 4,000 and 6,000–7,000 feet. Throughout much of Africa the traveller is impressed by the immensity of scale at which accordant summit levels fill the horizon. Viewed from the air it seems as if a giant dragline has notched the edge of a vast plateau at definite intervals. Africa with its wide flat or gently rolling surfaces and dramatic changes of slope is not unlike Brazil and Western Australia. There are also remarkable stratigraphical resemblances between these lands now far removed from each other. Other evidence, such as the distribution of late Carboniferous ice sheets, Dwyka tillites, *Glossopteris* flora, the apparent fit of South America into the bulge of West Africa, the varying position of the poles as shown by the differing axes of ancient magnetic rocks, the orientation of sand dunes in fossil deserts in Great Britain and the United States of America, the past distribution of corals and the geophysical evidence of faulting on the ocean floor off the western coast of North America, suggests that some form of continental drift or displacement of the continental blocks over thousands of miles may well have occurred.

Evidence has been put forward to show that the continents form rafts of light sial which move through a sea of denser sima. Wegener suggested that the continents could drift through the sima, and that these continental masses, once forming the composite continent of Pangaea, have split apart to form the separate land masses of the northern and southern hemisphere. Different gravitational forces caused a drift equatorwards and westwards, forcing the African and European land masses to move towards each other and the Americas to drive westwards. Du Toit working in South Africa was impressed by the correlation of the geologies of South Africa and South America, and supported the theory of continental drift. His evidence was based on a large number of reasons including the general similarity of the coastlines, the abruptly terminated plateaus or elevated erosional plains with disturbed or reversed drainage, submarine features such as the mid-Atlantic or mid-Indian ocean ridges, comparable fold systems passing out to sea at opposed shores and the crossing of fault or fold systems of specific ages.

The '*Driftists*' have been given new support by studies of the magnetization of ancient rocks which tends to show that the continents have changed position relative to the geo-magnetic poles at various times in the geological past, and have also moved with respect to one another. Thus between Europe and North America there has been a rotation of 25° or more corresponding to the opening of the North Atlantic.

Also operating from South Africa, L. C. King is a staunch supporter of the drift theory, which has been given a new look in his latest book.[1] He suggests that the former unity of Gondwanaland (South America, Africa, India, Western Australia and Antarctica) is amply demonstrated by the vast sequences of late Palaeozoic and Mesozoic continental sedimentaries which are to be found in all the southern continents. This series, thousands of feet thick and covering a great area in the same horizontal plane, could not have been derived from the areas in which they now occur. Similarly evidence of Carboniferous glaciation has been discovered widely in the whole of Gondwanaland, except in Antarctica where there is little or no evidence as yet.

In South Africa thousands of square miles are underlain by Dwyka tillites reaching a thickness of 2,000 feet in the Karoo. Evidence from striations and other sources suggests that the ice flowed from gathering grounds to north, north-east, and south of South Africa. King suggests, as glaciation was sequential from west to east, occurring from early to late Carboniferous times in South America, late Carboniferous in South Africa and India, and Permo-Carboniferous in Australia, that the southern continents have wandered across the south polar region.

He goes on to say that from Carboniferous to mid-Jurassic times the various parts comprising Gondwanaland drifted apart, but that the breaking occurred from the mid-Jurassic to the end of the Cretaceous period. The process of fracturing began with southerly incursions of the Jurassic sea into Tanganyika. During the first disruption, which was finished by early Cretaceous times, India, Madagascar and possibly Antarctica had been separated in one piece. The two halves then drifted into new positions where a further process of fragmentation took place and the western half composed of South America and Africa may have hived off by mid-Cretaceous times.

The importance of the Continental Drift theory lies partly in its explanation of the continental nature of the majority of Africa's rocks, partly that with the separation of the outer continental land masses Africa was left with an elevated and even surface. This latter fact forms the core of King's theory of *pediplanation* by which he endeavours to explain the distribution of extensive level surfaces separated from each other by steep changes of

[1] L. C. King, *The Morphology of the Earth*, London, Oliver & Boyd, 1962.

slope which are so characteristic of Africa. The old level surface is gradually reduced by scarp retreat consequent on the initiation of a new erosion cycle working its way upstream by headward erosion. The flat relict summit of a small plateau is thus separated from the encroaching plain by a scarp, palisaded at the top and then dropping steeply to a gently sloping pediment. Factors such as long continued erosion in a semi-arid climate and the horizontality of the rocks have aided the preservation of these erosion cycle units. The smooth sheet flow or the sudden flood of a seasonal climate have further emphasized the flatness of plateau level and the steep slopes of residuals. There have been four major erosion cycles, Gondwana, African, Victoria Falls and Congo. Planation during the Jurassic produced the oldest Gondwana surface now preserved on the highest parts of African watersheds as, for example, the high veld of South Africa. The breaking apart of Gondwanaland in Cretaceous times created new coastlines in the southern part of Africa, whereas in North Africa there was no disruption. The outward flexing of the eastern and western margins initiated a new erosion cycle and Africa was again reduced to a widespread pediplane now preserved over large areas of Africa as the African cycle. Tertiary uplift caused the southern plateau to be raised above 4,000 feet, the margins to be tilted and the rivers to seek new base levels by means of deep gorges, rapids and falls, while along the axis of uplift rift valleys opened in central and east Africa.

Strong Tertiary uplift had thus had a major influence on the drainage pattern of High Africa. The warped and tilted plateau has been fretted and seamed by rivers flowing slowly down the back of the plateau or cutting deeply into the edge of the plateau, particularly in the well-watered east, so reaching base level swiftly by means of nick-pointed fall and incised meander. Tertiary uplift initiated the Victoria Falls cycle to be replaced later by the Congo cycle, caused by zonal uparching.

Wellington[2] has put forward a different theory to account for the landscapes of Southern Africa, although he agrees with King that the landscape of today has resulted from the Gondwana surface of Karoo sediments and lavas. He maintains that the erosive agency has not been pediplanation but the continuous downcutting of rivers flowing to the Orange and the east coast. Tertiary earth movements and differential resistance to erosion of the Karoo and pre-Karoo rocks have affected the rate of downcutting. He accepts King's African cycle as being the clearest example of an erosion cycle, but doubts whether it existed in South Africa. He thinks that the

[2] J. H. Wellington, *Southern Africa*, Volume I, Cambridge, Cambridge University Press, 1955.

Jurassic opening of the Mozambique Channel and the separation of Madagascar was very important as it caused drainage to flow away from the Kalahari to the Mozambique Channel in mid-Cretaceous times.

Monica Cole suggests that King's concept offers the best interpretation of conditions in the eastern margins and Wellington's concept for the great plateau and diversified country of the Southern Transvaal. Thornbury's comment on King's concept is as follows: 'A most sweeping application of the pediplane concept has been made by King (1950) who has interpreted widely separated erosion surfaces in Africa, Asia, North America, Europe, South America and Australia as ancient pediplanes dating back as far as the Cretaceous.'[3] King's theory seems difficult to accept unless a steadily rising base level is postulated.

The effects of pediplanation are difficult or impossible to trace in areas where deposition or structural control are dominant. The African cycle sinks beneath the wide wastes of the Kalahari, while the later cycles are masked by the recent sedimentaries of the Mozambique Plain, or camouflaged by faulting and folding in areas such as the Cape mountains or the Limpopo and Zambezi troughs.

Low Africa is far larger in size and lower in altitude than High Africa. It contains great basins, some arid, some humid, extensive deserts from which arise the granitic plateau regions of Ahaggar and Tibesti, and the Atlas folded ranges. The latter are part of the Alpine system and therefore younger than the Cape mountains. They consist of a complex series of massifs, plateaux and depressions; the highest mountains with an Archaean core are the High Atlas with some peaks nearly 14,000 feet high. From here the ranges of the Middle, Riff and Tell Atlas splay out to the north and along the Mediterranean coast, and to the south are the Anti-Atlas and Saharan Atlas enclosing the high arid plateau of the shotts. The highlands of Ahaggar and Tibesti are island relics of ancient crystalline rocks intruded by granitic and volcanic rocks rising from the vast sweep of younger Palaeozoic, Cretaceous and Tertiary deposits, both stony and sandy in nature, comprising the Sahara desert extending from the Atlantic Ocean to the Red Sea.

There are three great depressions at the junction of the Sudan and Sahara, two of them, El Juf and Chad, being areas of inland drainage, the third the southern Sudan, being one of the main gathering grounds of the Nile. Because of aridity and the characteristic basin formation almost one-third of Africa has inland drainage.

[3] W. D. Thornbury, *Principles of Geomorphology*, London, Wiley, 1958, p. 296.

The *Nile Valley* cuts through the eastern edge of the desert, its sources being in the high altitude and high rainfall areas of the south. The river has evolved through the junction of a series of rivers in a north–south trending depression which formed part of an arm of the Mediterranean in Cretaceous times.

A section of the Sudan belt forms the northern half of *West Africa*, a much dissected ancient crystalline plateau mostly below 1,500 feet overlain by later sediments of various ages. Occasionally higher plateaux and volcanic highlands, the Futa Jallon, Jos, Cameroons and Adamawa, rise from the low plateau. The Niger and Benue have cut down into the plateau and flow in great faulted valleys to unite north of Onitsha. The *Niger*, rising in the Futa Jallon, flows for nearly a thousand miles in an easterly direction before turning south into the Bight of Benin. The lower Niger has captured the upper waters of a former proto-Niger that flowed north-east into a vast inland lake.

The *Congo basin* is an almost circular region astride the equator, its level floor being about 1,000 feet above sea level and the encircling rims rising to 3,000–4,000 feet in altitude. The floor is of thick Karoo sediments of Carboniferous to Jurassic age, the bed of a former lake, over which the river, the greatest in Africa, meanders sluggishly before breaking through the Crystal Mountains to reach the sea.

CLIMATE

The most important physical factor in Africa is probably climate, and in particular rainfall (Fig. 4). As the continent projects north and south of the equator to much the same latitude there is a certain degree of climatic similarity, with a vast curve of tropical continental climate half encircling much smaller patches of equatorial climate.[4] The equatorial type of climate is confined to parts of the Congo basin and West Africa, while the greatest extent of tropical continental climate is to be found south of the equator. On the outer fringe of the savanna are to be found deserts and steppes which by contrast reach their greatest extent in North Africa with only a narrow coastal strip of desert in south-west Africa. Mediterranean, highland and humid subtropical climatic zones cover a comparatively small

---

[4] It can be argued, however, that climates north and south of the equator are dissimilar. There is a larger extent of hot desert and Mediterranean climate in the north and the climate of the Ethiopian highlands is not matched south of the equator, while the high veld climate and the eastern margin climate of Natal and the Eastern Cape are not duplicated north of the equator. In particular, the narrower land mass in the south faces on to two oceans, in great contrast to the area north of the equator.

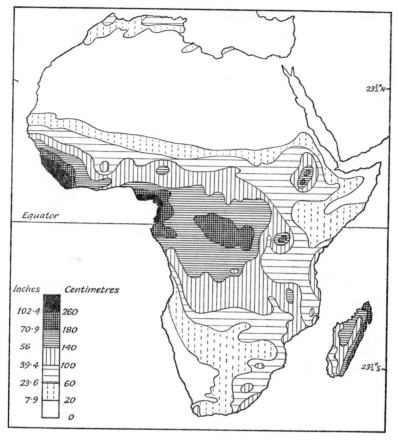

FIG. 4—Mean annual rainfall (after Worthington)

area. The bulk of Africa is thus tropical grading down to semi-desert and desert climatic types. The major climatic regions are separated by broad transitional zones as, except for features such as the Great Escarpment in South Africa, there are few major physical barriers to impede the movement of air masses.

There are a number of general factors that influence Africa's climates. The sun is always overhead somewhere in the Tropics so that insolation is generally high throughout the year. The equator cannot become really hot as the sun is overhead for two short periods and there is much cloud, whereas at the Tropics the sun is overhead for a long period each year and this with the longer daylight and clear skies causes maximum insolation at the solstices and therefore high temperatures. As Africa only projects about 35° north and south of the equator it is intensely tropical in its location and is therefore distinguished by its high temperatures, although relieved in parts by the high plateau.

The distribution of *pressure* and *air masses* is also of great importance in their effect on African climates. Sub-tropical anticyclones are found at about 30° north and south and are characterized by clear skies, low rainfall, subsiding air (particularly in the eastern parts of the anticyclones), trade winds and trade-wind inversion. These high-pressure cells are well marked in the southern hemisphere in the Atlantic and Indian oceans, but in the northern hemisphere the highs are complicated by the effect of greater areas of land and particularly the great width of the African continent north of the equator. This causes the thermal equator and the equatorial trough to lie at about 5° north of the equator. A significant factor is the preponderance of oceans in tropical latitudes, particularly off the coasts of south-east Africa. Various air masses—Tropical Continental and Superior, Equatorial, Sub-Tropical, Polar and Sub-Polar—influence the climate of Southern Africa. Sub-Tropical air masses, originating over the Atlantic and Indian oceans, control coastal weather for most of the year. A very cold Sub-Polar air mass may bring snow to the mountains and occasionally to the plateau.

Air masses over tropical Africa are of two main types. Equatorial or Tropical Maritime air is warm, moist and rainbearing and Tropical Continental air is dry and dusty owing to a long passage overland. Air moves from the subtropical anticyclonic belts of high pressure towards the equatorial low pressure, the junction between them forming the intertropical front or convergence (ITC) zone which is of great significance in African climates.[5] The passage of the front is related to the apparent movement of the sun between the tropics; the sweep is far greater on the east coast[6] than on the west coast.

In *January* during the southern summer the ITC loops far to the south over Southern Africa, the rainbearing winds in the south and east being the south-east and north-east monsoons respectively. Where these two air streams converge over Central Africa heavy rain is caused. In West Africa the ITC parallels the coast, and dry Saharan air flows southwards almost to

[5] Although T. Bergeron prefers to use the old term 'tropical front' rather than intertropical front, the problem of the existence of frontal discontinuities in the equatorial area is being debated actively, and it is agreed generally that frontal factors are of great importance in the tropics.

[6] Local variations due to relief and the presence of large lakes in East Africa cause significant changes in the behaviour of the intertropical fronts in this region.

the coast causing dry conditions over most of the region. In July with the northward movement of the sun the ITC has retreated wholly north of the equator and is situated at about 18° N. in an east–west direction across the Sahara. The great mass of maritime air in the southern hemisphere moves towards the ITC, changing its direction from south-east to south-west as it crosses the equator and causing a belt of heavy rainfall in West Africa and the Congo basin. The southern half of Africa is dry at this period except for the south-eastern tip of South Africa which has a Mediterranean type of climate. There are wide seasonal fluctuations in the behaviour and location of the intertropical front and a great deal depends on the depth of the air mass and the dimensions of the upper air.

Other factors affecting African climates are the influence of relief, latitude, ocean currents, distance from the sea and nature of vegetation. The effects of altitude and latitude on temperatures are well illustrated by the following figures of mean annual temperatures for Pretoria, Bulawayo, Nairobi, Kayes, Mombasa and Lagos.

	Pretoria	Bula-wayo	Nairobi	Kayes	Mom-basa	Lagos
Latitude	24° 45′ S.	20° 10′ S.	1° 16′ S.	14°24′N.	4° 2′ S.	6° 26′ N.
Altitude (feet)	4,471	4,470	5,450	197	50	25
Mean Annual Temperature ° F. (° C.)	63·5 (17·5)	66 (19)	63·2 (17·4)	84·9 (29·5)	78·5 (25·8)	80·5 (27)
Temperature Range ° F. (° C.)	20 (11)	15 (8·3)	6·7 (3·6)	19·2 (10·6)	6·5 (3·5)	5·6 (3)

The highest temperatures as shown by Kayes, with average temperatures for April and May of 94° and 96° F. (34° and 35° C.), occur in the dry cloudless areas of tropical Africa. The effect of *altitude* is clearly illustrated by temperatures for Nairobi and Mombasa, the latter being 16° F. (9° C.) warmer than Nairobi whose average temperature for the coolest month is 59° F. (15° C.) and which is 5,000 feet higher in altitude. Bulawayo and

Pretoria 20° and 24° S. illustrate the greater range of temperature with increasing latitude. The range of temperature is also affected by factors such as aspect, rainfall, relative humidity and distance from the sea. Therefore a station such as Lagos on the west coast of Africa, $6\frac{1}{2}$° north, has a mean annual range of only 5°–6° F. (3° C.) compared to a range of over 30° F. (16°–17° C.) in the Sahara, where maximum insolation by day and a very dry atmosphere, lack of cloud cover and great radiation by night cause a wide range of temperature.

The sharp *relief* of the tropics causes high *rainfall*, particularly where coastal hills and mountains lie athwart the rain-bearing winds, as in Sierra Leone and on Mount Cameroon where Debundscha receives nearly 400 inches of rainfall per annum. The rate of increase of rainfall with altitude may be high and over 10 inches per 100 feet compared to 1–3 inches in Britain. The leeward slopes of a mountain affected by subsiding air experience a considerable rain shadow, as for example in East Africa where the southern slopes of Mount Kilimanjaro receive over 60 inches of rainfall per annum and the northern leeward slopes less than 20 inches. On a major scale the effect is also marked in South Africa where the Great Escarpment may receive up to 80 inches and the adjacent plateau slopes only 30 inches; similarly the seaward edges of the Cape mountains over a 100 inches and the intermontane valleys less than 10 inches.

The link between tropical *vegetation* and climate is also of considerable interest, although there is little accurate information as to the direct effect of forest on rainfall. It is possible, however, that a coastal forest of up to 200 feet high would cause by its height and the turbulence it produces a considerable increase in rainfall. And equally well transpiration over a large area of equatorial rainforest will cause a high humidity and therefore high moisture content; for example, Atlantic air masses arriving in East Africa after travelling over the forests of the Congo basin have a very high humidity. In the United States of America Tannehill[7] thinks that the destruction of the vegetation cover has had an aggravating effect on the hot winds of the plains and has therefore produced a measurable deterioration in the climate of North America. In South Africa Brooks[7] thinks that, although no proof can be established of a reduction in rainfall, the destruction of the veld cover and appearance of large areas of bare soil has increased temperature contrasts and that the rain now falls in heavy instability showers and thunderstorms. The apparent southward spread of the Sahara in North Africa and the rapid encroachment of desert and Karoid vegetation in parts of South Africa give further emphasis to the problem as to whether

[7] *Veldtrust*, March 1955, p. 21.

man or changing climate has been the major factor in the alteration of vegetation in parts of Africa, a point that will be referred to again.

The climatic effect of *ocean currents* is significant. Near the equator warm currents flowing from east to west cause warm water on the western sides of oceans, whereas nearer the tropics relatively warm waters move polewards in the western parts of oceans and cool currents flow towards the equator on the eastern sides. Cool currents, the Canaries and Benguela, flowing equatorwards off the west coasts of North and South Africa respectively cause cool summer temperatures and foggy conditions along the coast. The cool water causes sea breezes to blow fog and mist inland. This fog forms easily over the sea, but is rapidly dispersed over the dry land. These areas receive very little rainfall. Warm ocean currents off the Guinea and south-east African coasts conversely cause high temperatures and comparatively heavy rainfall. Climatic statistics for Durban and Port Nolloth in South Africa illustrate these points.

	Jan.	Feb.	Mar.	Apr.	May	June	July	Aug.
Port Nolloth ° F. (° C.)	59·5 (15·3)	59·9 (15·5)	59·3 (15·2)	57·7 (14·4)	56·8 (13·9)	55·4 (12·9)	55·2 (12·8)	53·8 (12·2)
Durban ° F. (° C.)	76·3 (24·7)	76·8 (24·9)	74·8 (24)	71·8 (22·2)	67·8 (19·9)	64·8 (18·1)	64·3 (17·9)	65·8 (18·9)

	Sept.	Oct.	Nov.	Dec.	Year	Range	Rain-fall
Port Nolloth ° F. (° C.)	55·0 (12·7)	58·1 (14·5)	59·0 (15)	60·3 (15·6)	57·6 (14·3)	6·5 (3·5)	2·3
Durban ° F. (° C.)	67·6 (19·8)	69·5 (20·8)	72·0 (22·3)	74·6 (23·7)	70·5 (21·4)	12·5 (6·8)	42·7

Rainfall is of vital importance in any tropical region and especially in Africa which has nearly all of its great area in the tropics. The significance

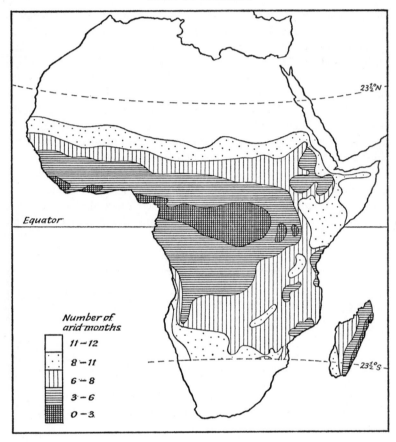

Number of
arid months

	11 – 12
	8 – 11
	6 – 8
	3 – 6
	0 – 3

FIG. 5—Length of dry season in intertropical Africa

of rainfall has long been recognized in African society and the rain maker
is an important figure. Of more interest to the African cultivator than the
total amount of rainfall is the time and onset of the first ploughing rains,
whether this season's rainfall will be spread out sufficiently to produce a
good harvest, or whether the rains will fail and drought cause destruction
of crops and animals. Annual and seasonal totals of rainfall are very variable
over most of Africa and are most marked where rainfall is low.

Tropical rainstorms are intense and rates of two inches an hour are not
unusual, so that most of the rainfall of tropical regions comes from a few
of these storms. The force and weight of a tropical deluge is impressive
and where soil is unprotected on a steep slope it is torn away and gullies

appear quickly. The number of rain days in the tropics is thus far less than in Britain; for example, totals of over 50 inches of rain fall only on about 100 days per annum in the Guinea zone compared to 200 days or more in Britain. The concentration of heavy storms in both period and place causes great variability of rainfall in time and amount and in annual and monthly totals. In semi-arid areas the variability may be up to 40%. Over most of the tropics droughts may occur in nearly any month. Over 90% of Africa has a dry season, the length of which increases with increasing distance from the equator, particularly in the northern hemisphere. The map (Fig. 5) showing the *length of dry season* is of major importance in understanding the distribution of vegetation, crops, livestock and man. One of Africa's most serious problems is lack of water during the dry season which may last over nine months.

It is essential to be able to forecast possible variations in rainfall with some degree of accuracy. Maps in the 1953–5 East African Royal Commission Report, which depict the percentage prospects of obtaining 20 inches and 30 inches of rainfall, indicate that most of East Africa is too marginal for crop production. The agronomist in Africa is more interested in water availability than amount. Evapotranspiration studies have been carried out in a number of countries by applying Thornthwaite's formula and it has been found that over most of Africa potential evapotranspiration exceeds the average annual rainfall.

Many attempts have been made to classify tropical climates. The Greeks selected latitude as their basis for dividing the world into climatic zones and Supan at the end of the nineteenth century adopted the annual isotherm of 68° F. (20° C.) as the outer limit of the tropics. This figure of 68° F. (20° C.) was also chosen by Köppen, but Austin Miller favours 70° F. (21° C.). Others have used a variety of geographical factors to define climatic regions. Herbertson, for example, combined configuration, climate and vegetation and Köppen used climate and vegetation together to draw up his major climatic types. Thornthwaite makes use of indices such as temperature and precipitation effectiveness. Making use of rainfall as a major criterion for climatic classification in the sub-humid and humid tropics, the majority of climatologists follow Köppen's divisions based on the main rainfall of the driest months: Af—tropical rain-forest, with no month below 2·4 inches mean rainfall; Aw—tropical savanna, with one or more months below 2·4 inches mean rainfall; Am—intermediate and often monsoonal, where a large annual rainfall can offset a dry period.

The main climatic types

The following paragraphs contain a brief description of the *main climatic types* in Africa (Fig. 6). The first is the *equatorial* type of climate which is found in a narrow strip on parts of the West African coast and the northern Congo basin. It is characterized by continuous heat, high humidity and well distributed rainfall. Mean monthly temperatures are usually 76°–80° F. (24°–27° C.) with very slight seasonal variation, the mean monthly temperature range being 5°–6° F. (3° C.) or less. The diurnal variation is greater and may be up to 15° F. (8° C.). Temperatures are not high, however, compared to semi-arid regions where temperatures of 110° F. (43° C.) are common. Monotonous temperatures, lack of air movement and high humidity cause sultry, enervating conditions. A narrow coastal strip experiences land and sea breezes and these are physiologically most beneficial as they reduce the sensible temperature considerably.

The small strips of equatorial type climate along the coast of West Africa have a monsoonal component in the rainfall, with a dry spell from December to February, when the region tends to be dominated by dry continental air. For the rest of the year tropical maritime air tends to control the coastal climate. The highest rainfall occurs where the coast is at right angles to the south-west winds and backed by high relief as in Liberia and the Cameroons. The Congo basin is more equatorial in location and more sheltered and tends to have a more equatorial climate than that of West Africa. As in parts of West Africa, there is a double maximum of rainfall which in the Congo occurs in April and October after the equinox. Mary Kingsley, who visited West Africa in 1893, describes graphically the intensity of equatorial rainfall: 'After we passed Cape Verde we ran into the West African wet season rain sheet. There ought to be some other word than rain for that sort of thing. . . . For weeks it came down on us that voyage in one swishing, rushing cataract of water.'

Tropical continental climates almost surround the equatorial climates and are characterized by their increasing seasonality of rainfall and increasing range of temperatures. Rainfall varies considerably from 60 inches or more on the equatorial margin to less than 20 inches on the edge of the steppe. There are marked seasons, a hot wet season with a rainfall maximum which occurs during the period of high sun and a cooler dry season when the sun is in the opposite hemisphere. The length of the dry season increases away from the equatorial margin and in northern Nigeria is about eight months long. Variability of rainfall is considerable, especially in East Africa, where in the drier part it approaches 100%.

The tropical continental climate is best developed in the great *Sudan*

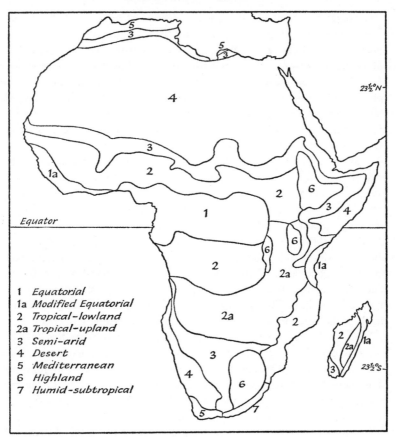

1 Equatorial
1a Modified Equatorial
2 Tropical-lowland
2a Tropical-upland
3 Semi-arid
4 Desert
5 Mediterranean
6 Highland
7 Humid-subtropical

FIG. 6—Climatic regions

belt north of the equator. Maximum rainfall occurs from May to October, when the belt of equatorial low pressure has moved northwards, the rest of the year being dry owing to stable, descending air associated with high pressure. Temperatures are high for most of the year and increase near the desert margin, Timbuktu and Kayes both having mean annual temperatures of nearly 85° F. (29° C.). Contrasts between the dry and wet seasons are great and Karl Kumm who crossed the Sudan in the rainy season in the early years of this century writes feelingly:

England is bad enough in the rains, but Central Africa in the rainy season is, I am tempted to say, beastly. . . . Africa is a land of extremes. In some places there is too much water and in others too little; it boasts the biggest swamp and the greatest desert. At one period of the year people seldom get dry, and at another time they die of thirst.

During the cool dry season the grass decays and the rivers dry up. With increasing heat and lack of moisture some of the deciduous trees and shrubs drop their leaves. But with the return of the rainfall in April the temperatures become lower, the vegetation revives and the fire-blackened, dusty trees put on fresh leaves and the grass springs up to a height of six feet and more. Rivers flow again and the hoes and mattocks break up the ground for replanting.

Near the equator the high plateau modifies temperatures considerably and the continental climate is generally far cooler than in the Sudan, as average annual temperatures show for Khartoum at $15\frac{1}{2}°$ N. with 83° F. (28° C.), and at Broken Hill $14\frac{1}{2}°$ S. with 68° F. (20° C.), but at an altitude of nearly 4,000 feet. East and south-east Africa have a great diversity of climate as they cover a latitudinal range of nearly 40° F. (22° C.) and altitude varies greatly from the coastal plain to the plateau at 4,000 feet, and the highlands of Kenya and Ethiopia which exceed 10,000 feet in places, and culminate in mountains more than 15,000 feet high and reaching the snow line. The southern subcontinent is also far narrower than north of the equator and therefore more open to oceanic influences.

Most of the plateau has a seasonal rainfall varying from less than 20 inches to over 60 inches, except for a narrow strip along the equator which has a double rainfall maximum with long rains from March to June and short rains from November to December. Elsewhere the bulk of the rain comes in summer; for example, in Rhodesia May to October is dry and only about 3% of the annual total occurs then.

Winter is a very pleasant period on the plateau of central Africa with maximum sun and cool nights (frost is not uncommon in the higher areas), low humidity and lack of rainfall. Sir Harry Johnstone in his *British Central Africa*[8] writes:

But the year is clearly divided into seasons of rain and drought. The rainy season generally begins at the end of the month of November and heavy rains fall in December. . . . April is a delightful month as it is in Europe, of alternate showers and sunshine. A little rain falls in May and an occasional shower in June. July is the height of the winter—cold, dry, sparkling. . . . October is quite the driest month and in the lowlands passes without a drop of rain, though in the highlands there may be an occasional thunderstorm. Towards the close of November (the first half being terribly hot and dry) the big rains recommence.

[8] London, Methuen, 1899.

Speke in his journey across East Africa experienced the cool temperatures and winds of the dry season:

> in the drier season they blow so cold that the sun's heat is not distressing; and in consequence of this, and the average altitude of the plateau, which is 3,000 feet, the general temperature of the atmosphere is very pleasant, as I found from experience; for I walked every inch of the journey dressed in thick woollen clothes and slept every night between blankets.

From October onwards maritime air masses invade the subcontinent, moving towards the low pressure of the south-migrating sun, and the period of maximum rainfall occurs in the southern summer. Rainfall tends to decrease steadily to the south and west, and on the west coast of Angola the low rainfall of less than 10 inches is caused partly by the northward-flowing Benguela current. Interior Angola, the northern part of Zambia, Rhodesia and Malawi receive abundant summer rain. Temperatures are higher on the plateau slopes, the Zambezi Valley and Mozambique coast and rainfall tends to be lower in the Zambezi Valley and other rain-shadow areas. The effect of the warm Mozambique current is seen in the average temperatures of 83° F. (28° C.) for November and December in the port of Mozambique compared to 73° F. (23° C.) for those months at Zomba in Malawi at an altitude of 3,200 feet.

Dry climates are difficult to define as aridity is a product of many factors such as high evaporation, high temperatures, low humidity, strong winds, type of soil and lack of vegetation cover. Because of the lack of water in a dry climate streams cannot originate in such an area, although permanent rivers such as the Nile, Niger and Orange, whose sources are in more humid regions, may cross a desert. According to Köppen's classification the boundary between humid and dry climates is the mean annual isohyet of $13\frac{1}{2}$ inches at 50° F. (10° C.), 22·3 inches at 70° F. (21° C.) and 26·7 inches at 80° F. (27° C.). It has been suggested that the northern boundary of the Sahara should be taken as 10 inches of winter rainfall, while the southern boundary in the hotter Sahel belt is taken as 15 inches of summer rainfall. Arid climates may be subdivided into desert and steppe, the boundary between them being arbitrary and according to Köppen should be half the amount of rainfall separating steppe from humid climates. Thornthwaite defines the desert boundary by a Precipitation Effectiveness Index of 16.

Towards the desert margin in the Sahara, the Namib and the Southern Kalahari and in East Africa a narrow belt of *tropical steppe climate* is found with high temperatures and 10 to 20 inches of rainfall which is extremely variable in amount and seasonal distribution. Rainfall is negligible in amount in the Sahara, many areas recording no rain at all in successive years.[9] Temperatures are also extreme, maximum temperatures being over 130° F. (54° C.) on occasions; much of the Sahara has a July mean of over 95° F. (35° C.), for example In-Salah with a July mean of 99·3° F. (37·5° C.) and no annual rainfall, compared to Timbuktu with a steppe climate and a July mean of 89·2° F. (32° C.) and 9 inches average annual rainfall. Minimum temperatures, however, may be below freezing point in winter, and stations such as Touggourt and Biskra on the northern fringe of the Sahara record more days with frost than do stations on the Mediterranean coast of Algeria 4° further north. Temperatures are much lower in the Namib which is affected by the cold Benguela current. Humidity is very low and sunshine at a maximum, the Eastern Sahara receiving 90% of possible sunshine each year.

Sand storms often occur and are extremely unpleasant. Denham encountered such a storm in crossing the Sahara:

the wind raised the fine sand with which the extensive desert was covered, so as to fill the atmosphere, and render the immense space before us impenetrable to the eye beyond a few yards. The summer clouds were entirely obscured and a suffocating and oppressive weight accompanied the flakes and masses of sand, which, I had almost said, we had to penetrate at every step.

The highland areas of East Africa above 8,000 feet and the high veld of Southern Africa above 4,500 feet have a *temperate* rather than tropical climate. Frost may occur for three to six months of the year on the high plateau of Southern Africa. Night minimum temperatures are low, for example Hanover at an altitude of 4,500 feet has a mean temperature of 42·9° F. (6° C.) for June and July. Rainfall is over 20 inches per annum in the east, but decreases rapidly to ten inches and less in the rain shadow areas of the south and west.

A narrow coastal strip in Natal and the south-eastern Cape has a *humid subtropical climate* with most rainfall in summer, but some in winter because of the eastward extension of Mediterranean climatic influence; Port Elizabeth, for example, has 1·7 and 1·9 inches in June and July whereas Pretoria has 0·2 and 0·1 inches in the same period. Along the coast temperatures are high and the annual range is low; Durban has an average annual temperature of 70·5° F. (21·4° C.) and a mean annual range of only 12·5° F. (6·8° C.) compared to a range of 26·5° F. (14·5° C.) at Hanover.

---

[9] The highland areas of Ahaggar and Tibesti may receive up to 10 inches rainfall a year, and streams flow seasonally.

The north-western and southern fringes of the continent have a Mediterranean type of climate, which is much restricted in range in the south-western Cape compared to North Africa. The climate is markedly seasonal, the rainfall maximum occurs in winter, brought by westerly depressions which cause a mild wet season, followed by a hot and dry summer when anticyclonic conditions prevail. Rainfall varies from about 10 inches to over 30 inches in North Africa, while in the mountains of the Cape Province rainfall may be over 100 inches per annum. Temperatures vary considerably according to location and the effect of cold currents; the Atlantic coast of Morocco affected by the cold Canaries current has winter temperatures which are lower than in Tunisia, for Mogador has a July mean of 68·2° F. (20° C.) and Tunis 77·7° F. (25·4° C.). The Mediterranean type of climate has long been renowned for its pleasant cool season weather and high degree of sunshine throughout the year. Sir Joseph Hooker, who visited Morocco in the spring of 1877, writes that 'nothing could be more agreeable than the climate of this season, the thermometer in the shade during the day varying from 60° to 66° F., and the air being delightfully clean and bracing'.

Climate and Man

The dire effects of climate on man in tropical Africa have been strongly stated in the past. Miss Semple wrongly says of South Africa that climatic monotony, operating alone, would have reduced South Africa to poverty of development and will unquestionably always avail to impoverish her national life. Minerals rather than climate, however, seem to have been the key to economic development in South Africa. Murray writing of tropical Africa before the Great War says, 'depression, moodiness, and acerbity of temper so characteristic of the tropics is mainly owing to the enervating influence of heat and fever'. Elsewhere the reader is warned both of the danger of standing still on certain soils which by overheating the feet may cause sunstroke and also not to risk exposure of the head to the vertical rays of the sun, so that frying betwixt earth and sky there seems no hope for the venturer into Africa's tropical hell. These examples illustrate the fallacy of prejudging an issue on insufficient evidence. Lord Hailey in his revised version of an African Survey is far more cautious and says that there are no simple or satisfactory standards by which tolerability of climate in Africa can be assessed. The third generation of white settlers in Kenya and Rhodesia seem to have suffered no ill effects from tropical climates. Conditions are admittedly more beneficial for them at high altitudes, but sojourn in the hot and humid lowlands of West Africa is

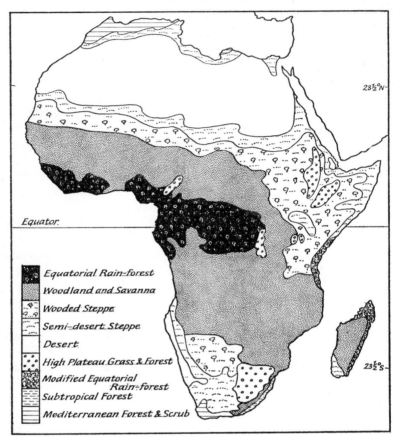

Equator.

Equatorial Rain-forest
Woodland and Savanna
Wooded Steppe
Semi-desert Steppe
Desert
High Plateau Grass & Forest
Modified Equatorial Rain-forest
Subtropical Forest
Mediterranean Forest & Scrub

$23\frac{1}{2}°N$

$23\frac{1}{2}°S$

FIG. 7—Vegetation

now no more arduous for a white person than in a temperate climate. Disease rather than climate has been man's enemy in the tropics.

VEGETATION

The generalized vegetation map of Africa (Fig. 7) reflects clearly in its broad outlines the effects of factors such as climate, relief, soil and man. Because of the depredations of slash-and-burn techniques of shifting cultivation and continuous burning of the vegetation for thousands of years by the nomadic pastoralist, the area of forest and woodland has been decreasing and savanna and steppe increasing in Africa over a very long period. The belt of *tropical rain-forest*, moist tropical forest, occurs discontinuously in equatorial latitudes with abundant precipitation and constantly high temperatures throughout the year, the greatest area being in the northern Congo Basin. This type of forest is evergreen and has several definite layers, the lower layers having abundant woody climbing plants.

It is difficult to describe accurately the great belt of vegetation in the seasonal rainfall zone that almost surrounds the rain-forest. A forest-savanna mosaic, or belt of derived savanna, is found bordering the moist tropical forest. Various types of drier and more open woodland and forest are characteristic of this broad zone, reflecting a dry season of varying length from three to over six months during which the vegetation is fired to burn off the dead grass of the dry season and discourage bush encroachment. The greatest extent of *dry forest and grass* forming a woodland-savanna mosaic is to be found south of the equator. On the plateaux of East and South-east Africa there are wide areas of *Brachystegia-isoberlinia*, or miombo woodland, with trees from 30 to 45 feet high, and short tussocky grass cover. The area of dry or Guinea forest is far smaller in West Africa owing to the higher density of population and greater prevalence of burning. *Brachystegia*, common to East Africa, is not found in West Africa. Trees are gnarled and fire-resistant and the grass cover is tall, grasses being up to 10 feet high and coarse; but, with a closed canopy of trees, grass is absent.

Where rainfall decreases below 30 inches and the dry season is six to seven months in duration, the grass cover becomes more continuous and trees more dispersed, as in East Africa and the Sudan belt of West Africa. In the *Sudan savanna* zone of West Africa grasses tend to be shorter and less coarse and therefore more useful for grazing than in parts of the Guinea zone further south. With a dry season of seven to eight months and rainfall below 20 to 25 inches a thorny scrub of semi-desert *steppe* replaces the woodland savanna. Conditions are too dry for the majority of perennial grasses and are somewhat inimical to tree growth so that acacias have developed protective devices such as thin leaves, thick bark and deep roots so as to conserve moisture. Characteristic of this steppe vegetation is the Karoo scrub of South Africa which is rapidly advancing into the temperate grasslands of the high veld. The long narrow zones of steppe or Sahel grade off gradually into the deserts, most of which carry some type of vegetation for a short period after rain. Ephemeral grasses and widely scattered patches of scrub are the only types of vegetation that can exist in this marginal environment.

In the relatively small areas of Africa that are above 5,000 feet in altitude *mountain forests and grassland* are to be found; the latter are most extensive on the high plateaux of south and east Africa. These grasslands probably do not represent a fire climax and are thus more temperate in character than the tropical grasslands. At the north-western extremities of the continent there are small areas of *Mediterranean type vegetation* with evergreen trees and shrubs adapted to a seasonal climate of warm dry savannas and wet winters. Evergreen forest is found in the Eastern Cape and Natal provinces of South Africa, while in North Africa there are scattered areas of cork-oak forest.

Forest and woodland were probably natural climaxes over much of Africa that is now savanna, in areas of over 30 inches rainfall per annum and a wet season of six months or more. Thousands of years of grass burning, with its seasonal swathes of fire-blackened vegetation, has cut into and reduced the area of forest drastically. Many African governments are attempting actively to halt the retreat of the forests and conserve the vegetation cover. When the tree cover is removed in areas of high rainfall the steep slopes are quickly stripped of their soil and the rivers develop an irregular régime with alternate flooding and long periods of little or no water. When cultivation is halted temporarily under a bush-fallow system there is regeneration of woody climbers followed by regrowth of tree stumps and then the seedlings of short-lived trees. But the succession is usually halted by cultivation before secondary forest can be re-established. Under a grass fallow in the derived-savanna zone, grasses such as *Imperata cylindrica* invade rapidly with regrowth. In areas of sandy soils a grass climax tends to replace a forest climax, as the sandy soils are less able than the heavier types to retain the water needed by forest trees. It is not a deterioration of climate but man's misuse of the soil that is primarily responsible for the apparent southward movement of the desert in West Africa and encroachment of Karoid types of vegetation in the grasslands of South Africa. All over Africa the cultivator demands more land and the pastoralist easy grazing and so through slash-and-burn techniques the area of forest and woodland is reduced each year.

The products of gathering and collecting

The great majority of Africa's population is dependent on agriculture for a living. Over much of Africa pastoralism is impracticable because of pests and diseases and the nature of the climate and vegetation, so that diets are mainly vegetarian and vegetation is of vital concern. Meat and fish are comparatively rare and milk seldom used. The African bush supplies a wide variety of roots, fruits, seeds and leaves which form an indispensable item in the peasant's diet. In Ghana, for example, there are

100 edible fruits and many different varieties of leguminous seeds and leaves are eaten. It is particularly during the hungry and dry season that wild roots, fruits and insects are appreciated and gathered daily by the women.

The main food is a kind of porridge or potage, pounded maize or cassava being common ingredients, enlivened by a spicy sauce or relish made from a variety of fruits, vegetables, seeds and leaves. In West Africa tomatoes, groundnuts, peppers, chillies and leaves from the baobab and shea butter tree are commonly used. The usual African diet is low in calories and quite inadequate at times, especially towards the end of the dry season. An ill-balanced diet lacking in proteins and vitamins causes malnutrition and inability to perform the heavy labour needed in tropical agriculture. However, the lack of a proper diet is due as much to custom and habit as to the poverty of the environment.

Because of the lack of fodder crops and animal feeding stuffs the grasses and bush of Africa provide the only source of forage for livestock. The pastoralist is thus vitally concerned with the amount of grass available and desirous of obtaining as early a flush of new grass as possible during the dry season when animals tend to avoid the tall tough grass. He is also concerned in preventing the spread of woodland which restricts the grazing available for his animals. The only weapon available is fire, and so each season the vegetation is fired and a pall of smoke settles over Africa.

Besides being an indispensable source of food and fodder Africa's forests and bush also supply a variety of other necessities such as clothing, shelter and economic products such as timber, cocoa, rubber and other tree crops. The oil palm (*Elais guineensis*) of Eastern Nigeria, whose natural habitat is western and central Africa, is the very staff of life to the majority of the people living in that region. Oil is obtained from the fibrous pericarp and is used for a number of purposes, including the preparation of food, while palm kernels are either exported or crushed locally to obtain oil for the margarine and soap industries. The sap is used for preparing palm wine, a heady brew that is drunk widely. The fibre is used for lighting fires, for paint brushes or calabash mending, the fronds for rafters and hut building, and the wood of the tree for many purposes. The Bemba of north-eastern Zambia use the ash from burnt trees and branches for enriching the soil; the bark forms ropes used for nets and snares or lashing building poles, while the trees supply material for huts, granaries, fences, beds, stools, churns and canoes.

The rain forests of West Africa contain hundreds of species of trees, but only about 20 of these are valuable for commercial purposes, the most important varieties being obeche, ebony, sapele, wawa and various types of mahogany. The *timber industry* of West Africa has been the most successfully developed of the various African timber-producing areas, because of proximity to Europe and the fact that the forests border the coast and are fairly accessible. The local market may be more important than the export market, for timber is used widely in the building industry and for firewood, because of the lack of other means of heating in many parts of Africa. Cutch, a tanning extract used for fishing nets and sails and obtained from mangrove bark, is produced in East Africa. Fibres are obtained from the raffia palm, oil palm, coconut and baobab and made into mats and ropes. Gum arabic is collected extensively in the Sudan from a number of different types of acacia trees. Both palm oil and rubber are obtained from wild and not plantation sources, although the percentage of palm oil from commercial plantations is increasing and most African rubber comes from *Hevea brasiliensis*, grown in Liberia and Nigeria.

A wide variety of trees and shrubs are used by native herbalists for medicinal purposes.

SOILS

Until recently knowledge of African soils (Fig. 8) was sketchy but a considerable amount of information has been amassed through the work of pedologists and soil scientists in recent years. There is a wide variation in the character of tropical soils, which are very different from temperate soils. It is increasingly apparent that most of Africa's soils are poor in quality, in fact far poorer than most temperate soils. The myth dies hard of tropical Africa's rich soils as supposedly evidenced by the luxuriant vegetation. However, the low yields of tropical agriculture are caused by backward farming methods as much as by poverty of soil. Although climate is of considerable importance in soil formation there are other factors as well. There is a close relationship between soil formation, the development of the landscape and systems of land use over considerable areas of tropical Africa. In East Africa, Milne has emphasized the fact that many soils have been formed over a very long period under past rather than present climatic conditions. Tropical soils are poor in mineral nutrients and humus; their chemical importance lies in the upper organic layers.

Mineral bases are quickly leached by heavy tropical rainfall aided by high temperatures and acidity, and are quickly lost as the organic layer is so thin. The superficial layer of humus is quickly decomposed by insects and micro-organisms. Generally, all that is left after humus and minerals have been removed are sand and clay particles. The sand consists of hard

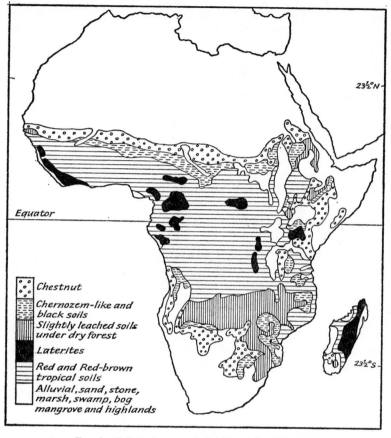

Fig. 8—Soils in intertropical Africa (after Kimble)

quartzite particles which tend to compact and develop a hard cement-like surface, while the clay degenerates into laterite, a dark brown or red brick-like layer that is completely infertile.

This hard compacted layer, known as *laterite*, occurs in hot tropical regions with high but seasonal rainfall, the depth at which laterization occurs tending to decrease with increasing seasonality. There is much speculation as to the cause of this layer of hard-pan but it may be caused by seasonal variations of rainfall and dry conditions. Laterite is characteristically found in level areas on plateaux or peneplanes, but may be transported and redeposited in undulating areas. The incidence of laterite in tropical soils may become serious, for example in Madagascar where it occupies most of the island. Once the thin top soil is removed in a savanna zone and extensive sheets of laterite exposed, cultivation is virtually impossible as laterite is a rocklike soil, and is of far more use in building or road construction than in agriculture.

The climactic soils of the humid tropical region with a seasonal climate and a rainfall of 60–100 inches a year are the ferralitic soils. These are strongly leached yellow-brown and red soils which have developed on crystalline and miscellaneous rocks, and are important as the zone of tropical tree crops such as coffee, cacao, bananas, oil palms and rubber. As these crops require forest conditions it is essential to maintain a tree cover. When the forest is completely cleared the shallow surface layer of soil tends to deteriorate rapidly and as the economy of coastal West Africa depends to a major extent on valuable tree crops such as coffee and cacao further indiscriminate clearance of the coastal forests will affect the economy adversely.

Brown tropical soils in areas with a rainfall of 40–60 inches, which are developed on crystalline rocks or volcanic ash, are often rich in organic matter and have a high mineral content. These fertile soils are formed on the plateaux and highlands of south-west Cameroon, East Africa and Ethiopia, and are used for the cultivation of many different kinds of cash crops including coffee, cacao, bananas, tea, pyrethrum, vegetables and food crops. Where well-managed, they are among the most productive in the tropics and maintain a dense population and zones of intensive cropping.

Ferruginous tropical soils are found fringing West Africa, in the Sudan, and south of the Congo Basin in Angola, East Africa, Mozambique and Malagasy. These soils often have a considerable mineral reserve but they are not very stable and are very vulnerable to erosion. They are quite deep in places but their structure breaks down fairly quickly when cultivated; therefore, as with the ferralitic soils, a vegetation cover should be retained as far as possible. These are the characteristic soils where bush fallowing is practised as a long established method of maintaining soil fertility. Alternative systems such as rotational cropping, increased use of fertilizers and mixed farming are essential to obviate the long fallow period and enable intensive farming to be carried on without a decline in soil fertility.

In the drier regions with a rainfall of 12–25 inches brown to red tropical soils are found. These soils are similar to the American brown and red soils and have some value for pastoralism and irrigation farming. Crops such as cotton and sorghum are typical of this zone. The structure of the soil tends to deteriorate quickly if irrigation is excessive or the permeability of the soil inadequate.

Reasonably fertile *red soils* have developed under a grass cover on the high veld of South Africa with a seasonal rainfall of over 20 inches per annum. These soils are used extensively for commercial cultivation of maize. *Red loams* have also developed on the terraced forelands of Natal, where there is a higher rainfall, and these are used for maize cultivation, dairying and the growing of wattle. The *loams and sandy loams* of the areas of Mediterranean rainfall in the south-western Cape tend to be poor in quality, shallow and with a low mineral content. They are used for wheat, vines and fruit. Wheat monoculture over a long period has further impoverished these soils. In North Africa brown-earths and rendzinas are found, the latter forming a dark loamy soil in limestone areas, and used extensively for the cultivation of vines and olives.

CHAPTER THREE

Economic and Social Geography

ECONOMIC GEOGRAPHY

THERE are great social, economic and political contrasts in Africa, with a very wide range in the social and economic scale from the Bushman and Pygmy hunters of the Kalahari and Congo forests to the nomadic Masai and Fulani pastoralists of East and West Africa, and from the sedentary Yoruba and Kikuyu cultivators to the sophisticated urban dwellers of towns and cities such as Cairo, Kinshasa, Ibadan, Nairobi and Johannesburg. Skyscrapers and crude shacks are found side by side in the cities, while in mining areas African tributors dig and pan tin and gold using age-old methods next to the most modern draglines that tear away the overburden to get at the mineral-bearing rock beneath. In agricultural areas plantations use the latest scientific methods to produce tea, pyrethrum, sisal and other crops, surrounded by a host of African peasant farmers practising a primitive form of polyculture. Economic development is very patchy and there are clusters of intense economic activity surrounded by vast areas of self-sufficiency (Fig. 9).

Throughout the continent the old ways of life are changing very rapidly under the impact of new ideas; the *communal spirit of the tribe* with its self-sufficient economy is decaying and being replaced by economic individualism, the growing of cash crops, mining, the rise of towns and new methods of communication by road, rail and air. A political, social and economic revolution has occurred during the last 25 years; from 1938 to 1956 the value of exports from tropical Africa increased nearly ten times, the colonial régime has largely disappeared and independent African countries are striving to overcome the problems caused by this revolution, problems such as the increasing pressure of population on the land and the difficulty of increasing productivity in a continent where the great bulk of the people depend on agriculture for a livelihood. The economy of Africa in the past was based on subsistence agriculture and self-sufficiency over vast areas, apart from the export of slaves, gold and a few other commodities. There was little difference in individual economic status between members of the group for wealth, in the form of cattle or crops, was usually confined to

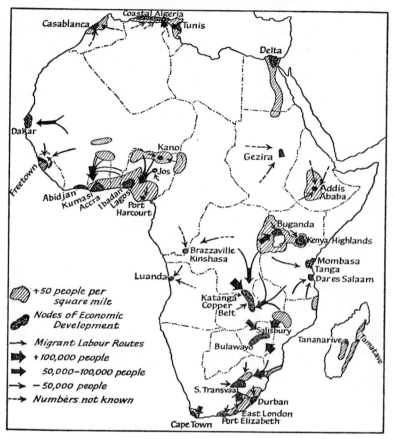

FIG. 9—Labour migration

the chief who was expected to entertain strangers, feed members of the tribes who harvested or helped in his fields and aid those in distress, much as the Saxon earl kept open house in his timber hall. Individual wealth was not encouraged or possible in a tribal society. This feature of communal help is still important in the family or kin group, those with money or position being expected to help less fortunate members of the group.

The effects of the *European exchange economy*, with its stress on individual enterprise, the clash of competing firms and swing of fluctuating prices, were seen at their worst during the great depression of the early 1930s when producers of tropical crops such as cocoa and coffee suffered great economic hardship. This type of capitalism is obviously alien to the old system of communal effort and subsistence economy and helps to explain the trend towards socialism in a country like Ghana. The state replaces the tribe, the president the paramount chief and communal endeavours by the citizens of the state are extolled, while the commercial operations of large individual firms are criticized.

Until the late nineteenth century European interest in Africa was largely confined to the coast and the slave trade, particularly in West Africa and Angola. Knowledge of the interior was very sketchy until the work of travellers and explorers like Mungo Park, Barth, Speke, Burton, Livingstone, Stanley and de Brazza opened up vast areas to European enterprise. The abolition of slavery was slowly achieved by the end of the nineteenth century and replaced by commercial activities. With the scramble for Africa and the division of the continent first into spheres of influence and later into colonial territories the indigenous inhabitants were soon affected by European economic methods. Each territory tended to develop on different economic and political lines according to the policy of its colonial ruler. In West Africa most of the agricultural economic development was undertaken by Africans who had produced oil palm products for much of the nineteenth century and cocoa in the Gold Coast at the end of the century. In contrast, Africans played a relatively minor role in East and Southern Africa where European settlement was encouraged and cash crops such as sisal, coffee, tea and tobacco were grown on European estates. It was only in areas such as Buganda that peasant production of cash crops was significant.

MINERALS

Throughout the continent *mining* was carried on largely by European enterprise, although gold, tin, copper and iron had been mined by Africans for a very long period. It was knowledge of these mine workings in West and Southern Africa that attracted European interest from the fifteenth century onwards. Over large areas in the Basement Complex rocks of the mineralized interior, particularly in Southern Africa, the extraction of minerals has been the major spur to economic growth. The modern phase of mining development probably began with the Namaqualand copper boom of the 1850s, to be followed more spectacularly by the discovery of diamonds on the Vaal in 1867 and at Kimberley in 1870, and gold in vast amounts on the Witwatersrand in the Transvaal in 1886. *Gold* to the value of nearly £5,000 million has been recovered to date from the Southern Transvaal and Orange Free State gold fields. South Africa's economy has been based primarily on gold for most of the twentieth century and it is only recently that secondary industry has outstripped the value of gold production. *Copper* has been the king pin of Northern Rhodesia's (now Zambia) economy for thirty years, its value now being over a £100 million per annum. In the former Belgian Congo minerals have provided over half the country's revenue.

Gold has been mined in West Africa for over a thousand years and was important to Portugal in the fifteenth and sixteenth centuries. The Iron Age came to West Africa over two thousand years ago, and tin was mined on the Jos plateau long before the Europeans came. The modern European phase of mining in West Africa dates from 1878 when the banket reefs were mined for gold at Tarkwa in the Gold Coast. Later *diamonds*, manganese and bauxite were mined in the Gold Coast, tin and coal in Nigeria and iron ore, chromite and diamonds in Sierra Leone. Minerals now form a considerable proportion of the total exports from Guinea, Sierra Leone, Ghana and Nigeria. East Africa is unfortunate as it has little in the way of minerals to compare with West and Southern Africa; diamonds in Tanzania and copper in Uganda have only become important recently.[1]

Investment in mining has been of great influence on the pace and scope of economic development, particularly in determining the growth of rail and road transport and mining towns. Mining employs about a million people in Africa as a whole and is the most highly developed section of the economy. *Exports of minerals* have provided a large source of foreign earnings and a major contribution to revenue in many African countries. Since the war African production of minerals has risen and her share of world trade in antimony, copper, manganese and tungsten has increased greatly. She now produces a considerable proportion of the world's output of antimony, asbestos, chromite, cobalt, copper and phosphates. Africa still dominates world trade in gold and diamonds as illustrated in the following table showing African production of selected minerals as a percentage of world (excluding U.S.S.R.) production of minerals.

	1937	*1959*
antimony	5	53
chromite	43	43
copper	19	29
gold	47	69
manganese	24	41
vanadium	42	20
asbestos	16	21
diamonds (gem)	97	99

[1] Kilembe copper reserves are small and will only last for another few years.

This post-war expansion has been caused by many factors. With heavy capital investment as in gold mining this expansion has been caused by improved mining techniques and decline in labour force, but in activities needing little capital outlay, as with diamond mining in West Africa, the increase has been caused mainly by a rise in the labour force. Mineral production tends, however, to be concentrated in Southern Africa and South Africa alone produces over 40% of the total mineral output.

Mining more than any other economic activity has provided the initial stimulus for development in many African territories, particularly in the field of communications. Previous economic development was hampered greatly by the fact that communications were almost non-existent over most of Africa because of formidable physical obstacles[2] and lack of economic incentive.

TRANSPORT AND COMMUNICATIONS

In the fly-free areas of West and North Africa the donkey and camel and in South Africa the ox could be used as transport animals, but over vast areas of tropical Africa the spread of the *tsetse fly* precluded the use of any other form of transport than *human porterage*. This is a wasteful and uneconomic means of transport, and costs and time taken are greatly reduced when alternative methods of communication by railway or road are introduced. Nevertheless, paths are still of great importance in Africa and a great quantity of goods is carried on the head. Bicycles are used extensively and it is a common sight to see man, woman, child and luggage all perched on one bicycle. The wheel was however not used, or unknown, until the coming of the Europeans.

The *ox waggon* was of great importance in the opening up of Southern Africa. In the Great Trek of 1836 hundreds of lumbering waggons drawn by long spans of oxen moved north from the Eastern Cape onto the waving grasslands of the tsetse-free high veld. The *camel*, introduced to North Africa some 2,000 years ago, was essential in the opening up of the trans-Sahara caravan trade. The rise of the slave and ivory trades in parts of East Africa occurred at the same time, as slaves acted as human caravans and carried the ivory to the coast. But at the end of the nineteenth century, with the decay of the slave trade and the growth of exports of tropical produce and minerals, it was essential to construct railways to bring these products out at reasonable cost, and also to establish political control over a country. Railway construction was given a great stimulus by the rapid

[2] See p. 14.

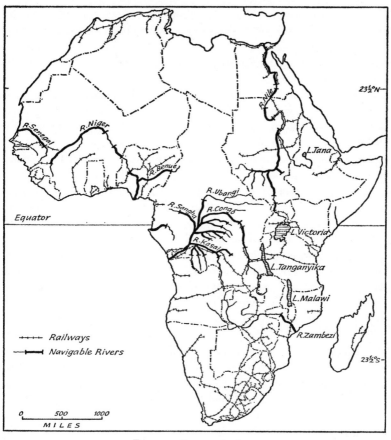

FIG. 10—Communications

occupation of territory after the Berlin Conference of 1884–5. Many railways were begun before the end of the nineteenth century (Fig. 10).

Railways

Railways were constructed in Africa for a variety of motives and not only for the exploitation of minerals. Strategic and political factors were important at first with the construction of lines in the French territories in West Africa, the Central Line in Tanganyika and the main line to the north in the Rhodesias. The latter forms part of Rhodes' much vaunted Cape to Cairo line. Rhodes the imperialist was a great believer in the value of Britain's civilizing role in Africa and railways were a vital factor in the spread of British control. The railway from Cape Town reached Kimberley in 1885 and was extended to Bulawayo in 1897 to meet the Beira–Salisbury line in 1902. The line to the north reached the Victoria Falls in 1905, the Copper Belt in 1909 and Katanga and Bukama in 1918. The link to Lobito Bay westward across Angola was only completed in 1931, while the Kamina Kabalo link of 1955 gave through rail connections from Cape Town to Lake Tanganyika, thence via Kigoma and the Central Tanganyika line to Dar es Salaam. Much of Rhodes's dream has come true and it is possible now to travel by rail, road and water from Cape to Cairo.

The use of certain routes was often controlled by political and national motives rather than economic reasoning. Examples are the long haul of copper from Zambia via Beira and Lourenço Marques to Europe and the United States of America rather than by the much shorter route through Lobito Bay, and the misuse of the Gambia river divorced from its natural hinterland in Senegal because of the accident of boundary delimitation.

Africa's railways system is rudimentary and single lines were usually built to serve the particular interests of the controlling country. In West Africa, for example, lines were not connected up to form a wider embryonic network as in Southern Africa. There was little or no overall planning of the route to be followed or appreciation of the size and potential of the region to be served. Railways are usually at right angles to the coast and were constructed quickly to reach their mineral or strategic objective with little knowledge of the terrain and with insufficient capital. Railways thus twist and wind unnecessarily with many curves and steep gradients. As most railways are single track and the turning loops inadequate, journeys are slow; it takes several days to cross Angola from Lobito Bay to the Congo border and the 136-mile journey from Freetown to Bo on the $2\frac{1}{2}$-foot gauge line in Sierra Leone takes ten hours.

Railways are still very important for the *bulk haulage* of products such as minerals, timber and crops; but since the 1920s, with the rapid growth of motor transport, roads have become of increasing significance in the opening up of new areas.

Road transport

Regarded formerly as only useful feeders to the railway line, *roads* have in a number of areas usurped the functions of the railway as the main carrier of goods and passengers. Commercial vehicles ply all over Africa and reach remote bush villages. The growth of motor transport in the last thirty years has had a greater effect on the lives of Africans than railways.

The mobility of the population has increased greatly and settlement has been attracted to the roads, villages and towns growing round the lorry park and roadside market with the familiar western appurtenances of filling stations, garages and repair shops.

A large number of individuals and small companies operate lorries and buses. A combined passenger and goods lorry is known as a mammy waggon, and this vehicle performs a useful service as a general carrier. The comparative value of motor as opposed to rail transport lies in its flexibility for it can take advantage of unusual and unexpected demands and peak loads, a common feature in Africa. On the other hand motor vehicles are affected by frequent breakdowns and accidents and goods are often damaged or suffer from rapid decay in either a hot and humid or hot and dry climate. Another problem is that owing to the patchy nature of economic development there is often no return load. There is no fixed schedule for passenger services and vehicles wait until they have obtained a full complement before proceeding on their journey.

Most African roads are of gravel, laterite being a common road-building material. These 'dirt' roads corrugate quickly unless constantly graded and may be impassable after heavy rain which tends to wash away the surface. The mileage of all-weather macadamized roads is increasing, but the expense of building a road that conforms to international standards is almost as high as the cost of a light railway. Bridging costs are increased by the nature of the river valleys, wide and sandy in places, deep and gorge-like in others. Adverse factors such as road corrugations, heavy rain, or alternatively clouds of dust and lack of maintenance cause high depreciation costs, so that the economic life of a motor vehicle in tropical Africa may be little more than two years. Names on the front of buses and lorries in West Africa are of considerable social interest and reflect a pragmatic attitude to life with its accidents, breakdowns and unexpected happenings.[3]

Waterways

Inland waterways are of limited value as the plateau-like form of the relief close to the coast prevents uninterrupted navigation for long distances into the interior of the continent. This is clearly marked in the case of the Congo system which provides nearly 10,000 miles of navigable waterway whose value is greatly reduced by the Crystal mountains which cause a break in the system near Kinshasa. Falls, rapids and low water limit navigation to particular stretches of water and seasons; for

[3] 'Still Head is Bone'; 'Life is War'; 'Six Feet at Last'; 'Why Worry'; 'Love is All'; 'Where To'; 'God moves in a Mysterious Way'; 'Time will Tell'.

example, the Benue is only navigable as far as the Cameroons for about two months in the year. Railway links often bypass the unnavigable stretches of a river, but the break of bulk entailed is costly and may limit the use of the rail and river system. Africa's major waterways are largely restricted to a belt approximately 10° north and south of the equator, except for the Nile.

The regularity, lack of bays, inlets and inhospitability of the African coast-line with its long stretches of desert, rocky or surfbound shore, greatly restricts the number of *ports and harbours* to a few natural harbours such as Freetown, Lobito Bay, Lourenço Marques and Nacala or artificial ports such as Takoradi and Cape Town. Lighterage and surf ports are still of considerable importance, although there has been progress in port development since 1945 as, for example, the construction of the new port of Tema in Ghana. A country's trade tends to be concentrated on one or two ports, usually the termini of railways to the interior, for example Dakar, Lagos, Lourenço Marques, Dar es Salaam and Mombasa.

Air transport

There has been a rapid growth in the use of *airways* since the last war. A network of airlines covers most of Africa with regular internal services and there are external services to many parts of the world. Most African countries have an internal air service and some are of a considerable size, as in South Africa. There are a large number of airfields, many of inter-national importance, such as Kano, Khartoum, Lagos and Nairobi. Besides passengers and mail an increasing amount of freight is carried, with com-modities such as dairy products, fish, meat, gold and diamonds, and selected manufactured products.

There are numerous problems hampering the growth of communications in Africa. Distances are great and many of the economically developed areas are far from the coast. Traffic density on many routes is low and long distances have to be travelled through unproductive regions, as, for example, the Central Tanzania railway line. Shortages of equipment and trained personnel after 1945 caused serious congestion, especially on the Rhodesian railways and in the port of Beira. Routes have not been planned on a broad regional basis and there is often little coordination of transport services within the various countries.

AGRICULTURE

Agriculture is the major economic activity in Africa, providing a liveli-hood for most of Africa's people; for example, 92% of Malawi's population is agricultural. It has been estimated that Africa has 15% of the world's arable and tree crops, 17% of its forested land, 18% of its pasture and yet produces only 4% of the world's agricultural output. This, however, represents 10% of the world's trade in agricultural products as Africa has a considerable share in certain commodities. Africa produces about 80% of the world's output of oil from palm products, 60–70% of the sisal and cocoa, 25% of the groundnuts, 15–20% of the coffee, but less than 10% of the world's total production of cotton and rubber.

Exports

The first agricultural export on any scale was *palm oil*, while cocoa was first exported from Ghana at the end of the nineteenth century. Since the war there has been a great increase in the volume of agricultural exports. Nigeria produces nearly half the total tonnage of palm kernels marketed, but only about a quarter of the oil. The Congo is important as an oil producer. Nigeria also features high on the list of African exporters of *groundnuts* with a production from 1955 to 1960 varying from ½ to 1 million tons. The main centre of groundnut production is in the savanna regions of West Africa.

The production of *cocoa* has also increased considerably. It is almost exclusively a peasant crop and is well suited to the high humidity, well distributed rainfall and consistently high temperatures of West Africa's rainfall belt. It also fits in with the prevailing system of bush fallowing. In Ghana cocoa is responsible for nearly two-thirds of the exports. The volume of *coffee* production had risen from 140,000 tons pre-war to nearly ¾ million tons by 1960. The biggest African producers are the Ivory Coast, Angola and Uganda. It is both a peasant and a plantation crop and nearly all the coffee is exported. *Cotton* is also a peasant and a plantation crop. In Egypt and the Sudan, with the exception of the Gezira, it is almost all produced on large privately owned estates. In Uganda, Tanzania and Nigeria it is entirely a peasant crop. Cotton accounts for over half the total exports of Egypt and the Sudan and over a third in Uganda. Yields of many of these cash crops, although still low by world standards, have been rising because of improved irrigation facilities, better seed, increased insect control, weeding, and the use of more fertilizer.

Subsistence farming

Subsistence farming, despite the growth of cash crop production, accounts for two-thirds to three-quarters of the cultivated land. Most

African countries are predominantly agricultural, exporting only a few products which are grown mostly on a small scale, for the bulk of agricultural produce is for local sale. It is difficult to estimate the value of African subsistence crops as there are few accurate statistics and much of the available information is only based on estimate.

The average African farm is very small, often under 3 acres in size, and usually consists of scattered strips of land of assorted shapes and sizes. A portion of the land is used to grow cash crops which may be tree crops such as cocoa or coffee or an annual crop such as groundnuts or cotton. The rest of the land is used for food crops, plantains or roots such as yams or cassava in the forest area, maize or millet in the savanna zone. Crops are usually interplanted and harvested when required.

Bush fallowing

The prevailing method of cultivation throughout Africa is a form of *bush fallowing* in which the vegetation is cut down with axe or cutlass and burned, the resultant ash supplying some phosphorus, potassium and calcium to the soil. The ground is then cultivated with the hoe and crops grown, using the stored-up fertility. When the fertility has declined and crop yields are too low, or weeds become uncontrollable, the clearings are abandoned and fertility is gradually restored by natural regeneration of the bush. On the good alluvial soils of a river valley like the Niger or the fertile volcanic soils of East Africa continuous cultivation can be carried on. On the poor soils of north-eastern Zambia, where the chitimene system of shifting cultivation was evolved, a fallow of up to twenty years may be necessary after only two to three years of cultivation.

When land is plentiful bush fallowing is a sensible adaptation to the exigencies of the tropical environment. With increasing pressure of population and new demands being made on the land, a long period of fallow is both impossible and uneconomic. The bush fallow system will only support a certain density of population and when this critical density has been surpassed the fallow period has to be reduced, with an inevitable decline in fertility and yield. This system, depending as it does on the rigidity of tribal custom, is not adaptable enough to cater for the needs of increasing population and the demand for a higher standard of living.

Natural and economic hazards

Declining fertility of the soil is a serious factor over much of the continent and *soil erosion* is a major problem in areas with a variable rainfall and a bare soil. Hill and plateau slopes in countries as far apart as Ethiopia and Basutoland are scarred by deep gullies and the thin soil may be removed completely by sheetwash to expose bedrock. *Pests and diseases* are additional hazards for the African farmer. Queleas and locusts may destroy crops completely over a wide area and small boys are detailed to act as guards during the ripening and harvesting season to ward off attacks by predators. A disease such as rust or smut may ruin a cereal crop and swollen shoot a tree crop like cocoa. Storage difficulties are serious in a hot climate which causes foodstuffs to decay or dry up rapidly, and every homestead has its bins or huts for the storage of grain and food crops. The advantage of a root crop like cassava is that it can be left in the ground until needed.

There are also *economic problems* of transport, marketing, variable quality and fluctuating prices. Over large parts of Africa cash crops still have to be carried on the head before reaching a railway or road. In areas far from a port, such as Zambia, only a valuable crop like tobacco can absorb the high costs of transport. Marketing boards have assisted greatly in problems of marketing, variable prices and quality. They have been in existence since the war and by paying an assured price to peasant producers have given a measure of stability to agriculture and shielded farmers from the vagaries of world prices. Prior to independence in British West Africa, the marketing boards built up large reserve funds based on the high prices paid for tropical commodities after the war. These funds have been of major assistance to economic development in Nigeria and Ghana. France provides an assured market for agricultural products from her ex-colonies, often at prices above the prevailing world price.

It is little wonder that the yields of African crops are low considering the limitations imposed by poor soil, uncertain rainfall, pests and plant diseases, and backward farming methods with little or no use of fertilizer, a lack of crop rotation and uneconomic methods of holding land. Whereas in temperate regions soil fertility can be restored by using manure and fertilizers and by practising crop rotation, manure is not available or of little use over much of tropical Africa, and fertilizers are too expensive for the average farmer. Green manuring and composting have been partially successful and a cover crop such as elephant grass has had great success in Rhodesia. Communal forms of land tenure persist over most of Africa, but are being modified by pressure of population and realization of the economic value of land. Among most African societies, however, land is not regarded as a negotiable asset, but as being held in common for the benefit of the tribe, the chief usually being the custodian of the tribal land. There are strong religious ties between the land, kin groups, ancestors and tribal gods, the latter being responsible for the happiness and welfare of the tribe, by ensuring a good harvest and so the prosperity of the group, for example. African farmers are thus reluctant to break the sacred ties with the land and settle permanently on new land away from the ancestral area.

Cattle

Pastoralism also faces great hazards in Africa, not the least of which is the widespread occurrence of tsetse fly which limits greatly the keeping of cattle in the forested regions of West Africa and the Congo basin (Fig. 11). The main *cattle* areas are to be found in the drier savannas and pastoralism tends to dominate land use in these areas, whereas crop cultivation is more important in the better watered areas which are usually infested by tsetse fly. One of the major problems in Africa is how to combine cropping and cattle-keeping. Mixed farming has met with little success apart from areas on the fringe of the tsetse-infested zone such as northern Nigeria and the highlands of Kenya. There are a few tsetse-resistant cattle such as the Ndama in West Africa, but over most of Africa pastoralism and bush fallowing are separate activities and almost mutually exclusive.

Goats are widespread throughout tropical Africa, but sheep are to be found mainly outside the tropics. Although the African continent has one-sixth of the world's sheep and nearly a third of the world's goats, and about one-seventh of the world's cattle, there is far less edible weight than in temperate countries. Meat, hides and skins are important products, but productivity is low. Tropical Africa had about 92 million cattle in 1960 with a potential beef production of one million tons, which represents about one-sixth of the comparable yield in Western Europe. Replacement rates are also low and disease causes considerable losses. Many areas are over-grazed and the stocking rate is too high, causing destruction of the grasses and accelerating soil erosion. Tropical grasses have little nutritive value during the dry season and water becomes scarce so that animals lose much weight. Cattle are still regarded as valuable more for social and religious than economic reasons and emphasis is on quantity rather than quality.

Cattle numbers should be adjusted to the carrying capacity of the grazing area and systems of management and proper control of burning should be introduced. Other measures to improve the condition of the cattle include a better distribution of watering points to obviate long treks in search of water and undue concentration on available supplies, and also the growing of fodder crops and planting of fodder trees to eke out reserves of food during the dry season. Effective vaccines have been developed for livestock diseases such as rinderpest, anthrax, blue tongue, horsesickness, Rift

84

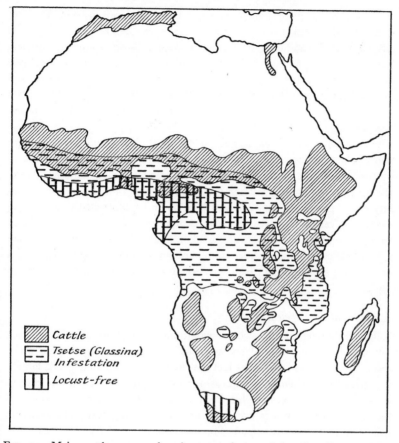

FIG. 11—Major cattle areas and main areas of tsetse infestation (from F.A. O. *African Survey*)

Valley fever and heartwater. Tick-borne diseases are controlled by regular spraying or dipping. The control of the tsetse fly is undertaken by selective bush clearing, the use of insecticides, injecting and dipping cattle and other measures where economically justifiable, as the fly can only be kept down by close agricultural settlement and pastoralism.

Serious consideration is being given to the economics of *wild animal farming* in the savanna areas of Africa. It is thought that larger stocks of, e.g., kob and other species of buck can live in the drier areas of Tanzania and Kenya than the existing herds of cattle and produce a greater weight of protein per acre without irrevocably damaging the habitat.

However, adequate transport, slaughtering and refrigeration facilities and a nearby market are essential before large amounts of meat can be disposed of. In addition, over large areas of Tanzania there is Moslem prejudice against animals killed by unbelievers. In the 32,000 square mile Ngorongoro Conservation Area some 10,000 Masai pastoralists and 100,000 cattle coexist satisfactorily with large herds of game. The Masai are not protein hungry and in contrast to nearby cultivators do not poach the game.

Experiments have been carried out in South Africa and Rhodesia with controlled shooting of wild life so as to provide an annual crop of meat and yet allow the stock to increase. In Uganda's Queen Elizabeth Park the number of hippopotami had increased too much for the carrying capacity of the grass, but they are now kept down by controlled shooting and provide a valuable source of meat. Attempts have also been made at various periods to domesticate wild animals. In the former Belgian Congo African elephants have been successfully tamed and the eland can be tamed quite easily, but little success has been achieved with the African buffalo.

Agricultural development

The technological and urban revolutions have forged ahead in Africa but social and agricultural change is slow. Agriculture lags behind other economic activities despite the fact that Africa's future still depends largely on agriculture. Schemes for increasing productivity are thus of the greatest importance. Urban growth has been a marked feature since the war and the drift to the towns is occurring on an increasing scale, although the urban percentage is still small. The towns seem to offer greater economic advancement and better educational and recreational facilities than the impoverished rural areas. Young people dislike the grind and uncertainty of subsistence farming, and the boredom and lack of economic opportunity in the villages. The agronomist and sociologist are thus faced with a dual problem of increasing agricultural productivity and stemming the drift to the towns.

A great deal of valuable scientific information on grasses, crops, soils and climate has been collected and collated at *agricultural experimental stations*. Efforts are being made to bridge the gulf between the scientific methods and high yields of the best farms and estates and the backward farming methods and uncertain yields of the average African peasant. Plans for agricultural development range from the self-help type of scheme set on foot by a village council or enthusiastic agricultural officer, to the costly multi-purpose hydro-electric irrigation project such as the Kariba Gorge or Volta River Dam Schemes. Such plans may also include the reclamation of land as in Rwanda and Burundi and Tanzania or the consolidation of fragmented holdings as in Kenya.

A great deal of preliminary surveying has to be carried out and information collected about soils, rainfall, diseases, pests and the profitability of crops before a *scheme* can be commenced. Too many schemes, such as the Tanganyika Groundnuts Scheme, have failed in the past for want of accurate ecological, economic and sociological information. Schemes may be too ambitious and the jump from hoe cultivation to mechanized farming too rapid. Mechanization has a considerable future in African farming, but machines used indiscriminately can do more harm than good; for example, the tractor-drawn plough exposes far more ground to the threat of soil erosion than does the hoe. Greater attention should therefore be paid to improving existing hand tools and animal-drawn implements, and by the provision of a domestic water supply, pumps, hand carts and other simple equipment to reduce the amount of unnecessary hard work performed by the men and particularly the women of Africa. In the past, Africans have displayed a marked ability in adapting imported crops such as maize, tobacco and cocoa to their own use and, given advice and financial aid from outside, should succeed in adopting new and scientific farming techniques which will help to overcome the present imbalance between population and resources.

INDUSTRIALIZATION

Dismayed by the seemingly impossible task of improving agriculture, a number of African countries have attempted to seek a short cut to economic prosperity through industrialization. The obstacles are many, capital is very difficult to obtain, power, machinery, communications and housing have to be provided *ab initio*. Labour, hitherto unused to continuous factory work, has to be supervised and trained for a long period, an expensive procedure. The state is usually obliged to provide this economic infrastructure.

The growth of *manufacturing industry* in Africa is fairly recent and before the war only Egypt, South Africa and Southern Rhodesia had begun to industrialize. This is illustrated by the following table showing the scale of manufacturing industries in these countries before and after the war.[4]

---

[4] Note size of establishment is different in each country, Egypt minimum of 10 or more workers, Southern Rhodesia 5 or more, South Africa 3 or more.

Countries	Number of establishments		Number of persons employed (thousands)	
	1937–8	1953–4	1937–8	1953–4
Egypt	2,404	3,651	151	251
S. Rhodesia	299	714	18	70
S. Africa	8,713	14,257	269	704

From 1950 to 1956 the volume of manufacturing grew by 100% or more in the Congo, Angola, Southern Rhodesia and Mozambique and to a smaller extent in Egypt and North Africa. In most African countries manufacturing is still characterized by the small scale of production and small number of factories, the predominance of consumer and durable industries and the lack of intricate manufacturing processes. South Africa is by far the most advanced industrially of African countries and has a higher proportion of workers engaged in heavy industries. *Wages* are still generally too low, which restricts purchasing power and hence the size of the market, but the compulsory raising of wages in many industrial and urban areas has had the effect of increasing purchasing power and productivity. The potential *market* in many African countries is great and there is an increasing demand for a wide variety of manufactured goods.

Sources of power

An important aspect of industrialization in Africa is the provision of power. The growth of secondary industry in Uganda, for example, was dependent on the provision of power from the Owen Falls Scheme which began supplying electricity in 1956. Until recently the generation of electricity in Africa has been based on coal and oil. The largest known coal deposits are in South Africa, with reserves of 75,000 million tons but most African countries lack exploitable coal. Large deposits of petroleum and natural gas have recently been found in the Sahara. However, with the building of great multi-purpose dams at Kariba on the river Zambezi and Owen Falls on the Victoria Nile and projects such as the Volta River Scheme, Aswan High Dam and the Inga Scheme on the lower Congo (the largest hydro-electric power scheme in the world but held up by political unrest) an increasing proportion of power generated will come from hydro and not thermal electricity. There has been a very rapid growth in the demand for electricity since the war because of the development of towns and industries.

SOCIAL GEOGRAPHY

POPULATION

It is essential for a modern country to have accurate information about the size, composition and distribution of its population as an indispensable aid to social and economic planning. Until recently, however, most African countries have had very little accurate demographic information and what information was available was often only an estimate. Although South Africa had its first complete *census* in 1904, it was not until 1931 that this was done in Nigeria, Ghana, Sierra Leone and the Gambia, and for most of Africa complete counts were not made until after 1948. These early censuses are also often inaccurate and of doubtful value for comparative purposes. Even when a census was held there were many difficulties to overcome before accurate information could be obtained. Only the simplest of questions can be asked of a largely illiterate and conservative population, and there is a marked reluctance to being counted for fear of taxation or conscription. In East Africa the results of the 1948 census showed a wide gap between the census total and the estimate.

There is also very little information about characteristics of population such as sex ratio, age range and vital statistics of births and deaths. The sex ratio in most countries tends to be low by world standards and varies greatly, being highest where there is a marked degree of urbanization and economic development, as in the coastal regions of West Africa and the mining and industrial areas of Southern Africa. The sex ratio of South Africa's adult population is 105, that of Malawi only 68·4. Africa is characteristic of young countries with a high proportion of its population under 15 years of age. Information on births and deaths is also very limited and even in South Africa registration of births and deaths amongst the bulk of the African population is not enforced.

Until recently it has not been possible to hazard more than a guess for the total population of Africa, but with the post-war batch of census material a more accurate figure can be postulated: Africa had an estimated total population in 1962 of about 269 million with a density of about 23 people per square mile, compared with a population of 140 million in 1920. This density is small compared to Europe and North America, so that Africa has the lowest density of population of all the continents except Australia.[5] The *map of population density* (Fig. 12) reveals striking contrasts between adjacent areas of high and low density of population in the same

[5] Although W. A. Hance (in *The Geography of Modern Africa*, New York, Columbia University Press, 1964, p. 52) estimates that nearly one-third of Africa is overpopulated.

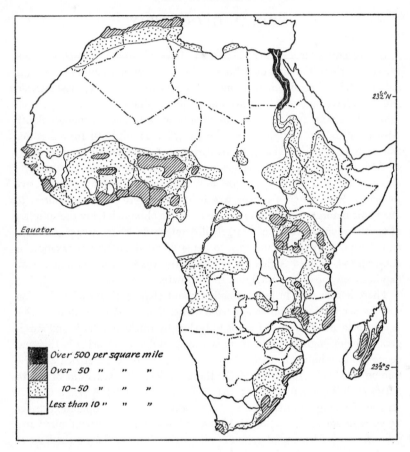

FIG. 12—Density of rural population

environment, giving the appearance of islands or clusters of population concentration surrounded by a sea of slight density. Parts of the Eastern Region of Nigeria have a population density of over a thousand people per square mile next to large areas in the Middle Belt or the Cameroons with less than 50 per square mile.

AREAS OF HIGH DENSITY

There are vast areas with a density of under 5 per square mile, particularly in arid and semi-arid areas such as the Sahara and Kalahari and forested regions such as the Cameroons and northern part of the Congo basin. Most of the continent has less than 50 per square mile, and there are only five major areas with a density of over 50 per square mile. The first occurs in the coastal belt of North-West Africa, known as the Maghreb, where the Mediterranean climate, a rainfall of over 20 inches per annum and long contact with Mediterranean culture and economy has encouraged a fairly high density of population. To the south, increasingly meagre rainfall and desert-like conditions permit only nomadic pastoralism. The Nile Valley has long been an area of dense population with permanent, irrigation cultivation of a high order, and the historical legacy of an ancient civilization and later influence of Islam explain a density that is more comparable to Asia than Africa. West Africa has a well distributed rainfall and constant high temperatures in the southern forest region, with tree and root crops, while the Sudan savanna belt is a zone of contact with seasonal rainfall permitting the growth of annual crops and rearing of livestock; both these regions have seen the rise of strong states such as Ashanti, Dahomey, Gao, Songhai and the Fulani emirates and a marked density of population. In East Africa favoured regions such as the shores of Lake Victoria and the Rift Valley Highlands, with a well distributed rainfall and lack of tsetse fly, encourage crop cultivation, the rearing of livestock, and powerful Bantu states such as Bunyoro, Buganda, Rwanda and Burundi organized by an Hamitic aristocracy.[6] In South Africa factors which help to explain the high density of population are mining, the industrial development in the Southern Transvaal and the influence of European settlement and higher rainfall along parts of the south and east coast.

Factors affecting distribution

When considering the map of population distribution in Africa as a whole, it might appear that *physical factors* of relief, climate, drainage, vegetation and soil are paramount in their influence on the location of settlement. However, a closer analysis of the rainfall, vegetation and soil maps with maps of population density fails to reveal such a close identity and other reasons must be sought in an endeavour to understand the pattern of population distribution. Areas of heavy rainfall are not necessarily areas of dense population, and indeed the reverse often occurs, as in the Kano-Katsina region of Nigeria which has an average annual rainfall of approximately 30 inches and population density of over 100 per square mile, compared to large areas of the Middle Belt with a rainfall of over 50 inches per annum and density of less than 50 per square mile. But even

[6] In the latter states the power of the Tutsi minority has declined greatly.

in regions of aridity a high population density may be found, as in parts of the Nile, Niger and Orange River valleys where permanent water is available in adequate quantities and the population sufficiently advanced in agricultural technique to use it for extensive irrigation. In many parts of the tropical world highland areas attract population because of a cooler climate and higher rainfall, but a number of Africa's highland areas in Ethiopia, Kenya, Malawi and Lesotho may be too high and cold for permanent African settlement, which is only found at lower and warmer altitudes.

However, in East Africa as a whole individuals such as Gillman[7] and reports such as that of the admirable East Africa Royal Commission have shown how important is the *availability of water* for population distribution. Sukumaland, south of Lake Victoria, has a surprisingly high density of population in an area with an average annual rainfall of less than 30 inches but where water is readily available at the base of granitic tors, and where the light sandy soils, as in northern Nigeria, can be easily worked and crops such as cotton grown. There is also a very close correlation between heavy rainfall and dense population on the southern slopes of Mounts Kilimanjaro and Meru, while the northern rainshadow slopes have only a slight population.

African *river valleys* often have a markedly low density of population because of their unhealthy nature with malaria, sleeping sickness, river blindness and other diseases and considerable fluctuations of water level with severe flooding at times. This is in marked contrast to the dense population of major river valleys in Asia and Europe.

Disease is also a major factor affecting the composition and distribution of population. A formidable list of tropical and temperate diseases may cause death, ill health, lack of energy and low agricultural productivity. The major endemic diseases are found, such as pneumonia, tuberculosis and venereal diseases. The most serious are malaria, sleeping sickness, leprosy, tuberculosis, hookworm, filariasis, bilharziasis, malnutrition, blindness and diseases of the respiratory tract and the bowel. Malaria is slowly being brought under control, although it is complicated by the fact that the main carrier, the *Anopheles Gambiae* mosquito, breeds freely in water. Mention has been made of the tsetse fly which causes sleeping sickness in man and nagana[8] in animals. The diet of the African peasant, particularly in the forest areas, is deficient in animal protein because of inability to keep cattle in tsetse-infected forest areas. Protein deficiency is

[7] *Geographical Review*, **26**, 1936, pp. 353–75, and other articles.
[8] Animal trypanosomiasis.

most common in young children and causes kwashiorkor which is often fatal. Despite the toll taken by diseases in Africa, advances made by medical science in the control and eradication of many diseases have caused a decline in the death rate, so that with a high birth rate the population of Africa is expanding rapidly in many areas.

Historical and contemporary economic, social and political factors may exert an overriding influence on population. Refuge zones such as dense forest or broken country may attract a large population which has sought sanctuary there in times of danger. It is difficult otherwise to explain the anomalously high population density of Iboland in Nigeria, as it has a meagre environment with a poor, sandy, leached soil and very heavy rainfall, producing little other than oil palms. The dense forests, now largely cleared for oil palms, probably sheltered Negro peoples fleeing from slave raiders, just as in the Middle Belt of Nigeria pagan villages are perched on rocky outcrops, guarded by steep slopes and cactus hedges against the incursions of Fulani horsemen. In Basutoland the palisaded upper slopes and steep lower slopes of mesas aided greatly the survival of Basuto clans against attacks by Bantu, Boer and British forces. These flat topped mesas, often large in area and with adequate supplies of water and grass to withstand a long siege, acted as fine natural fortresses somewhat akin to the ambas in Ethiopia. Sparsely populated zones in a potentially favourable environment often represent shatter belts or zones between recently warring tribes and states, as for example between Buganda, Busoga, Bunyoro, Ankole or between the Yoruba states, Dahomey, Benin and the Fulani emirates.

In Southern Tanganyika Arab *slave raids* during the nineteenth century and later opposition to German rule caused great loss of life, the abandonment of settlement, encroachment of bush and tsetse infestation over a wide area. Lack of population here is thus due rather to the initial ravages of slavery and warfare than the direct effects of pests and diseases. Strong states such as the lacustrine Bantu states in East Africa and the Sudanese states of West Africa made good use of the physical environment and supported a dense population. Over a million people live in the region round Kano and practise intensive cultivation in a region which has light sandy soil and high water table, but which is potentially little more favourable than underpopulated areas to the east. In proto-historical times the same environment in the better watered eastern parts of Southern Africa supported in turn a very low density of Bushman hunters, larger numbers of Hottentot pastoralists and a far greater density of agricultural-pastoral Bantu who made far better use of the environment. The same region now

93

supports well over ten times the original population at a very much higher living standard. An obvious factor of great importance is the ability of the population in a given region to support increasing numbers through the growth of technological and social skills.

Social and religious factors anchoring people to areas where family and ancestors have lived for a long time may make them reluctant to leave their own overcrowded areas for adjacent empty and more fertile regions and thereby break home ties. This is a serious problem now in Northern Ghana and parts of Nigeria. The African attitude to children is also of importance, as children are regarded as assets, particularly girls who command a good bride price. Women are rarely unmarried and they produce large families, which is also an expression of the necessity for survival in pre-European days, when the chief was anxious to gain fighting men and increase the strength of the tribe. Africans are, however, great travellers and migrants and in East and South Africa a complex series of migrations occurred until the coming of the Europeans which tended to stabilize the population pattern somewhat prematurely. The Zulu wars of the 1820s caused the devastation of large areas of South Africa and the northward movement of the Ndebele and other tribes and the settlement of the Nguni in areas as far afield as the Upper Zambezi Valley and Lake Malawi. This was a reversal of the previous trend of southward migrations of the Bantu people. During earlier ages various waves of migration and dispersion have carried groups of Bushmen, Hottentot, Negroes, Hamites and Bantu all over Africa. Many have entered Africa by way of the Eastern Horn and Ethiopian highlands or Nile Valley and moved slowly west across the Sudan savanna, or south via the Great Lakes and high plateau of East Africa.

The *economic revolution* of the twentieth century has brought profound social changes to Africa and the population pattern has reacted correspondingly with increased density in the economically developing areas. Lines of communication by rail and particularly by road have obviously attracted settlement and caused a knotting of population at nodal points such as crossroads or markets. In West Africa lorry parks often form the economic centre of the small town replacing the chief's or ruler's headquarters as the raison d'être for the town. In the savanna zone many towns such as Kano and Timbuktu sprang up at the termination of a caravan route, and particularly, as in the case of Kano, a change of mode of transport from camel to donkey aided greatly in its growth. Many of the East African towns originated as railway centres and ports, and two of the largest East African towns, Mombasa and Dar es Salaam, are both railway termini and ports.

Smaller towns such as Fort Portal, Mbeya and Iringa rely exclusively on road communications.

Agricultural development and the growing of cash crops by African peasant farmers in Southern Ghana, Buganda and on the slopes of Mount Kilimanjaro and by European settlers in East and Southern Africa have caused a great increase of population in areas that may have been relatively sparsely populated before. The *mining* of gold, diamonds, copper, tin and other minerals has caused a rapid growth of settlement in areas almost empty of population. For example, the only inhabitants of the high ridge of White Waters or Witwatersrand in the Southern Transvaal early in 1886 were a few Boer farmers and wandering prospectors, whereas with the discovery of vast exploitable reserves of gold later in the year thousands of people flocked to the site, and now the city of Johannesburg has over a million inhabitants. In Zambia a waste of bush near the Katanga border was practically uninhabited in the 1920s, but with the later exploitation of the great reserves of sulphide copper ores in the Copper Belt, a number of mining towns such as Ndola and Kitwe sprang up almost overnight and this region now has a population of over $\frac{1}{4}$ million people. In West Africa towns such as Bukuru and Jos have developed on the Jos plateau in Nigeria based on tin mining, while Obuasi and Tarkwa in Ghana owe their origin to the mining of gold.

URBANIZATION

Almost everywhere the European invasion of Africa and the subsequent economic and political development has been accompanied by urban growth. Urbanization has accelerated since the last war with increased economic and political development and the drift of people from rural areas who see little hope of economic betterment in agriculture. Ports and capitals of newly emergent countries have increased considerably in size as trade has expanded and the prestige and expense of independence has resulted in a building boom. Although the urban percentage of Africa's population is slight, with less than 10% of the people in settlements of over 5,000, compared to 80% in Western Europe, the rate of urbanization in parts of Africa is as high as in Western Europe. Many towns have trebled in size in the last twenty years; for example, Nairobi in 1937 had a population of 61,000, and in 1960 250,000, while Bamako in 1941 had 22,000 people and in 1958 100,000.

The most urbanized areas of Africa are in the north-west, Egypt, West and South Africa; West Africa has two-thirds of the towns in tropical Africa.

95

A significant factor is that the majority of towns of over 30,000 are ports, while three-quarters of the thirty-two African countries abutting on the sea have their largest towns on the coast. There are also a number of important inland river and lake ports, the largest of which, such as Bamako, Kinshasa and Khartoum, have become the major commercial centres and capitals of their countries. Other African towns are situated at significant junctions of land routes as in Northern Nigeria and Rhodesia with the towns of Kano and Bulawayo. Mining activities are also responsible for a number of towns including the Copper Belt towns of Zambia and the great complex of towns based originally on gold in the Transvaal and Orange Free State.

Except Cairo, Alexandria, Algiers, Casablanca, Dakar, Lagos, Ibadan, Kinshasa, Nairobi, Addis Ababa, Salisbury and towns in South Africa, the towns are generally small. The majority are new and reflect European influence to a considerable degree with a rectangular pattern, zonal layout and separation into different quarters and groups of people. This is particularly marked in Southern Africa where ethnic groups tend to be segregated in clearly marked zones. Countries with a more complex economic structure tend to have a larger number of non-Africans employed in non-agricultural occupations, which is reflected in the greater racial diversity of their towns.

It is difficult to distinguish between a small town and a large village over much of Africa, as in most towns the percentage of people dependent on agriculture is still high. A small administrative centre with a population of a few hundred people may possess more urban functions than a large village with many times the population. Compared to the industrial towns of Europe, most towns in Africa are dependent on commerce and administration rather than industry.

Temporary migration

In addition to the permanent movement to the towns, there is also temporary migration to areas of economic growth. Over much of Africa a dual economy has developed with the direction of the economy in European hands and unskilled labour performed by African migrant workers. Migrant labour has developed on a vast scale because of the long distances between areas of dense population and centres of economic activity. The exploitation of minerals in the ancient rocks of the plateau interior has taken place in underpopulated areas, so that migrant workers may travel hundreds of miles to seek work in the copper mines of Zambia and the Katanga and the gold mines of South Africa, the sisal belt of Tanzania, the cotton and coffee areas of Buganda and the cocoa areas of West Africa. Movement has occurred from overcrowded areas such as Lesotho, Rwanda and Burundi, Eastern Nigeria and Northern Ghana. The absence of the majority of adult males for long periods has an obvious effect on the society and economy of the rural areas. Family and tribal life is disrupted and agriculture neglected. It was calculated in the former Belgian Congo that not more than 10% of the adult males should be absent at any one period to maintain a healthy rural society and economy. Migrant labour may be a useful temporary bridge between labour supply and demand areas, but it is an uneconomic use of labour as it tends to promote inefficiency and a high rate of turnover.

European influence

The effects of differing European economic and political control on population distribution is also marked. The creation of law and order and the provision of medical and health services has lowered the death rate and this, combined with a high birth rate, has caused a general growth of population, which increased by over 70% from 1920 to 1950 in Africa south of the Sahara.

Where European settlement has been extensive Africans have been confined to reservations as in South Africa, Rhodesia and Kenya and acute local pressure of population has arisen. In West Africa greater economic development has taken place in the former British territories and large numbers of people have migrated from former French territories into Ghana and Nigeria. Fear of forced labour has also caused many thousands of Africans to flee from Mozambique and Angola into Zambia and Malawi.

Finally it must be remembered how little is known about African population, and how difficult it is to plan any comprehensive schemes for economic development on the basis of the present statistical information. The available evidence seems to indicate a fairly high rate of increase of population varying between about 1·9% for Tanzania and 2·8% for Zambia. In some areas of the world with a high living standard an increasing population is an asset, but this is usually not the case in Africa where living standards are low and the market small. There are some fifty African countries, many with a meagre economic base, so that the problems of economic viability are serious. Under these circumstances some form of closer economic and regional planning would seem essential, but in the light of political events since 1962 this is less and less likely.

CLASSIFICATION OF POPULATION

There is a great diversity of peoples on the African continent, each group differing considerably in social, economic and political organization and ethnic composition. One of the major problems facing the geographer is to decide on suitable criteria for classifying these peoples. A population distribution map indicates the areal extent of distribution, but obviously does not differentiate between the various groups. Attempts have been made to group people according to *physical anthropological traits* of skin colour, hair, eyes, stature, skull shape, nose, lips, etc., but this is unsatisfactory as the distribution of different criteria does not coincide. Prolonged migration and conquest have caused such an intermixture of people that it is difficult to distinguish between them. For the most part the lack of any major geographical barriers to movement, except for areas such as the Nile swamps and parts of the Sahara, has encouraged the amalgamation rather than the separation of ethnic stocks.[9]

A more accurate criterion than physical factors for classifying Africa's peoples may be *linguistic*. Therefore, by using a number of criteria including language, and perhaps culture, one may classify the various ethnic groups into those of European and Asian origin, Hamites, Semites, Malgache, Negroes (or West Africans), Bantu, Nilotes, Bushmen and Hottentots, and various intermixtures of the above groups such as the Cape Coloureds and Half Hamites.

European origin

People of European origin numbered some 6 million in 1959 of whom a half lived in South Africa, one-third in North Africa and the rest were scattered about the highlands of tropical Africa with significant numbers in Southern Rhodesia and Angola. The following table shows the total population and ethnic composition of selected countries by percentage in 1959:

Country	African %	European %	Asian %	Total population (thousands)
Kenya	96·0	0·9	3·0	6,450
Uganda	98·6	0·2	1·2	6,517
Tanganyika	98·5	0·2	1·3	9,238

[9] G. P. Murdock (in his book *Africa: Its Peoples, their Culture and History*, New York, McGraw-Hill, 1959, p. 7), says that Africa has been inhabited by representatives of only five races: the Bushmanoid, Caucasoid, Mongoloid, Negroid and Pygmoid.

Country	African %	European %	Asian %	Total populatton (thousands)
N. Rhodesia	96·7	3·1	0·2	2,360
S. Rhodesia	92·0	7·1	—	3,000
S. Africa	68·2	19·4	3·0	15,841

The Europeans of South Africa are over 3 million in number and, including those in Zambia and Rhodesia, Angola and Mozambique and South-West Africa, the total is now nearly 4 million. The Afrikaners of South Africa are descended from the original Dutch settlers with admixtures of French and German blood. They have evolved as an African race with a distinctive culture. Those of British origin form some 40% of the white population of South Africa, but are still a separate group and have strong ties with England. They tend to dominate the commercial, mining and industrial economy of South Africa.

The large European population of North Africa has diminished greatly in numbers as Morocco and Tunisia and finally Algeria gained their independence. There were about 1 million European settlers of French and Italian extraction in Algeria who opposed bitterly de Gaulle's policy of creating an independent Algeria. The best land had been alienated to Europeans who largely controlled the export of wine, fruit, tobacco, olives and cereals. Nearly three-quarters of the Algerian farmers owned only 14% of the land, but with independence and the flight of many of the European estate owners their land is being apportioned among the landless Berbers.

Although the total number of Europeans in Africa is small their influence in the last five hundred years has been very great. Through their efforts many millions of slaves were transported to the Americas where Negroes now predominate numerically in many areas. They introduced Christianity to Africa and in the past eighty years brought in the last colonial phase that is now coming to an end.

Asian origin

The next largest group of non-African origin, the Asians, came from the Indian sub-continent, and now over ¾ million of them are found in East Africa and the province of Natal in South Africa. About ½ million Asians live in South Africa, 80% of them Hindu and the majority in Natal where they were introduced originally from India to serve on the sugar estates as indentured labourers. They no longer work on the estates and have become market gardeners, small shopkeepers and factory workers, but are restricted in occupation because of the civilized labour policy adopted in South Africa. Kenya has the largest number of Asians in East Africa, the larger numbers being associated with a greater degree of economic development than in Uganda and Tanzania. There has been long historical contact between India and East Africa and Indian traders have operated along the coast of East Africa for many hundreds of years. It was only with the colonial occupation of the late nineteenth century, however, that they moved into the interior. Asians now play a dominant role in the commercial life of East Africa and are also important as artisans. With the retreat of colonialism and the political ascendancy of the Africans the economic importance of the Asians will diminish.

Cape Coloured and Swahili

There are a number of small ethnic groups such as the Cape Coloured and the Swahili of East Africa that have evolved in the coastal areas of Africa through intermixture of indigenous peoples and foreigners who have arrived by sea from Europe and the Far East. The Cape Coloureds are a mixed race with Hottentot, Malay, European and Bantu blood and as a result differ greatly in appearance. They live mainly in the Western Cape in South Africa and provide the bulk of the unskilled and semi-skilled work on farms and factories. They now number about $1\frac{1}{2}$ million and are increasing rapidly, as the birth rate is nearly 50 per 1,000 and the death rate is decreasing.

Immigrants from Arabia, Persia and India have intermingled with the Bantu of East Africa and produced a group of people speaking Swahili, which has now become a lingua franca in East Africa. It is a distinctive language with a mixture of Arabic and Bantu terms. The coastal Swahili have been affected greatly by the Arab influence, and their culture and religion is Islam.

Hamitic origin

The Hamites are a Caucasian people who probably invaded Africa from Southern Arabia or the Horn of Africa, and settled north of the equator and along the shores of North Africa. In North Africa they are known as the Berbers and form the majority of the population in Morocco, Algeria and Tunisia, the Semitic influence increasing farther east. They have given their name to this region which is also called Barbary. The North African Hamites are tall, light brown in skin colour, with wavy hair, thin straight noses and thin lips. They are mainly settled agriculturalists living in villages, but practising a certain amount of transhumance in the Atlas mountains. In the eighth and succeeding centuries they were conquered by the Arabs who imposed their language and religion to a great extent on the Berbers.

Most of the Egyptian peasants, the fellaheen, are also of Hamitic origin, whereas in the upper and middle classes in the towns there are many foreign elements. Many of the nomadic camelmen of the Sahara, the Moors and the Tuareg are largely Hamitic, but the Tibu of the Eastern Sahara centred on Tibesti are darker than the Tuareg and have a greater admixture of Negro blood. In the oases the settled agriculturalists are largely Negroes, the descendants of those enslaved by the Moors and Tuareg. Greater than the ethnic distinction is that between the nomadic pastoralists of the desert and the agriculturalists of the oases who form over half the population of the Sahara. They produce irrigation crops of cereals, fruit and vegetables which grow under the shade of the date palm, the date being the most important source of food and medium of exchange.

Hamitic influence is also strong in the Western Sudan along the fringes of contact with the great Negro block of West Africa. It is probably strongest amongst the Fulani who number about 6 million and are scattered over most of the savanna belt from Senegal to the Sudan. Some are nomadic pastoralists, the Cow Fulani, who have retained Hamitic traits more strongly than the sedentary and town Fulani who have been integrated more with the Negroes, particularly in the Hausa states of northern Nigeria. Hamitic influence is considerable among the Mandingan and Songhai peoples who founded the powerful medieval states of Mali and Songhai.

The Hamites, Amharas, Galla, Somali, Danakil, Boran and others form the great majority of the people of the semi-arid and arid regions of northern East Africa and Somaliland and the better watered highlands of Ethiopia. Hamitic blood becomes more diluted farther south, but the combination of Negro and Hamite has produced the Bantu of East and Southern Africa and a different admixture of these basic strains has produced Nilotic peoples such as the Acholi, Lango and Luo in East Africa. A Hamitic aristocracy also controls the well organized Lake Victoria states such as Buganda and Bunyoro. The Hamites probably brought with them a knowledge of iron working, cattle keeping and sheep rearing and with tribes such as the Masai constitute a wedge of pastoralism between the lacustrine Bantu on the shores of Lake Victoria and those inhabiting the highlands and coastal areas of Kenya.

Semitic origin

The Semites are physically similar to the Hamites, the main distinction being linguistic. The Arabs are Semitic and in North Africa form an increasing proportion of the population in the arid regions farther east. Under the impulse of Islam they invaded the coastal areas of North Africa and, particularly during the Hillalian invasion of the eleventh century, destroyed much of the Berber culture. The Arabs are mainly nomadic pastoralists living in tents and eking out a precarious existence based on their herds of camels, cattle and sheep. In East Africa Arabs arrived as traders and sea-borne invaders rather than land-based nomadic pastoralists. Perhaps their most significant contributions to the society and culture of Africa has been the spread of Islam as a religious and cultural force and with it the Arabic language. Their influence is most strongly felt in the savanna zone to the south of the Sahara where Islam has been a powerful force in the growth of the Sudanese states. In the Sudan today there are two strongly contrasted zones; the north which has a population of about 8 million which is Moslem in religion, speaks Arabic and is the most advanced socially and economically; and the south with about 3 million inhabitants predominantly Negro or Nilotic in ethnic origin, speaking a variety of languages and with very different customs from those of the north, and pagan rather than Moslem in religion.

The Hamito-Semitic influence on Africa has been profound and the Hamites have invaded the continent in greater numbers and over a longer period than the Semites; thus they have influenced the ethnic composition of Africa's peoples to a major extent and introduced the arts of pastoralism and iron-working. The Semitic influence has been no less important in the north with the tremendous impact of the Moslem religion and culture and the Arabic language.

Malagasy

Malagasy is like the African continent in miniature with a high plateau and steep escarpment to the east, but a broader western coastal plain. Its society and economy has been influenced considerably by contact with Indonesia. It has been suggested that the bulk of the Malgache people are of Malayan origin, but it seems probable that only about one-fifth of the population is Malayan and that the rest are Africans. The total population of over 5 million is at present small, but it is increasing rapidly and pressing heavily on the limited amount of fertile soil. The most numerous and advanced people are the Merina. Economically the influence of the Far East is seen in the production of rice, which is the main crop on the island. The cattle complex from Africa has invaded Malagasy and cattle are regarded as social and religious as well as economic assets.

Negro groups

West Africa is inhabited largely by Negroes, who number nearly 50 million. They are to be found in a number of different geographical regions, but most characteristically in the coastal and forest zone.

The West African Negro tends to be tall with a brown to black skin colour, black woolly hair, flat broad nose, protruding upper jaw and thick lips. The main distinction, however, between the Negro peoples of West Africa and the Bantu of East and South Africa is linguistic rather than ethnic. A very large number of mutually unintelligible languages of the Sudanic group are spoken in the forest areas of West Africa, while in the north, Hausa, as is the case with Swahili in East Africa, tends to be a lingua franca.

Some of the main groups in the *south* are the Ibo and Yoruba of Nigeria, the Akan of Ghana and the Ivory Coast, the Temne and Mende of Sierra Leone and the Wolofs of Senegal. Some five million *Ibo* live in a poorly endowed region of Eastern Nigeria whose high rainfall and highly leached acidic soils only the oil palm will tolerate. The Ibo have a loose social organization with small communities, the largest unit being a group of villages with a common meeting place. In pre-European days there was little tradition of urban life, although recently the towns of Port Harcourt, Enugu, Aba and Onitsha have been growing rapidly. The *Yoruba*, nearly 5 million strong, are a great contrast to the Ibo and are composed of a number of linguistic groups. They have a degree of urbanization remarkable in Africa, with six towns of over a 100,000 inhabitants, Ibadan being the largest town in tropical Africa. There is an ancient tradition of urban life going back for a thousand years. Towns had to be fortified during the wars of the pre-colonial era, but walls and gateways have largely disappeared by now. The gridiron pattern of European administrative commercial and residential quarters has been grafted on to the old core sprawling haphazardly around the Oba's palace and market place. Buildings are rectangular with thick mud walls and steeply pitched thatched or corrugated iron roofs to counteract the heavy rainfall. Rooms face onto an inner courtyard.

In the *north* a fusion of *Hausa and Fulani* has created the remarkable series of Hausa states *cum* Fulani emirates. Kano is the largest of these emirates, the town of Kano now having a population of 200,000. Seen from one of the minarets of the main mosque the city stretches out to a flat horizon, with a sea of flat mud roofs and thick mud walls intersected by winding paths and studded with borrow pits, palms and thorn trees. The centres of Kano are the great market place and the Emir's palace. Surrounding the city and within a radius of some thirty miles live over 1 million energetic Hausa farmers, practising intensive agriculture and producing a variety of food and cash crops, of which groundnuts are by far the most important.

Economic specialization and trade had reached a far higher level of development in West than in East Africa in the pre-colonial era. Cloth dyers and weavers, leather workers and tanners, metal workers and blacksmiths, potters, calabash carvers, makers of grass mats and many other *craftsmen* played and still play a vigorous part in the local economy of West Africa. Much of the *sculpture* of West Africa is of great artistic merit. It is strongly associated with religion, much of it being symbolical and stylized, for example the carved dance masks in which the dancers impersonate ancestors or spirits. Being a well forested region with little stone most of the sculpture and carving is in wood.

The standard of sculpture is declining with the decay of tribal religion and the carvings no longer reflect the customs and mythology of a passing age. There are two different styles in sculpture, the Moslem art of the Western Sudan being formalized and geometric compared to the more exuberant and varied art of the Guinea coast. West African art dates back over 2,000 years to the Nok culture with its vigorously executed terracotta figures, its tradition can be traced to the splendid bronzes and terracottas of Ife and bronzes and ivories of Benin which reached their height in the fifteenth and sixteenth centuries A.D.

Bantu

South of a line from the eastern frontier of Nigeria to the south of the Uele river to Lake Kyoga, along the eastern shore of Lake Victoria and across Tanzania, and stretching southwards for thousands of miles to the eastern Cape Province in South Africa is a great region inhabited by over 60 million Bantu who form Africa's largest ethnic group. Physically the Bantu are similar to the Negro and there is a great variety of types from the near Hamite to the near Negro among the Bantu. The major criterion distinguishing Bantu from Negro is linguistic. Compared with West Africa the Bantu of East Africa had little in the way of economic specialization and no tradition of urbanization except amongst the Swahili and Arabs of the coastal belt in pre-European days. Population distribution in East Africa is more patchy and the population is far smaller than in West Africa.

Pastoralism is far more important than in West Africa, although the incidence of tsetse fly limits cattle keeping over large areas of Tanzania.

Perhaps the most advanced Bantu group in East Africa is the *Kikuyu*. They inhabit the Kenya Highlands north of Nairobi and are a great contrast to the almost feudal structure of the Buganda kingdom. They are a forest people who expanded from the present Fort Hall area southwards into Kiambu. They have a complicated social organization with two main patterns, the one being based on the extended family and closely linked with land ownership, the other being territorial and based on the control by councils of senior elders of ridges between parallel streams flowing to the Athi, Thika, Tana and Uaso Nyiro rivers. It is in Kikuyuland that serious conflicts have arisen with the white settlers over land ownership, culminating in the recent Mau Mau rebellion. The Kikuyu are the most politically active group in East Africa and with the growth of land consolidation may become the most progressive of East Africa's people.

The great majority of the inhabitants of the *Congo basin* are Bantu, with Sudanese Negroes on the northern fringe and Pygmy hunters in the Ituri and other forest regions. Because of the abundant and well distributed rainfall crop cultivation is predominant and pastoralism of minor importance. The Congo is thus ecologically and agriculturally more akin to West Africa than to East and South Africa. Also the population has settled in the Congo basin for a longer period and been less subjected to migration and movement than in Southern Africa. There are a number of important tribes, some of them with a strong social and political organization such as the kingdoms of Congo, Bushongo and Lunda. The Bushongo and Baluba people produce sculpture of a high order, particularly in their carving of wood, and their art is somewhat similar to that of West Africa.

The *Southern Bantu* inhabit the high plateau and coastal regions of Southern Africa comprising the political territories of Angola, Mozambique, Rhodesia, Malawi, South-West Africa and South Africa. They are both agricultural and pastoral, some tribes, such as the Zulu and Ndebele, being more pastoral and warlike than others. The Bantu trekked southwards probably from the region of the great lakes in a series of migrations over a period of hundreds of years, pushing Bushmen and Hottentots before them. The Bantu are a vigorous people who withstood the European onslaught well and did not suffer the virtual extinction of the Hottentot and Bushman groups. There are great differences physically and socially among the Bantu tribes and between individuals of the same tribe. Compared to the Negro, prognathism tends to be less marked, the nose is more prominent and skin colour lighter.

There are many different tribes, but they can be grouped in a number of different ways. In South and South-West Africa the southern Bantu are divided into four main ethnic groups: Nguni, Sotho, Shangana-Tonga and Venda. The *Nguni* group, which includes the Xhosa and Zulu tribes, inhabits the curving swathe of plateau foreland between the Great Escarpment and the sea in the Ciskei, Transkei and Zululand. The *Sotho* group is found on the plateau and includes the Southern Sotho of Basutoland and adjacent regions, the Northern Sotho of central and northern Transvaal and the Tswana of Botswana and northern Transvaal. The *Venda* also live in the Zoutpansberg of the northern Transvaal. The *Shangana-Tonga* inhabit the coastal region both north and south of the Mozambique border.

The Bantu *social organization* is tribal, the basic unit being the family or homestead which is usually dispersed in a seemingly haphazard way but with a location dictated by the choice of the individual or the chief and having suitable access to natural resources of water, soil and grass. Only in Botswana where there is a scarcity of water is settlement concentrated around springs. Among the coastal tribes huts tend to be beehive in shape and composed of a wattle framework covered with grass. On the plateau huts are more cylindrical or rondavel in shape with mud walls and steep conical thatch roof. The reason for this may be the greater scarcity of thatching grass and timber compared to the better watered coastal region. The square or rectangular form of hut is becoming more popular under European influence.

Huts were built formerly in a circular pattern with the cattle kraal in the middle and the chief's or headman's hut in the centre surrounded by those of his wives and children, the whole encircled by a palisade or zariba of thorns and sticks. Amongst the predominantly pastoral tribes the cattle kraal occupied the most important position in the enclosure.

At the head of the tribe is a chief, or of a confederation of tribes such as the Basuto, Xhosa, Zulu and Swazi, a paramount chief. The chief and his council and a hierarchy of sub-chiefs and headmen rule the tribe. The chief has varied powers, but is usually held in check by his advisory council, though strong personalities like Chaka or Moshesh pursued their own course of action without hindrance.

Over much of Southern Africa, however, the tribal organization has decayed greatly. The *economic revolution* wrought by the discoveries of diamonds and gold in the period 1870–90 and of other minerals such as coal, iron ore, copper, manganese, asbestos and tin caused a great demand for labour in the fast developing mining and agricultural industries and recently in manufacturing industry. The bulk of the unskilled work in these industries is performed by Africans, the majority of whom are temporary migrants. An increasing number of Africans are becoming permanent dwellers in the towns and European areas and now little more than one-third of the twelve million Bantu live permanently in the Reserves. These were demarcated as tribal preserves during the nineteenth century and now form the so-called Bantu homelands, which are being developed as part of the government's overall scheme of apartheid that aims at separate development of the various ethnic groups in South Africa.

The rapidity of economic growth has had disturbing *social effects* and both Africans and Europeans have drifted to the towns in large numbers. A spectacular rise in the value of land and the increasing necessity for the white man to perform skilled work or enter the professions meant a decreasing opportunity for the poorly equipped Afrikaner to find suitable work, especially as it was beneath his dignity to perform manual work. By the 1920s a large poor white class had arisen which could not fit into the economic framework of either European or African society. The problem of the poor white in South Africa has largely been solved by the growth of secondary industry and the application of the civilized labour policy to safeguard the interests of the Afrikaner. The problem of the African migrant worker has still to be solved.

Bushmen

The Bushmen formerly occupied much of Southern Africa, but have now been driven by Boer and Bantu invaders into the fastnesses of the Kalahari. They have small heads, are short in stature and with a yellow-brown skin colour, peppercorn hair and broad faces with flat noses. They are divided into hunting bands each about fifty strong with a well defined territory. They are skilled hunters and collectors using bows and arrows, pits and traps. They are nomadic and live in rough shelters of branches, somewhat similar to those erected by the Australian aborigine. Bushmen were adept at depicting lifelike hunting scenes on the walls of caves and rock shelters, practising a form of art akin to the Neolithic cave paintings of Southern France and Spain.

Hottentots

The Hottentots also migrated southwards a long time ago and are reputed by some writers to be a cross between Bushmen and Hamites. They are similar in appearance to but taller than the Bushmen with whom they were confused by the early Dutch visitors and settlers at the Cape.

CHAPTER FOUR

Political Geography

POLITICAL geography is today the most dynamic aspect of the geography of Africa. Since 1950 vast regions of Africa have become independent, and in 1963 only comparatively small areas in the south were still under external control. The last phase of the colonial era in Africa, during which the present boundaries were drawn up, has lasted little more than seventy years, from 1890 to 1960, a very short period in the history and proto-history of Africa but a time of great significance in the political and economic sphere. Despite the artificiality and rigidity of the ex-colonial boundaries, the newly independent African states are at present fitting their political institutions to these boundaries and no major boundary changes have occurred as yet.

A DEFINITION OF POLITICAL GEOGRAPHY

Political geography is concerned firstly with the political relationships between the state and the various groups within the state; with state and internal administrative boundaries; the homogeneity of the population and, in so far as their distribution is significant for human geography, with concepts such as nationality, religion and language, and their influence on the structure of the state; the degree of loyalty expressed towards the state by various sections of the population and in the problems of minority groups. The second premise of political geography is the resource basis of the state and the effect of factors such as location, size, shape and com-munications on the power potential and cohesion of the state; the composi-tion, degree of skill and technological development of the population; the distribution and variety of natural resources and the use made of them for the benefit of the state. Thirdly, political geography is concerned with the state's extra-territorial relationships as expressed through trade, commerce, economic and political alliances and mutual control with other states of a physical unit such as a river basin. Political geography thus covers a wider field than the state alone and analyses the relationship between a country and its territorial dependencies, or the structures of supra-national organi-zations such as the United Nations Organization and the Communist bloc, or the regional organizations such as the East African Common Services Organization, or the strength and extent of a concept such as Pan-Africanism.

In general the evolution of African states must be viewed against a background of both physical and human factors which may be interrelated. There are few barriers to movement in Africa, except perhaps for the sudd and equatorial forest, but there are a number of refuge areas afforded by desert, dense forest and highly dissected relief. The Sudan savanna encouraged movement across Africa, and the plateau grasslands of the east and south migration down the length of Africa. The Sahara was not an insuperable barrier and fruitful contact with the Mediterranean and the resources of the savanna and forest belts aided the growth of powerful Sudanese states such as Gao, Songhai and Mali. The clash of warring peoples and the cruelty of slave raiding caused the flight of broken groups of people to seek refuge in the tropical forests, the semi-deserts such as the Kalahari or the precipitous highlands of Basutoland (now Lesotho) and Ethiopia. The sanctuary afforded by the high Ethiopian plateau enabled the Ethiopian Coptic church to survive the attacks of Islam.

Tropical diseases, particularly malaria and other fevers, long delayed European exploration of the interior of West Africa. Pests such as the tsetse fly preclude the keeping of cattle and horses over much of the continent, and formerly restricted Fulani raids to the Sudan and Guinea belts and the access of Boer ox-waggons to the veld of the south. Climate influences man mainly through agriculture, and in large areas of Africa highly variable rainfall prohibits the cultivation of crops. The supposedly debilitating influence of climate on man's energy and the power of the state has not been proved.

The influence of man with his skill and technology on the growth of the state is profound. In pre-colonial Africa the agricultural economy was almost entirely on a shifting subsistence basis, while minerals were exploited sporadically and on a small scale, although the gold of West Africa and the lower Zambezi basin did give rise to widespread commercial links. The European invasion of Africa led to a rapid and systematic exploitation of minerals such as gold, diamonds and copper, and the swift growth of youthful states with an embryonic communications pattern, of towns and the cultivation of a wide range of cash crops. The revolutionary changes brought about by the impact of Western technology is nowhere more clearly manifested than in South Africa, where a complex modern state arose in a period of about thirty years in a region that was previously inhabited by a handful of pastoralists.

In this chapter some of Africa's traditional political systems will be studied in broad outline, rather than in detail, followed by an account of the onset of colonialism, and the types of boundaries that were drawn up at the end of the nineteenth century; then a study of differing colonial policies and finally an assessment of the contemporary political scene. The multiplicity of political units emerging in Africa makes it necessary to treat them in major regional groupings rather than separately as individual units.

PRE-COLONIAL POLITICAL INSTITUTIONS

There are two important types of African society, the one based on the system of kinship, the other on the system of chieftainship. A *kin group* consists of those males who trace their descent from a common ancestor. The kin group is a unit holding land for the benefit of the group, and the head of the kin group has to make sacrifices to the ancestors for the welfare of the group. When a number of kin groups came together in a wider association for some common purpose, such as warfare, the system of *chieftainship* tended to evolve. Chieftainship and the tribal system have developed over the greater part of Bantu Africa. These migratory warring Bantu tribes have evolved further towards the creation of a state than the kin groups.

In pre-colonial Africa there was a marked interrelation between traditional laws and customs, shifting agriculture and the physical environment. Communal effort was needed to clear the land, build huts and make paths and tracks. Possible lines of action were discussed and decisions ratified at the tribal assembly by chiefs, elders and council, in much the same way that the Germanic tribes and Anglo-Saxons discussed affairs at the folk-moots. Life was a hard struggle and required communal effort to ensure the growth and welfare of the community. There was no place for the uncertainty of individual effort and perplexity of divided voice.

The forces, natural and spiritual, influencing tribal society were incomprehensible to its members, so that tangible symbols in the form of myths and ritual had to be adopted and strictly adhered to if the tribe were to prosper. These mystical values reflected basic physical and human elements such as the soil, rainfall, family and fertility. The head of the tribe, be he king, paramount chief or chief, had a mystical relationship with his people, as he was the guardian of the distinctive tribal mores which symbolized the unity of the tribe.

As African societies were largely based on a self-sufficient economy, wealth did not accrue in the form of commercial capital, so that the right to levy tribute, taxes and labour was an important aspect of chiefly power. Perhaps the most important characteristic of the African state is the right to employ organized force in times of danger and stress. In the hands of an unscrupulous ruler such force would be used oppressively, but in the hands of a wise ruler it would be employed for the good of the group. However, it was rare for the head of an African state to become too autocratic, as there were a number of checks on his authority. The paramount chief had to delegate responsibility to a hierarchy of chiefs and headmen in charge of smaller administrative divisions. If the chief became too powerful and tried to abuse his authority the subordinate chiefs would break away. There were also councils and courts to act as a check on central authority. Just as the chief had rights, so also he had obligations to his people, to protect them from attack, to interpret tribal laws and customs and to be responsible for their welfare.

Political institutions vary greatly throughout Africa. Sophisticated and complex states have evolved over a long period along the shores of North Africa, in the Nile Valley and the Sudan savanna belt. In West Africa there are marked contrasts between the loose political organization of the Ibo and the centralized Yoruba kingdoms headed by the Obas, and between the scattered pagan tribes of the Middle Belt and the alliance of the Akan-speaking peoples who formed the Ashanti Confederacy. In East Africa there are many types of political organization. The Nilotic tribes have a decentralized form of political structure, while among the lacustrine Bantu society was highly evolved under the impact of Hamitic invasions. Further south the tribal structure headed by a chief became stronger, and in Southern Africa the political system was very centralized, with a paramount chief controlling a number of groups through a hierarchy of chiefs and headmen. In a number of tribes, particularly amongst the Zulu, organization for war was based on the enrolment of all the young men of the tribe into regiments.

A more detailed account of the *Zulu political system* may help to illustrate some of the above points. Under the influence of Shaka and Dingaan the various Zulu clans were placed on a military basis, regiments being directly under the paramount chief and living in barracks round the Great Kraal. Hundreds of clans were thus forged into the Zulu nation, headed by the paramount chief to whom they owed allegiance, military service, labour and gifts of grain, beer and cattle, for which in turn he was expected to feed and aid his people generously. The paramount chief exercised powers of authority. Directly controlled by him were the chiefs under whom were the indunas or sub-chiefs and in turn the heads of lineage groups. Tribal and ward capitals were the centres of social life. The state featured largely in the life of the people and, as the young men were gathered at the Great Kraal, the women and girls had to help with cattle herding and milking, normally a man's work, as well as with the cultivation of crops.

Frontiers and boundaries

Frontiers were of importance in pre-colonial Africa and among the hunters and collectors, such as the Bushmen, groups wandered within their own strictly defined territory. Travel beyond these bounds was often very difficult, as strangers were treated with great suspicion. It was only when states extended over large areas, as in the Sudan belt, and were influenced by Islam that a traveller like Ibn Batuta was able to travel far and wide and cross many boundaries. The nineteenth-century explorers were able to cross Africa only with the aid of guides, the consent of the various peoples through whose territory they passed, with superior technical means and sometimes with the use of force.

Onset of colonization

It was only towards the end of the nineteenth century that *colonial boundaries* were delimited and the whole continent rapidly partitioned. European exploration and colonization began with the Portuguese in West Africa, Angola, Mozambique and the East African coast in the fifteenth and sixteenth centuries. This was followed by the Dutch settlement at the Cape in 1652 and later English and French penetration of parts of West, East and North Africa. The appearance of the Germans and Belgians on the African scene towards the end of the nineteenth century caused the scramble for African territory by European powers and made it imperative to define rival spheres of influence. Prior to this, European exploitation had been confined almost entirely to the coast, except in the south, and for various reasons Europeans had not penetrated into the interior. Thus, even as late as the mid-nineteenth century, much of the interior of Africa was unknown to Europe, as illustrated by Arrowsmith's map of 1842.

The process of boundary delimitation was very rapid, and in the twenty years from about 1890 to 1910 most of Africa's boundaries had been drawn (Fig. 13), a process that took many hundreds of years in Europe. Frontiers and marcher zones, which had previously separated groups and European spheres of influence, were hastily converted into boundaries, while many new lines of division were drawn across the map regardless of population distribution or routes.

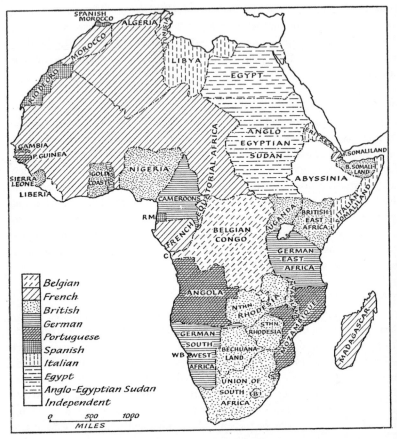

FIG. 13—Africa: political pattern, 1914

RM = Rio Muni.   B = Basutoland.
C   = Cabinda.   S = Swaziland.
WB = Walvis Bay.

The British, advancing far into the interior of *West Africa* from small coastal colonies in the Gambia, Sierra Leone, Gold Coast and the Niger Delta, were able to annex much of the savanna belt in Nigeria, the bulk of the new areas being incorporated as Protectorates. The French operating from Senegal north of the Guinea forest zone were able to drive swiftly and far across the Sudan belt and in a few years had occupied vast areas of West Africa and the Sahara, so linking up with their North African posses-sions. They also joined up with coastal footholds in Guinea, the Ivory Coast and Dahomey. The Germans, pursuing the colonial game with great vigour, were able to acquire Togoland and the Cameroons in West Africa, German East Africa and German South-West Africa, which was interpreted as constituting a threat to British and French possessions in Africa.

The political mosaic of West Africa was constructed quickly with British, German, French and Portuguese enclaves and the republic of Liberia jostling each other along the coast, backed by a French controlled Sudan savanna zone. The Anglo-French agreements of 1898–9 gave the Sudan to Britain, but *French* North, Equatorial and West Africa were joined up, forming a vast contiguous block of territory. To the south the *Congo basin* was incorporated by Leopold into what was virtually an immense private estate. To the east Uganda and *British East Africa* represented an eastern sphere of British influence that was prevented from linking up with the British possessions in southern Africa by the intermediate strip of German East Africa.

The *Portuguese* dream of a trans-African empire was shattered by the northward thrust of Rhodes and the creation of Northern and Southern Rhodesia astride the Zambezi. War between the Boers and the British, which ended in 1902, saw the incorporation of the states of *South Africa* under British control, and later unification as the Union of South Africa. The history of colonial boundary delimitation in Africa has only been treated in broad terms as the details have already been discussed in Chapter 1.

African boundaries drawn up hastily at European conference tables towards the end of the nineteenth century often represented the temporary limits of military occupation, as in Nigeria, or the efforts of explorers and agents who rapidly traversed the country and persuaded local chiefs to sign treaties in favour of the metropolitan country represented by the agent, as in Nyasaland and East Africa. German claims were put forward to territory as far north as the Tana river in East Africa, partly because of Dr Karl Peters' work as an agent. German claims were, however, countered by Kirk, Johnston and others. Johnston had gone out to East Africa on a natural history expedition and obtained a concession of land from a local chief at Taveta, which proved to be the only concrete piece of evidence to substantiate British claims to the hinterland.

The great bulk of Africa's boundaries are *artificial* and based on astronomical or mathematical lines. This is a feature of newly settled countries, as for example the boundary between the United States of America and Canada, or the boundaries between the Australian states. Africa's artificial boundaries reflect ignorance of the geography of inner Africa and also the rapid occupation of the interior so as to forestall rival claims to territory. There was little available knowledge of the distribution of resources and population, and many boundaries were drawn across potentially valuable areas. In Africa boundaries based on *relief* only account for about a quarter of the total, compared to the boundaries of a country such as France where relief is an important criterion for boundary delimitation.

Astronomical lines, based on parallels or meridians, are usually found in uninhabited areas, or those lacking definite physical features, as in the Sahara or Kalahari. Some mathematical lines made use of relief or factors of human occupation, as for example the boundary between Togoland and Ghana. Relief features—mountains, rivers and watersheds—were often chosen as boundaries because they were easily recognizable. In addition, watersheds were often used, as nineteenth-century exploration attempted to elucidate Africa's major drainage problems. Towards the end of the century the puzzle of the Nile and Niger drainage had largely been solved.

Many *river boundaries* reflect the desire of colonial powers to have access to a navigable river, as, for example, the curious Caprivi strip linking German South-West Africa with the Zambezi. The Congo Free State was given access to the Atlantic via the mouth of the Congo, thereby separating Angola from the Cabinda enclave.

African states of the early twentieth century did not grow organically and through the efforts of the African people themselves. The size and shape of the colonial state was determined from without and boundaries imposed for the most part with little knowledge or consent of the indigenous people. Major *problems* arise largely where colonial boundaries cut across ethnic or linguistic groups, and divide them between different administrations. This is particularly noticeable in West Africa where political boundaries run mainly north and south counter to the main east–west trending physical and vegetational belts and zones of population. Where the boundaries of Nigeria, Dahomey, Togo and Ghana are compressed into a short coastal strip of little more than a hundred miles wide there are many examples of boundaries cutting across ethnic groups. Perhaps the most publicized example here is of the Ewe separated by the Ghana—Togo boundary. Similar instances occur among the Angoni, divided by the Rhodesia—Malawi border, or the Somali parcelled out among four different administrations, those of Ethiopia, French Somaliland, Somalia and Kenya.

Some of the difficulties arising from colonial boundaries have been minimized by the effect of *divergent policies* introduced by the different colonial powers, which have endeavoured to make viable political and economic units of their possessions. Countries like Nigeria, Senegal and Kenya have developed markedly separate identities, and in their new-found national pride are determined to prevent a change in their boundaries. Conversely, there have been moves in East Africa recently towards a wider grouping of these territories under a form of federation or closer association, and for a short period the two Rhodesias and Nyasaland (now Malawi) were linked in an ill-starred federation. In a number of areas the effect of *warfare or slave raiding* caused the breakup of groups of people, who thereby lost their identity, or who fled to the shelter of dissected or dense forest. Nigeria's Middle Belt harbours many of these tribal remnants divorced from their original territorial base. On a wider scale, too, *movements of people* have occurred over a very long period up and down the length and breadth of Africa. Colonial boundaries have netted or crystallized their movements on a continental scale, a process that has happened on a smaller scale throughout Africa's history. In some instances treaties have been drawn up at the instance or appeal of a powerful chief, such as Moshesh in Basutoland, whereby tribal groups have been absorbed without much loss of territory or people, as in Hausaland or Barotseland. On occasions *tribal influence* has even been extended, and in Buganda with its powerful political organization the chiefs were used to administer areas of Uganda beyond the confines of Buganda. A great deal also depends on the goodwill of the rival colonial powers as to the harmful effects of boundaries on the frontier society. The Somali on either side of the British and French Somaliland borders were allowed to graze and water their cattle freely, whereas friction has arisen on the ill-defined Somali-Ethiopian border, as the Somali are not allowed by Ethiopia to use their traditional grazing grounds and waterholes.

Although the problems of external boundaries are numerous, just as serious may be the *problems of internal boundaries* in a large and well populated state with diverse ethnic groups. In Ghana there are now over a hundred native states, many of whose boundaries were not properly defined at the beginning of this century and are still in dispute. Nigeria's regional boundaries, established some forty to fifty years ago, have caused considerable trouble recently with the growth and the conversion of Nigeria from a unitary to a federal state. There are many boundary problems, some involving comparatively large areas. The Minorities Commission was called on to consider three major and a number of minor claims. The major claims were those of the Western Region to Ilorin-Kabba Province from the Northern Region, and the Eastern Region's claim to Aboh and Asaba divisions and Western Ijaw from the Western Region. But these claims have little historical or geographical support, and the Commission recommended a change only as a result of a plebiscite.

Other *economic problems* due to boundary delimitation have arisen where newly established patterns of communication have cut across older trade routes, or where a valuable mineralized region has been divided unintentionally by a boundary. After the ratification of boundaries the next stage of exploitation and establishment of law and order depended essentially on the rapid growth of communications. As mentioned in Chapter 3 Africa's rivers have not provided easy access to the interior, and pests such as the tsetse fly have precluded the use of animals and wheeled transport throughout great areas of the continent, so that railways were indispensable and quickly built. Over much of Africa *railways* were constructed for *strategic and political* aims at first, economic considerations being of secondary importance. As in the continents of North America and Australia, and in Siberia, so in Africa railways made possible the effective political occupation of the continent by tying the outlying regions to the core, and the interior to the coast.

The awkward partition of Africa among rival spheres of influence caused the building of many uneconomic railways which, had they not been instruments of political planning, would have followed a more logical economic route, as for example the French and British railways in West Africa and the German-built Central Railway in Tanganyika. In Southern Africa railways were pushed hastily inland from coastal outlets to the mineralized regions of the interior plateau of the Northern Cape, Transvaal, Rhodesias and the Katanga. The boundary between the Congolese Republic and Zambia cuts across the great mineral resources of this region and acts as a kind of economic watershed causing rail traffic to flow in devious ways, dictated by political and not economic considerations. Until recently most of the copper, cobalt, tin and other minerals were exported via Rhodesian railways to Beira and now to Lourenço Marques, or by means of a tortuous and costly rail and river route through Port Franqui on the river Kasai and Matadi on the river Congo. A shorter and more economic outlet is through Lobito Bay in Angola.

The breakup of the Central African Federation emphasizes some of the problems faced by Zambia, now dependent on external communications and power, formerly common services with Rhodesia. The main rail outlets lie through Rhodesia, Angola and Mozambique, and power comes from the Kariba generators in Rhodesia. Zambian politicians are thus vitally concerned in the future of Rhodesia and the Portuguese territories, and are considering seriously the construction of an alternative rail link to the Central Railway in Tanzania.

The *Gambia* is a flagrant example of the loss of a useful river route. The long sinuous outline of the river, and a short distance on either bank for an east–west distance of less than two hundred miles, was confined within British boundaries. The hinterland became French and later Senegalese, thereby depriving the Gambians of their hinterland and the Senegalese of a useful waterway. The position was made worse by the French construction of a parallel railway from Dakar to Kayes and Koulikoro on the Niger river. A similar instance occurs with the severing (by the Nigeria—Dahomey boundary) of the coastal lagoon route from Lagos to Cotonou, and the consequent disruption of waterborne traffic. *River basins* are physical units and if used for irrigation, navigation or the generation of hydro-electricity they must be considered as single units. Only by investing separate political control in an inter-state body such as the Tennessee Valley Authority can the rational development of the resources of a river basin be achieved. Egypt is utterly dependent on the Nile waters, so is vitally concerned in the political geography of the whole Nile Valley.

Although railways have cut across ethnic groups and disrupted traditional trade routes, diverging the flow of traffic in a new direction, nevertheless the economic patterns set up within the political framework of the new African state will not only survive the change from colonial rule to independence, but will grow stronger as the state seeks to broaden and strengthen its economic base.

DIFFERING COLONIAL POLICIES

The next section is concerned with the political effect of differing colonial policies on various parts of Africa until the onset of independence, that is over a period of approximately seventy years, from 1886 to 1956. With the rapid spread of colonial rule at the turn of the century the first step was to establish law and order, and organize political stability as rapidly, efficiently and cheaply as possible. With the spread of colonial rule over Africa many variations in administration occurred, not only between the different controlling powers, but also between individual territories. Two main points of view emerged which can be characterized broadly as *identity* and *differentiation*. The former envisaged the growth of African political institutions similar to those of Europe, while the latter as in South Africa, considered that African political systems should evolve separately from European.

British colonial policy

British policy in Africa has been far more empirical than the French and has tended to encourage local government. The principle of Indirect Rule, which was adopted in many of the British territories, was applied first in India and then adapted by Lugard to Northern Nigeria where the powerful Fulani emirates proved admirable media. Where a strong form of indigenous authority was found, as in Ashanti, Northern Nigeria and Buganda, with paramount chiefs, kings or emirs, the British authorities tended to leave local rule and customary power in their hands subject to the minimum of control by British colonial officers. A system of Native Authorities developed based on local chiefs, and charged with the maintenance of law and order, tax collection and other duties. Where there was little or no central authority attempts were made by the British with far less success to set up local institutions applying the same principle of *Indirect Rule*. It was thought that the use of Indirect Rule would enable the disrupting influence of Western European culture to be introduced gradually without the decay of tribal society. The pattern of Native Authority rule has changed, however, and proved insufficiently flexible to cope with the increasing range of social services and the complexity of government at the local level. There are other problems, too, of fitting educated commoners, resentful of traditional chiefly rule, into the N.A. framework, and of administering the complex large urban communities which have mushroomed in size since the 1930s. The N.A.s have thus been converted into elected local authorities in order to handle the increasing range of administration at the district level.

Throughout Africa traditional authorities have found it extremely difficult to interpret fairly the demands of the colonial power on the one side, and of their own people on the other. Chiefs, bedevilled by the duality of their role, have been pulled both ways by these conflicting demands. Under the withering blasts of nationalism the chiefs have lost many of their original functions, but have tended to retain their ritual significance, especially if the tribal group has been strong enough to reassert its identity against the attacks of nationalism and Pan-Africanism.

The British in West Africa have not been faced with the same problem, as in *East or Southern Africa*, of reconciling immigrant white and Asian interests with the political aspirations of the African majority. The problem was temporarily resolved in *Southern Rhodesia* in 1923 in favour of the 32,600 Europeans, who were given the choice of incorporation within the Union of South Africa, or responsible government; they chose the latter. But the Kenya whites, eager to obtain power, were baulked when the Devonshire White Paper of 1923 stated that interests of the Africans must be paramount, although the white settlers had been more perturbed about Asian than African political influence.

The Asian group in *Kenya* is the largest in East Africa, and in 1960 was estimated to number about 175,000. Most of the original Asians employed on the Kenya—Uganda railway remained after it was completed and now form an intermediate class between Africans and Europeans. They fulfil a useful function as artisans and clerks and tend to dominate the commercial life of the country.

White settler influence was dominant in the 1930s in Kenya, but has since declined with the growth of African political representation. Much of the bitterness which erupted in the 1950s with the Emergency, when a section of the Kikuyu tribe revolted against the administration, was engendered by the alienation of land in the so-called White Highlands. This was particularly so in the eastern section of the alienated area where crowded Kikuyu look out on scantily populated, prosperous European farms. One of the main issues in Kenyan politics is still the quarrel over possession of land after independence. The British government's land reservation policy has created many problems for the future African government in Kenya to solve.

In *Tanganyika* the white community was much smaller, in 1960 only one-third of Kenya's 65,000 whites, and more widely scattered. Their influence on the political scene was, therefore, far less significant than in Southern Rhodesia or Kenya. A watching brief by the League of Nations and later the United Nations Organization over Britain's role as trustee also encouraged the more rapid growth of African majority rule. In *Uganda* the number of Europeans has always been far smaller than in Kenya or Tanganyika, and there were very few settlers. The influence of Colonial Office rule has been correspondingly stronger. The movement towards independence in Uganda has, therefore, taken a different form, and has been complicated by the desire of Buganda to develop separately and in political isolation from the rest of Uganda.

The British have attempted to solve the critical problem of balancing the political claims of the various groups and sections in East Africa and the Rhodesias and Nyasaland by a system of *partnership* with communal and not common representation, and by linking their territories in a closer form of economic or political association. The policy of partnership, and of endeavouring to build a multi-racial state in which each group could play an effective political role, has not been successful. It has, in fact, tended to increase African suspicion and friction between the groups. Elsewhere Africans have come to power swiftly on the wave of common representation and universal suffrage, and they feel that the device of communal representation is merely another ruse to perpetuate white rule. Sir Edward

Twining, when governor of Tanganyika (1949–58), tried to fashion a multi-racial system of government, and divided the country into ten great constituencies, each to be represented by three members, one from each of the racial groups, African, Asian and European. However, the scheme proved unworkable, as at the first elections only those candidates backed by the Tanganyika African National Union, or TANU, the main African political party, were successful.

Closer association

There has been little in the way of *closer association* in the ex-British teritories of West Africa, largely owing to geographical separation and great distances between the different countries, except for a somewhat artificial community of interest created by the Second World War, with its post-war survival in the West African Inter-Territorial Conference. But East Africa, with compact shape and the three contiguous territories of Kenya, Uganda and Tanganyika, has far more common interest than West Africa. There have been various attempts to create closer political links. In 1929, for example, the Hilton Young Commission recommended an eventual union of the three territories, but this was rejected by the House of Commons Joint Select Committee. Periodic conferences of the Governors of Kenya, Uganda and Tanganyika were held between the wars, and during the Second World War the East African Production Supply Council was set up to deal with joint economic problems. Partly because of experience gained during the war, and because of the apparent advantages of closer association, it was decided to explore the possibilities of achieving economic rather than political federation. As a result the East African High Commission and East African Central Assembly were set up in 1948.

The *High Commission* was responsible for the administration of some twenty-three common services such as railways, harbours, posts, telegraphs, customs and excise and inter-territorial research, without trespassing into the realm of politics. Revenue was obtained from Britain, Zanzibar and the three East African governments, and the annual budget was £5–6 million. The work of the High Commission is now being continued as the *East African Common Services Organization*, partly because so much vital research work is being done on a wide variety of subjects. An economic federation has thus proved successful where a political federation could not be achieved in view of African mistrust of white settler influence. It is hoped that a political federation may eventually develop from the Common Services Organization, as was suggested at the Addis Ababa Conference in June 1963.

The three *Central African territories* of Zambia (formerly Northern Rhodesia), Rhodesia (formerly Southern Rhodesia) and Malawi (formerly Nyasaland) are approximately equal in size of population but differ greatly in area, natural resources and degree of economic development. Rhodesia has a white population of some 225,000 who dominate the political scene, while Zambia with 75,000 whites and Malawi with only 10,000, had for a long time been controlled by the Colonial Office but are now independent African states; Zambia with a valuable mineral industry is the most prosperous, Malawi with a poorly developed agricultural and migrant labour economy is the poorest, while Rhodesia with its range of mining, farming and industry has the most balanced economy. The Rhodesia—Nyasaland Royal Commission Report of 1938 rejected the idea of closer association, as the differences between the three territories were so great. However, the post-war economic boom and seeming advantage of a common market and common services, coupled with a rapid rise in the white population of the Rhodesias which had numbered little more than 100,000 in 1946, amongst other factors, led to the establishment of the *Federation of the Rhodesias and Nyasaland in 1953*. Political representation, in contrast to South Africa's policy of segregation, was based on the system of partnership, which has been interpreted in various ways by the British government, the different federal governments, and by Africans and Europeans. Partnership represents a kind of balance between the types of administration that evolved under the Colonial Office in the two northern territories, and the system of parallel development in Rhodesia.

To the *European* partnership meant a continuation of European economic, social and political standards for a long period, and the gradual acceptance of the African into European society, provided he measured up to these standards. To the *African* it meant a far more rapid growth than hitherto, leading to political and economic equality with the European, and thus a dominant voice in the control of affairs in the Rhodesias and Nyasaland. The Europeans envisaged the palliative of partnership as having only a slow effect, and in the first Federal Assembly of 35 members only 6 were Africans, although there were also 3 Europeans charged with special responsibilities for African interests. From 1953 to 1960 there was little increase in African representation, and the system of electing representatives to both the federal and unit assemblies became too complicated for the average voter to understand. Federation, which was imposed on the Africans despite widespread opposition, became increasingly to be regarded by them as a device to keep the European in power.

The *Republic of South Africa* is racially a microcosm, with nearly 11 million Africans, 3 million whites, $1\frac{1}{2}$ million coloured people of mixed origin and $\frac{1}{2}$ million Asians. The white group, of whom 60% are Afrikaners and less than 40% are of English origin, own nearly 90% of the land and 60% of the national income. They are skilled workers, managers, entrepreneurs, and professional men and dominate the economic, social and political scene. The Africans, who form the bulk of the population, are unskilled or semi-skilled workers on the farms, in the mines and factories, and have no political representation in the white parliament. White South Africa's political philosophy of segregation has evolved over a long period, and has been expressed clearly and unashamedly in contrast to the ambiguous policy of partnership that was put forward in the Rhodesias.

Apartheid

The policy of segregation, or apartheid, stresses separate development for the various racial groups in South Africa, and is an attempt to safeguard the European way of life and high standards of living from non-European competition. It is a policy directed by fear of economic competition and political domination by the non-whites, who outnumber the whites by four to one, and are increasing more rapidly in proportion than the white group.

The *mineral revolution*, caused by the discoveries of diamonds and gold in 1870 and 1886, and the recent industrial revolution has caused increasing integration of all the racial groups in South Africa, and a massive invasion by Africans of the European urban and industrial sectors. In 1936 less than half the African population were estimated to be in the European areas, but by 1960 this fraction had increased to two-thirds. The policy of separate development is economically impracticable, and if enforced vigorously as a policy of total apartheid, which the Dutch Reformed Church advocates, would mean economic ruin. The scheme of separate development is only applicable fully to the native reserves, which now occupy only about 12% of South Africa.

Apartheid has its roots in history, and has been official South African policy for fifty years. In 1948 the Nationalist Party came to power and implemented apartheid more rigorously and rapidly than the United Party which had held office previously. A series of Acts was passed,[1] one of the most important being the Group Areas Act, which laid down for each racial group the areas where they might own property. The Bantu Education Act

[1] Prohibition of Mixed Marriages Act, 1949; Suppression of Communism Act, 1950; Group Areas Act, 1950; Separate Representation of Voters Act, 1951; Bantu Authorities Act, 1951; Native Building Workers Act, 1951; Bantu Education Act, 1954.

of 1954 was also a significant step in the stage of implementing separate development, as it laid down that African education should be based on African ways of life as these were so different from European standards. But the main plank in the Nationalist plan for African development is the building up of the reserves as a national home or Bantu Homeland, where the African can become self-governing, and can pursue any economic activity he desires. Africans away from the reserves will be regarded as migrant workers with no political rights.

The *Tomlinson Commission*, which was set up to examine the possibility of implementing these plans, carried out an exhaustive survey of the reserves. It suggested that at the end of the twentieth century some two-thirds of the estimated African population of 21 million could be accommodated there if certain recommendations were carried out. Seven African territories were proposed, including the High Commission Territories, which were to be economically viable, based on agricultural and industrial development and the expenditure of over £100 million. Other points put forward in the Report were that towns and industries should be established, and that individual land tenure and white capital should be encouraged in the reserves. Government reaction was strong and Dr Verwoerd, then Minister of Native Affairs, said that white capital would not be allowed, that industries were to be set up on the edge of the reserves and that tribalism and chiefly power were to be strengthened.

The political situation of the *three High Commission Territories*—Basutoland (independent since October 1966, and renamed Lesotho), Bechuanaland (independent since September 1966, and renamed Botswana) and Swaziland—is also difficult, particularly that of Lesotho, which is embedded within the Republic. Botswana and Swaziland have independent access to the sea through Rhodesia and Mozambique. The Protectorates are economically dependent on South Africa, which takes much of their exports and employs the bulk of their labour. The South African government has laid claim to the Protectorates on various occasions and, as stated, the Tomlinson Commission Report envisaged their incorporation into the Bantu Homelands. The Act of Union in 1910 stated that the wishes of the inhabitants of the High Commission Territories were to be considered and that the British Parliament was to have every opportunity of discussion before any proposed transfer of these territories to the Union of South Africa should take place.

However, with the decline of British control (particularly since Botswana's and Lesotho's independence) the situation has changed rapidly and moderate political parties have come to power in all three countries which are prepared to cooperate more closely with South Africa in the future.

French colonial policy

French colonial policy differed considerably from British policy and was far more logical and orderly. Initially, with territories like Senegal, a policy of identity and of economic and political affiliation with France was carried out. Attempts were made to model local institutions on those of France, to select Senegalese representatives to the French parliament, and encourage the Senegalese to become French in culture. With the rapid growth of an extensive French colonial empire in West and Equatorial Africa, the policy of close political association with France had to be modified. In addition the policy of assimilation failed in most of the French colonial empire because of the vigour of indigenous religions and cultures in areas such as North Africa and South-east Asia. Separate development occurred, which was anti-assimilation to some extent, although the policy of assimilation was not lost sight of. The 1946 Constitution reflects both points of view and a policy of modified identity, the Associated States and Territories being allied with France in a Union. France was represented equally with all overseas territories in the Assembly of the French Union. There were also 32 African representatives from Africa south of the Sahara out of a total of 627 deputies in the French National Assembly. Africans were given a majority in the Territorial Assemblies.

French colonial policy, therefore, has two main characteristics, firstly the granting of a considerable measure of representation in French parliamentary institutions and French citizenship to the colonial people, and, secondly the retention of a definite measure of political control by France herself. The ex-French territories have thus tended to develop in *direct association* with France, whereas the ex-British colonies have pursued an independent course, within or without the British Commonwealth, according to their inclination.

Belgian colonial policy

The Belgians with little previous experience of colonial rule were somewhat reluctant to outline their colonial policy in definite terms. Responsibility for governing the Congo Free State was forced on the Belgian government in 1908. Faced with the problem of governing a great area, much of it consisting of the low-lying and unhealthy forests of the Congo basin and surrounded by a plateau rim of older and higher rocks containing potentially valuable minerals, the Belgian authorities decided to concentrate on economic rather than political development. The copper, gold, tin and diamond deposits of the Katanga, Kasai and other areas were rapidly exploited and the government, aided greatly by private capital, played a prominent part in the initiation and growth of industrial enterprises on a massive scale.

Attempts were made to reconstruct African social and political institutions. Customary chiefs were recognized, but over much of the Belgian Congo it proved well-nigh impossible to find chiefs of a sufficient calibre to perform the complex tasks of colonial government. In many areas Belgian administrative officers took over the duties of the chiefs and were in control of the larger towns. Thus a form of *paternalistic rule* developed with the accent on economic growth and a rising living standard, but not on political training with a view to Congolese self-government. No African was appointed to the ranks of the higher administrative service. However, the rapid economic development and urbanization encouraged the appearance of an African middle class of *evolués* who were anxious to adopt a European way of life. In 1956 there was very little sign of the organized political movement of an anti-colonial and anti-Belgian nature that was soon to sweep the country so precipitately and unpreparedly into self-government. In the mid-1950s the colonial policy of the Belgians was based on the assumption that political control would be held firmly in Belgian hands for a long period.

Portuguese colonial policy

Although there is a superficial similarity between French and Portuguese colonial policy there are important differences between the French and Portuguese views of identity with the metropolitan country. In the French territories the criteria whereby an African could be granted French citizenship were far less restricted than in Angola, Mozambique and Portuguese Guinea. These territories had been acquired at the end of the fifteenth century and Portugal has ruled them for nearly five hundred years. Her colonial policy and views on assimilation of Africans antedate the republican views of France, and rest on the assumption that Portuguese citizenship should only be granted to those Africans who abjure tribal customs and become Christians. A very small proportion of the African inhabitants of Portuguese Guinea, Angola and Mozambique have been able or willing to qualify for civilized status. In 1950 out of a total population of nearly 6 million in Mozambique only about 4,300 Africans had become assimilated, and in Angola the figure was slightly higher with about 30,000 *assimilados* in a total population of just over 4 million. There has been a marked differentiation of Portuguese policy towards the few who acquired the privileges of

Portuguese citizenship, and the great majority who were regarded as backward and needing firm control with direction of labour. However, the old *assimilado* system has now been abolished and Africans no longer have to achieve a particular level of education and Portuguese culture before becoming citizens, but can decide whether to live under Portuguese or tribal law.

Portuguese colonial policy, except for a brief period of colonial autonomy and consequent financial chaos early in this century, has for long emphasized the unity of Portugal and the colonies, now regarded as overseas provinces of the mother country. Angola and Mozambique return three deputies each to the National Assembly, although they may not necessarily be local residents. Portugal exercises a close control over Angola and Mozambique, and their budgets have to be approved by the Minister for the Overseas Territories. Angola is under a Governor-General, while there is a government council of officials and elected representatives, and an executive council in Mozambique. In October 1963 the local legislature was granted wider powers so that local interests should be more widely represented. Centuries of neglect and exploitation have meant economic and political stagnation until recently. It says little for the much vaunted policy of assimilation and identity with Portugal that over such a long period so few of the indigenous inhabitants should have reached the status of Portuguese citizenship.

The political significance of Angola and Mozambique lies partly in their control, through the ports of Beira, Lourenço Marques and Lobito Bay, of the valuable mineral and agricultural exports from landlocked Rhodesia, Zambia, Transvaal and the Katanga, and imports of manufactured goods and raw materials required for these areas. Much of the revenue of Angola and Mozambique is derived from handling this lucrative transit trade. Apart from this these countries are under-developed and very sparsely populated. Angola has an area of nearly ½ million square miles and a population of only 4 million. Exports consist largely of agricultural products—coffee, cotton, sisal, maize, cashew nuts, copra and sugar—except for the export of diamonds from Angola. Both countries have a great economic potential, but large areas, particularly in the interior, await development. The location of the Angolan capital at Luanda, on the coast instead of in the interior, reflects this lack of internal growth. The political future of these countries is of vital interest to Rhodesia and Zambia and of considerable concern to the Congolese Republic and the Republic of South Africa.

Spanish colonial policy

Spanish colonial policy is more difficult to define, but was developing on similar lines to those of Portugal until 1963 when it was decided to grant autonomy to the provinces of Fernando Po and Rio Muni. Since 1959 these territories have been regarded as provinces of Spain. However, Spain's colonial possessions are now but a shadow of their former extent and Spain's interest has lain far more in the Americas than in Africa, where she has never held much territory. Spain only retains the small possessions of Ifni, Spanish West Africa and Spanish Guinea[2] with a total area of about 136,000 square miles, much of it sparsely populated desert and a population of about ¼ million. She has relinquished her control of Spanish Morocco.

INDEPENDENT AFRICA

Since Lord Hailey's statement in 1956 that 'it is the singular fate of Africa that so many of its countries should be subject to the political control of one or another of the European powers', there has been a revolutionary change, and by 1961 some 23 countries formerly under French or British rule had become independent. Now only Southern Africa remains under white control, and even here Africans and not Europeans have a political majority in Malawi and Zambia. The causes of this political *volte-face* can be seen clearly in the ten-year period after the Second World War, and may be sought, too, at an earlier date. The long-term reason lies in the British policy of eventual self-government, although if it had not been for the shattering effects of the Second World War the process would have been far more gradual.

Until the late 1930s British colonial policy, a kind of belated Victorianism, aimed at achieving political stability, but leaving the colonies to finance schemes of economic development from their own resources. The depression of the 1930s, riots in the British West Indies and the publication of the Moyne Report caused a change of policy. It was realized that a strong economic base had to be built up before social services could be expanded and the foundation laid for eventual self-government. The Colonial Development and Welfare Fund was created for this purpose, and over £100 million spent on development schemes.

The defeat of European forces in the Far East by the Japanese caused a great loss of prestige. African troops serving in the Far East and North African theatres of war gained knowledge of a free world beyond Africa,

[2] Spanish Guinea includes the islands of Fernando Po and Annobon, Corisco Island and the mainland territory of Rio Muni.

and a realization of the white man's vulnerability. Colonialism has become a word of contempt and world opinion has turned against the idea of colonial exploitation, while anti-colonial views are expressed constantly at the United Nations. The principle of self-determination for Europe's subject states was affirmed at the Treaty of Versailles, and this principle was extended to African colonies after 1945. The British grant of self-government to India, Pakistan, Burma, Ceylon and other territories fired African nationalist leaders. However, the process leading up to African political control might have been much slower had it not been for leaders like Dr Nkrumah in *Ghana* who forced the pace and wrested the reins of power from more moderate politicians such as Dr Danquah and the traditional chiefly rulers. Ghana's success in obtaining independence in 1957 caused the fires of African nationalism to become a holocaust that swept east and south. Nigeria, Sierra Leone and the French West and Equatorial African territories and the Belgian Congo, Tanganyika and Uganda obtained their independence within a few years after 1957.

Britain has rapidly granted political responsibility to her overseas territories. The change from an official to an elected Legislative Council with an elected African majority and then to a ministerial system of government and an entirely African elected parliament based on universal suffrage, has occurred within a short space of time. The French Union, which was highly centralized, could not survive the impact of the war in Algeria and in 1958 was replaced by the French Community. As a result of President de Gaulle's dramatic offer in 1958 of independence, or self-government within the *French Community*, all the ex-French territories, except Guinea, favoured French association. They are now grouped, except for Guinea and Mali,[3] and including Rwanda, in the Organisation Commune Africaine et Malgache, known as OCAM. Matters of common policy include foreign affairs, defence, financial and economic policy, justice, higher education, transport and telecommunication. The French President is also President of the Community, and as such is represented in each state of the Community by a High Commissioner.

An attempt at closer association of Senegal, Soudan, Dahomey and Upper Volta in the *Federation of Mali* was unsuccessful, the various countries apparently lacking sufficient common interest and wishing to pursue separate courses. Similar attempts to link Ghana and Guinea have proved somewhat abortive, and Guinea now seems to be seeking greater French support after enlisting the aid of Communist countries.

[3] Nevertheless, all the former French colonies are closely linked to France by a number of economic, financial and technical agreements.

In March 1956 both *Morocco* and *Tunisia* achieved independence, but Algerian independence was delayed by the long period of civil war, and it was only on 4 July 1962 that Algeria finally obtained her political freedom under the Evian Agreement. The attempt to integrate *Algeria* completely with France failed and the bulk of the white community, formerly a million strong, have left the country.

Farther to the east along the North African coast, *Libya*, under Italian control for just over thirty years, became the first independent state to be created by the United Nations Organization at the end of 1951. The United States and Britain provide financial aid and are granted military facilities, as Libya is of considerable strategic significance.

British influence in *Egypt*, which lasted for about seventy years, ended with the declaration of a republic in 1953, headed first by General Neguib and later by Colonel Nasser. The formation of the *United Arab Republic* in 1958, comprising Egypt and Syria linked with the Yemen, as the United Arab States, raises the question of Arab as opposed to African nationalism, and the possibility of Arab unity. This has been a dream of Arab politicians since the successful revolt against the Turks in the First World War. An even wider link is the Arab League consisting of all North African states and the Sudan. The factors of geographical background, historical process and cultural, linguistic and religious ties all tend to foster Arab unity in a way that is not found in Africa south of the Sahara.

Climatic conditions, with a Mediterranean to steppe and desert climate, summer drought and scanty rainfall, are characteristic of the North African coast, and evoke a somewhat similar response of extensive pastoralism and cultivation in the non-irrigated areas, and intensive irrigation agriculture in the great oasis of the Lower Nile valley. Discoveries of petroleum in the Middle East and the Sahara have caused a revolution in the society and economy of the Arabic-speaking world. The delineation of colonial boundaries enclosing large oil resources has made spectacular differences to the economy of a state such as Kuwait, and will affect increasingly the prosperity of Algeria and Libya. Algeria's extensive Saharan area is the envy of her neighbours, and was a cause of the prolongation of the war with France. Despite the appeal of Arab unity it is difficult to foresee any form of closer association of Egypt with North-West Africa, because of the large area of desert and semi-desert separating these regions, and also because of the Maghreb's links with France. Equally a link between Egypt and the Sudan appears unlikely in view of the fact that Egypt's cultural and historical ties lie far more with the Middle Eastern countries than with the Sudan, despite Egypt's absolute dependence on the waters of the Nile. President Nasser has, however, made overtures to a number of African countries, and has rivalled Dr Nkrumah as a champion of Pan-Africanism.

Egypt's union with Syria has been somewhat uneasy and suffered temporary setbacks, as Egypt has tended to dominate the union. There is a great difference in size of population, Egypt having 25 million people, compared to Syria's 4 million. Whereas Egypt is confronted with grave problems of a mounting population, land shortage and poverty, Syria has land to spare and a rising living standard. Aid has been received from Communist countries, notably for the High Dam project in Egypt, and United Arab Republic foreign policy has at times been anti-Western.

Thus, over the North African political scene there is a thin veil of Arab unity, easily pierced by closer scrutiny, which reveals external associations with France in the Maghreb, with Britain and the U.S.A. in Libya and in the past with the Communist bloc in Egypt.

The *new African countries* are faced by many grave problems such as the great dependence on inefficient forms of agriculture, widespread poverty, important differences of language and custom and the arbitrary nature of the old colonial boundaries. Animosity between rival groups or the struggle for power between military and civil authorities may endanger the continued existence of the state. Under these circumstances opposition to the ruling party is considered as dangerous to the welfare of the state. Many African politicians consider that a British-type democracy based on a strong opposition party can only succeed in a homogeneous society; and therefore if an African state (with its many different groups of people) is to prosper politically it must be under the control of a single or mass party.

Europe has exported the idea of nationalism and the nation state to the rest of the world and in the new Africa there is much talk of *nationalism*, patriotism and the need for a common purpose and pride in the state. The burning question is how to achieve this purpose in the face of strong regional and tribal differences. Can this be done swiftly, arbitrarily and vociferously through the use of mass propaganda and the party machine, the whipping up of emotions and the chanting of Uhuru, attacks on colonialism and neo-colonialism and the use of colourful personalities in politics; or gradually and less spectacularly by means of economic, educational and cultural programmes? It is possible, as in Denmark, to build up a healthy pride in community and state by studying folk history, customs and literature, as long as this does not degenerate on the one hand into a decaying, folksy cult, and on the other hand to an overweening and rabid pride in a meagre history which breeds contempt for the other groups in the state. Africa has for long adopted foreign crops, customs and ideas, and recently many African countries have shown an increasing interest in their past cultures and history by promoting the study of such subjects as pre-history, linguistics and social anthropology.

A closer examination of a West African country such as *Nigeria* may help the onlooker to understand more clearly the political problems confronting a newly emergent African state and thus hazard a forecast as to future political trends. Nigeria, with an area of 366,000 square miles and a popula-tion of 36 million, is the most populous state in Africa. There are marked differences between the four regions of the North, East, West and Mid-West and also within the various regions themselves. The *North* with its 18 million people[4] has the largest area, consisting largely of savanna with a pastoral, cultivating and mining economy, a patchwork of strong Moslem emirates and numerous small pagan tribes, and is larger than many an individual African country. The *East* is the poorest of the regions with an Ibo core, dense population and a meagre economy based on palm oil grown on sandy, leached soils in a region of dense rainfall, and now looking to petroleum and the growth of secondary industry to bolster up its economy. Yet here the movement for independence was strongest, led by Dr Azikiwe, and there were no strong traditional forces to delay it. The *West* is more like Ghana in size of population and a prosperous economy based on cocoa, rubber, timber and other products. It has a Yoruba core and tradition of state and city life in centres such as Ibadan, Ife and Abeokuta. The Mid-West is the smallest of the four regions and is centred on Benin. It has Ibo and Yoruba minorities, and was excised from the Western Region.

Within the regions themselves there are considerable ethnic and linguistic differences and therefore, in view of Nigeria's large and great ethnic and regional diversity, it is perhaps understandable that a federal, rather than a unitary form of government, should have developed.

Ghana's well populated and prosperous coastal and forest region with a variety of products such as cocoa, rubber, timber, gold, diamonds and manganese, has become the core of the state; the Northern territories are backward and isolated, lacking a railway and suitable cash crops, and contri-bute little to the political and economic life of the state. Ghana's unde-veloped north is in strong contrast to the powerful Northern Region of Nigeria. Nigeria has, therefore, a number of potential cores to the state, and

---

[4] According to the latest census (1964) the North has a population of nearly 30 million. This high figure has great political significance, as the North is able to dominate the South by sheer power of votes. However, the results of this census are somewhat suspect and the Eastern Region has not accepted it.

once the problems of regional differentiation have been overcome, its large size, population and varied resources should prove sources of strength which, if wisely handled, could enable Nigeria to become a major political force in Africa.

Regional and tribal *political parties* have developed, the Northern Peoples' Congress in the North, the National Council of Nigerian Citizens in the East, and the Action Group in the West. A coalition of the N.P.C. and N.C.N.C. rules the country, and the Action Group have provided the main opposition party. The checks and balances that should have developed within a regional and federal framework have not proved strong enough to rule out the possibility of a mass party and the decay of democracy. In-effective and corrupt rule in the Western Region led to riots and interference by the federal authorities in the government of the Western Region. This is viewed by some as the prelude to one-party rule, that is a central govern-ment dominated by the N.P.C., despite the great differences between the traditionalist Moslem political forces of the North, where the women have no vote, and the more modern and Christo-pagan elements of the south. Some Nigerian politicians feel that the tremendous political problems faced by an under-developed country do not warrant the luxury of opposi-tion, and that the resources of the country would best be exploited by a one-party system of government, which might give Nigeria the necessary political stability. Conversely some Northern politicians have talked of secession from Nigeria and a link-up with other Moslem states such as the Sudan. This would be extremely difficult to achieve as the North has very close economic ties with the South, and if secession occurred it would have to face the same economic problems of an ocean outlet that is proving so hazardous for countries like Mali and Niger.

In *East Africa* the cultural differences between the various African, Asian and European groups are wider than in West Africa. In Kenya there is mistrust between the Bantu cultivators, represented by such tribes as the Kikuyu and Kamba, and the Hamitic pastoralists, and between individual African groups. Tension has also developed between the Africans and the small Asian commercial and artisan group, within which in turn there are wide differences in outlook, religion and customs. The position is further complicated by the presence of the compact white settler group. With the rise to power of the Africans, however, the political influence of the white and Asian communities has been greatly reduced, and the major issue is between the main African political parties, the Kenya African National Union (KANU) and the Kenya African Democratic Union (KADU). KANU's main source of power comes from the Kikuyu and Luo cultivators; that of

KADU from a number of small pastoral tribes who fear KANU's domination.[5] KANU presses for a unitary state, KADU for a federal state with a considerable devolution of power for the regions. A compromise has been reached and a regional framework adopted, the new regional boundaries cutting across many of the original provincial boundaries. The land reservation policy introduced by the British government has engendered a bitter quarrel over land ownership, and this is perpetuated in the different forms of government proposed by KANU and KADU.

On the north-eastern frontier some 200,000 Somalis included in Kenya are anxious to join their compatriots in independent Somalia in the Eastern Horn of Africa. Kenya African politicians are not willing to accede to their requests, which would mean a change in the boundary and loss of a considerable area to Kenya. Despite the fact that this is a semi-arid region sparsely populated by wandering pastoralists, the yielding of territory to another country would be a blow to national pride that the Kenya Africans will not allow to occur.

In *Uganda* the major field of contention before independence was the relationship between the rest of the country and the conservative Buganda, almost isolationist, and yet wishing to dominate the political scene, although with only a quarter of Uganda's population. Buganda has a thriving economy based on coffee and cotton in the well distributed double maximum rainfall region on the northern shores of Lake Victoria. The rest of the country consists of the smaller kingdoms of Bunyoro, Toro and Ankole, the densely populated region of Kigezi, the prosperous Eastern Region, somewhat similar to Buganda, and the seasonal rainfall area of the pastoral north which includes the scantily populated and backward Karamoja. There are considerable differences in ethnic composition, economy and outlook between the Bantu lakeside cultivators and the pastoralists of Northern Uganda. However, skilful bargaining between Mr Obote, the present Prime Minister of Uganda, and the Kabaka and his Lukiko ironed out the differences, and the country gained independence with an apparent unanimity of voice, if not of purpose.

In *Tanganyika*,[6] despite the great barriers to political unity of great distances between the economically prosperous and densely populated areas and the empty heart of the country, the proliferation of tribes and languages,

---

[5] In the elections of May 1963 KANU and allied independent candidates took 71 of the 112 seats in the house of representatives, 20 of the 38 seats in the senate and gained control of three of the six regional assemblies. Since Independence in December 1963 tension has lessened between KANU and KADU and KANU is now the dominant party.

[6] Now known as Tanzania since the union with Zanzibar in December 1963.

the lack of common ties and of educational and political experience, independence came more quickly than in Kenya and Uganda. The reason for this lies in the triumph of man over the difficult human and physical background, and particularly the triumph of Nyerere who forged the weapon of the Tanganyika African National Union, which upset Governor Twining's attempt to establish a multi-racial state. Tanzania, more than Uganda, faces awkward physical problems of aridity and variability of rainfall, and soil poverty which restricts agricultural development. The country has, however, a considerable mineral potential, and in 1964 produced about £8 million worth of minerals, diamonds being the most valuable at £6¾ million.

In the *Eastern Horn of Africa* the two major political units are Ethiopia on the high plateau and Somalia on the arid lowlands. For most of its history Ethiopia has been inviolate in its mountain stronghold except for a very short period of Italian rule. Ethiopia, therefore, exercises much political influence in ex-colonial Africa—Addis Ababa has been selected by the United Nations as the headquarters of the African Economic Commission. Two-thirds of the 3 million Somalis live in Somalia, most of the remainder in The Ogaden. They are nomadic pastoralists and split up into a number of clans who frequently dispute the possession of scant resources of grazing and water. There are a number of boundary disputes, particularly with Ethiopia. Unity will be difficult to achieve and the new state is also faced with critical economic problems.

The closely integrated political and economic framework, in which the Belgian administration and large companies acted as partners in the Belgian Congo, was disrupted by the hasty withdrawal of the Belgians in 1960. Very few Congolese had been trained to take administrative responsibility, so that confusion and anarchy reigned until United Nations aid was sought to help restore the situation. The mineralized areas of the South and West with a wide range of valuable mineral exports of copper, tin, gold, diamonds, cobalt, manganese, zinc and uranium, coupled with agricultural exports of oil palm products, cotton, coffee and tea, cause the *Congolese Republic* to be potentially one of the richest countries in Africa. But before political stability can be achieved problems must be solved of the unwieldy size and peripheral nature of the capital and economically developed regions, the unhealthy nature of the steamy Congo basin which forms the heart of the country, and the wide ethnic diversity and scattered distribution of the population. Protracted negotiations have brought the Katanga, the richest region which aspires to independence, into the political fold of the Congolese Republic, but the Katanga and other peripheral regions, difficult to govern from Kinshasa, a thousand miles away, are liable to try and pursue an independent course of their own again in the future. A loosely federated state will probably develop in which the federal units are based on tribal groups. The former mandated territories of Ruanda and Urundi have become independent countries, now called Rwanda and Burundi. After ten years the attempt to build a multi-racial state comprising the Rhodesias and Nyasaland has failed, and the *Federation* has broken up. The main reason for this was mounting African opposition to a federation imposed on them by the white minority and the British government. To the Africans federation has meant a device for ensuring the maintenance of white superiority. The Monckton Commission referred to the pathological hatred of Federation amongst the Africans, but suggested that it should be continued because of its great economic advantages. Large sums of private capital have been invested, the national income has increased considerably, and federal projects such as the Kariba Dam have been constructed since Federation came into being in 1953. Some economists criticized this assumption and maintain that Nyasaland would have benefited more in association with Northern Rhodesia alone than with the two Rhodesias in a federation. The African sees the expanding towns and industries of the white highlands of Rhodesia, and knows where the economic benefits have gone.

The core of *Rhodesia* lies in the white highlands above 4,000 feet in altitude, and stretches in a wide semi-circle from Bulawayo, via Salisbury to Umtali. This is the railway belt containing the main towns and industries, the tobacco belt and the chief gold, asbestos and chrome mines. Similarly the railway belt of Zambia, from south of Lusaka to the Copperbelt, constitutes the economic and political core of the country. These well developed zones, with the capitals of Salisbury and Lusaka, are bordered by great, poverty-stricken, peripheral areas containing the bulk of the African people who supply migrant labour for the prosperous European areas. Perhaps if more of the economic and political gains consequent on federation had been dispensed more widely in the African areas federation might have succeeded.

The violent opposition of Dr Banda and the nationalist movement in Malawi compelled the British government to agree to its secession from the Federation. African majority rule in *Zambia* has enabled the other northern partner to achieve the same objective. Faced with the prospect of African domination in Rhodesia the white voters returned Mr Smith and the right wing Rhodesian Front to power, pledged to maintain white supremacy. The ideals of partnership in Central Africa have proved abortive, and the political dynamite of democracy with universal suffrage has again meant the rise to power of the common man, the African, except in Rhodesia where the white minority declared independence in November 1965.

The gates of *South Africa's* protective cordon of apartheid are opened periodically to allow the African to work in the white preserve. Apart from this, there is very little contact between the various racial groups, who are confined to their own spheres; there is very little social and economic mobility and no political mobility. This has led inevitably to increased tension and embittered race relations, as the subservient non-white groups are not allowed to reap the full rewards of their labour for the white group. The Nationalist government, employing the maxim 'divide and rule', are pushing ahead with their policy of setting up quasi-independent areas, the *Bantu Homelands*, for the various African groups. By the end of 1962 there were six territorial authorities for the Xhosa in the Ciskei and the Transkei, the Tswana in the Western Areas, and the Tsonga, Bavenda and many of the Northern Sotho people in the Transvaal. According to the original plan such bodies have still to be established for the Southern Sotho, Zulu and Swazi groups. Despite the opposition of the Tembu chief, Sabatha Dalindyebo, and disturbances in Pondoland, the Transkei has been given a measure of independence under Paramount Chief Matanzima. The authority of the chiefs in the other reserves has been bolstered up by the Nationalist government in an attempt to restore tribalism.

Meanwhile the prosperity of South Africa depends to a great extent on the increasing economic integration of the various groups. Economically the policy of total apartheid is unworkable and politically it is not acceptable to the Nationalist government. The attitude of the Commonwealth to apartheid has been stated clearly and has resulted in the withdrawal of South Africa from the Commonwealth. The Union of South Africa has at last become a Republic, a triumph for Afrikaner nationalism, but one more stage in South Africa's isolation from the rest of the world. How long will the Afrikaners be able to retain control over the non-white majority in South Africa; will they be able to turn the tide of African nationalism, or will the non-white group soon seek a solution through revolution? Some South African economists and industrialists maintain that the recent economic boom in South Africa will cause restrictive legislation such as the civilized labour policy to be modified, as skilled labour is in such short supply and many non-whites are performing skilled work.

PART II: THE REGIONS OF AFRICA

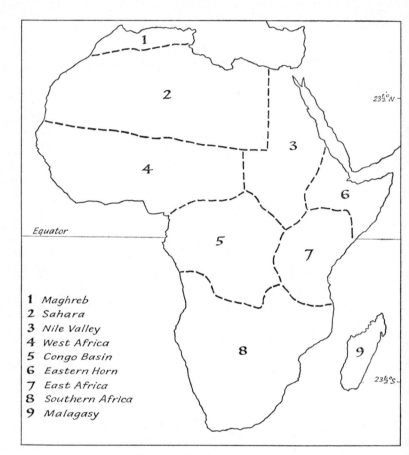

1 Maghreb
2 Sahara
3 Nile Valley
4 West Africa
5 Congo Basin
6 Eastern Horn
7 East Africa
8 Southern Africa
9 Malagasy

FIG. 14—Geographical regions

CHAPTER FIVE

The Maghreb

THIS region consists of the north-western mountainous belt of Africa, a counterpart of the Cape folded zone and yet more akin physically to southern Europe than the rest of the African continent. It consists principally of the Atlas ranges which form a zone of mountains and plateaux stretching for 1,500 miles from the Atlantic coast to the Gulf of Gabes. To the south and east is the Sahara. The Arabs called this region Geziret-el-Maghreb or the Island in the West. Bordered by sea and desert it forms an island of comparatively fertile and well watered country that has for long attracted invaders. The region has been divided into the three political units of *Morocco*, *Algeria* and *Tunisia*, and treated on a more detailed regional basis within the political boundaries of each of these countries.

The total area, excluding the Saharan parts of Algeria and Tunisia, is about 300,000 square miles and the population about 26 million.

HISTORY

The Mediterranean has exerted a strong influence on the history of the Maghreb. A constant stream of invaders has come by sea and land, usually from the east to settle in and trade with this attractive region. Its situation has also been of great significance, with Tunis controlling the Sicilian gap between the basins of the eastern and western Mediterranean and Ceuta and Tangier the Straits of Gibraltar and western entrance to the Mediterranean. Despite the stamp of diverse invasions and cultures the basic Berber stock and culture has remained and, although Arab culture and religion have pervaded the area, Berber dialects are still widely spoken, particularly in the mountain areas of Morocco and Algeria. Tunisia has been most affected by Arab influence, partly because it lies a thousand miles further east than Morocco, and partly because relief is more subdued than in Morocco and Algeria.

The first permanent settlements were established by the Phoenicians in about the tenth century B.C. The greatest of these was Carthage, which attempted for four hundred years to maintain her trading monopoly against Greek and later Roman attacks. After the destruction of Carthage Roman influence spread with a string of coastal towns, extension of cultivation and an export trade in olive oil, wine, corn, hides and skins and a luxury trade in animals. Fruit trees were probably introduced by the Romans and were adopted by the Berbers. Latin was widely spoken and Christianity was adopted. Roman influence declined after four hundred years of prosperity and the Maghreb passed under Vandal and Byzantine control until the Arab invasion of the eighth century A.D. More serious was the effect of the Hillalian invasion when the Arabs swarmed like a plague of locusts from the Nile Valley in the eleventh century, destroying towns and settled agricultural areas so thoroughly that recovery in some areas is not complete after nearly a thousand years.

At times Berber domains have spread from North Africa into Spain where brilliant Moorish cultures arose. The Almohad empire with its capital at Seville was the strongest of these. On its ruins arose three kingdoms that correspond approximately to modern Morocco, Algeria and Tunisia. In the fourteenth century the Merinid dynasty dominated North Africa and its powerful corsair fleet ravaged North African waters. The corsairs became particularly powerful during the seventeenth and eighteenth centuries, making Christian shipping unsafe. Disruption of European trade in the Mediterranean was one of the main reasons for interference by France and other European powers in North Africa and the suppression of the Barbary pirates. French influence became important in Algeria during the 1830s, control being established by about 1850 and later extending south into the Sahara. In Tunisia a protectorate was declared in 1881. The Spanish obtained a foothold in northern Morocco in 1880, while the French assumed increasing control over the rest of the country at the beginning of the twentieth century.

Since 1945 increasing local nationalism has enabled the region to achieve independence after a long period of colonial rule, and there are now three independent countries of Morocco, Algeria and Tunisia, still with strong French links.

STRUCTURE AND RELIEF

The Atlas ranges are separated from each other by a series of wide plateaux, depressions and deep valleys trending west-south-west to north-north-east (Fig. 15). Hercynian movements caused the uplift of the central plateau of the Atlas and the Moroccan Meseta but Tertiary folding produced the present relief, aided by recent glaciation. In the north of Morocco, the first range of the Atlas mountains are the Riff, a continuation of the Sierra Nevada and separated from the main body of the Atlas by the

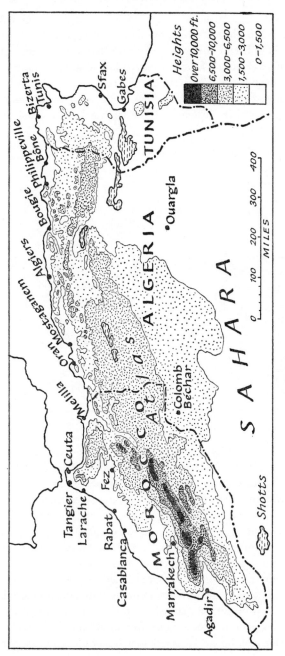

FIG 15—The Maghreb: relief and main towns

Taza gap. The High Atlas are a formidable barrier, the core of Archaean rocks forming the highest peaks, and a wall of rugged highland, mostly over 9,000 feet in height and with many peaks over 12,000 feet, extends for nearly 300 miles (Plate 1). Much of the Middle Atlas is a limestone plateau dissected by deep gorges and serving as a great aquifer which gives rise to Morocco's largest rivers. The Anti-Atlas (with an average altitude of 5,000 feet) are much lower in altitude than the High Atlas and consist of schists

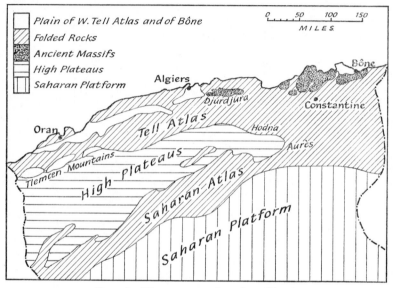

FIG. 16—The structure of northern Algeria

and quartzites flanked by dolomitic limestones. To the south is the Wadi Dra and the Sahara. North-west of the Middle Atlas is the Moroccan Meseta which breaks down into a narrow coastal plain.

Further east the Tell Atlas extends for nearly a thousand miles into Algeria and Tunisia to culminate in Cape Blanco (Fig. 16). It consists of a number of mountain masses and plateaux separated by plains and river valleys into a number of distinct belts. The coastal mountains, best developed in the Djurdjura massif, largely of Mesozoic limestone, are separated from the interior mountains by the Oran Bougie depression. From Tlemcen to Frenda there is an extensive dolomitic limestone plateau similar to the Middle Atlas and forming a marked scarp to the north. South of the Tell Atlas is a high plateau, 3,000–4,000 feet in altitude, cold and arid with inland drainage and the salt lakes and marshes of the Shotts, water-filled in winter, but dried up and salt-encrusted in the summer. Before the Sahara is reached there is a further range of the Saharan Atlas which terminates to the east in Cape Bon. Contact between the Saharan Atlas and the desert may be abrupt as in the case of the great faulted depression of the Chott Melhrir, which is partly below sea level and situated at the base of the Aures Massif.

The valley of the Medjerda in Tunisia separates the well forested sandstone belt of the northern lower Tell from a more barren limestone region of the higher Tell to the south. This breaks down into a zone of hills and plains of interior drainage further south, a continuation of the salt marshes and lakes in Algeria. To the east is a broad coastal plain. South of the Chott el Djerid and Gabes is a low plateau composed of Mesozoic rocks.

CLIMATE

The region has a transitional Mediterranean type of climate ranging from a humid coastal type to arid and semi-arid types, with a spring maximum of rainfall in the interior and a dry hot summer (Fig. 17). Because of the very diversified relief there are many variations in temperature and precipitation. In Morocco the climate is derived more from the Atlantic than the Mediterranean. There is a marked effect of the cold Canaries current on the climate of the coast. Temperatures are equable, low in summer and comparatively high in winter, rainfall is low and fogs occur often. The following statistics for Mogador, Marrakech and Tunis illustrate the effects of coastal and interior location.

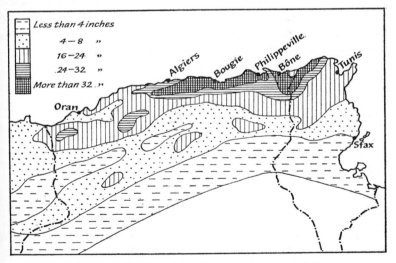

FIG. 17—Algeria and Tunisia: mean annual rainfall

	Altitude (feet)	Distance from sea (miles)	Jan. mean temp.	July mean temp.	Temp. range
Mogador	33	—	51° F. (10·5° C.)	68·2° F. (20·1° C.)	11·5° F. (6·3° C.)
Marrakech	1,542	80	51·6° F. (10·8° C.)	82° F. (27·8° C.)	33·7° F. (18·6° C.)
Tunis	141	—	48·4° F. (9·3° C.)	77·7° F. (25·4° C.)	30·2° F. (16·6° C.)

Along the Algerian coast mean summer temperatures are from 75–80° F. (24–27° C.) and in winter 43–46° F. (6–8° C.). In the interior altitude and distance from the sea may cause low winter temperatures; for example Geryville at 4,280 feet has a mean January temperature of less than 39° F. (4° C.), comparable with Oxford. Summers are, however, hot with mean temperatures above 80° F. (27° C.).

In winter low pressure forms over the Mediterranean and westerly winds are dominant. Depressions passing to the north bring considerable rainfall to the mountain ranges of Morocco and Algeria and the coastal strip west of Algiers. The Riff Mountains in Morocco have over 30 inches a year, but rainfall decreases rapidly to the south and Mogador has only 13 inches. In Algeria and Tunisia rainfall also decreases rapidly from over 30 inches in a narrow belt from Algiers to Bizerta to below 16 inches on the high plateau and less than 8 inches south of the Saharan Atlas. On the higher mountains snow may lie for a considerable period, especially in Morocco, and the snow-melt waters of spring are useful for cultivation.

Rainfall (in inches)

	J	F	M	A	M	J	J	A	S	O	N	D
Mogador	2·2	1·5	2·2	0·7	0·6	0·1	0	0·2	1·3	2·4	2·0	13·2

1·6

	J	F	M	A	M	J	J	A	S	O	N	D
Algiers	4·0	2·6	3·3	2·0	1·7	0·7	0·1	1·2	3·4	4·1	4·0	27·4

5·7

The above figures of average monthly rainfall for Mogador and Algiers illustrate the fact that summer is a dry period, Mogador only receiving 1·6 inches out of an annual total of 13·2 inches during the six months from the beginning of April to the end of September. In winter high pressure pre-vails over the Sahara and the easterly and north-easterly winds are dry and descending. The sirocco, a very dry and hot southerly wind, is probably the most damaging wind, causing crops to wilt.

VEGETATION

The vegetation pattern reflects the varying influence of soil, rainfall, altitude and man. In the well watered areas there are limited stands of forest with Aleppo pine, holm oak and thuya at lower altitudes and cedars at higher altitudes. The cork oak is the most characteristic tree and in the eastern part of the Maghreb needs over 25 inches rainfall. In the High Atlas of Morocco there are a number of altitudinal zones with distinctive types of vegetation. The lower calcareous foothills have a degenerate brushwood of olive, sumac and dwarf palm, junipers and carob trees. From about 3,000 to 5,000 feet is the zone of thuya woods, conifers that regenerate quickly, with wide areas of scrub, followed by red junipers from 5,000 to 8,000 feet and evergreen oak and thurifers above this. Widespread de-forestation has occurred over a long period and with consequent erosion of the mountain slopes it has proved difficult to re-establish a forest cover which is replaced by bush and scrub. The French have attempted to re-afforest in a number of areas with eucalyptus and pines.

With a rainfall of less than 15 inches on the high plateau of Algeria there are few trees or shrubs and there is a steppe vegetation of drought-resistant, perennial grasses such as the halfa, which grows in large tufts and, as esparto, is exported for paper making. Near the salt pans are found holo-phytes, such as rice grass, sea rushes and sea lavender, which provide good grazing during the dry season. The desert margin bears close against the Saharan Atlas with its forest of Aleppo pine, holm oak and red juniper on the lower northern slopes and cedar and mixed deciduous oak forest on the higher slopes.

POPULATION AND AGRICULTURE

A significant proportion of the population is of Berber and Hamitic stock, generally of medium height, long-headed, thin, straight-nosed and with black hair. There is also a fair tall type that would pass unnoticed in Northern and Western Europe. In the rural areas they are usually cultiva-tors living in small towns and villages with a well developed family and clan system. Fruit trees such as oranges, lemons, figs, peaches and apricots, olive groves and vineyards are widespread (Fig. 18). Cereals are grown in scattered areas, but yields are poor as the fields are only visited at sowing

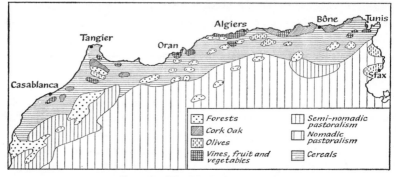

FIG. 18—The Maghreb: land use

and harvesting time, the rest of the year being spent in a semi-nomadic existence seeking pasturage.

Every district tends to have its weekly market where livestock products and dates are exchanged for oil, grain and manufactured products. Permanent cultivators usually live in low windowless houses (flat-roofed or tiled) which cling to the contour on steep slopes or mountain and plateau top. Villages are often sited in inaccessible defensive positions and are hard to identify against a background of rock and stony ground. Prolonged invasion, and particularly the Arab invasion of the eleventh century, caused the Berber people to retreat to the highlands and away from the coastal region. The semi-nomadic cereal cultivators usually live in temporary huts or poorer houses built of sun-dried brick or stone.

The nomadic herdsmen are mainly Arabs. Those from the desert may move great distances with their flocks of sheep and goats and herds of camels in search of pasture and water. Those in the Atlas mountains practise a more local form of transhumance, moving into the mountains with the onset of summer and when the cereal crop has been harvested returning to the lower areas for winter. Nomadism has decreased considerably under French influence as the better land has been taken up for permanent cultivation. Nomads live in tents made of long strips of wool, camel hair or goat skins often pitched over a low mud or stone wall.

Proximity to Europe has encouraged a large foreign element among the population. Prominent are the Moriscos or Moors who left Spain at the reconquest, the Jews, some of whom came in before the Arab invasion, and recently the French, Spanish and Italians, who have settled in the towns or fertile rural areas. In 1962 Europeans in Algeria owned about 6 million acres, some the best in the country, and produced most of the export crops.

The majority of the population of the Barbary states is still agricultural and rural, although the urban percentage is increasing rapidly. Whereas the rural population is Berber or Arab, the towns are inhabited by a mixed Arab-Berber population with a large foreign element of Moriscos, Turks, Jews, French, Spanish and Italians. Until recently, for example, over half the population of Oran and Algiers was of French origin. There tends to be a greater distinction between town and country than in Europe, although this is becoming blurred with the rapid post-war urban growth. Rural poverty has driven large numbers in Morocco to migrate to Casablanca, which has quadrupled in population since 1945 and now has a million people. Large numbers of Algerians have also migrated to France to seek temporary work and the foreign element in the towns has declined greatly since independence.

The towns of the interior tend to be predominantly Arab or Oriental in style, the houses backing onto narrow streets and lanes and with busy souks or market places, and craftsmen occupying their different quarters as they did in medieval Europe. The most prominent buildings are usually the mosque and the palace for the sultan or Kaid. The town is often surrounded by a defensive wall. European urban planning influence is strongest in the ports and coastal towns where the wide streets and regular layout contrast with the indigenous sprawling pattern of the old town.

MOROCCO

The kingdom of Morocco has an area of about 180,000 square miles and an estimated population in 1963 of $12\frac{2}{3}$ million people. In 1954 there were nearly $\frac{1}{2}$ million Europeans, but this number has been reduced greatly by the movement of about 150,000 back to France following on Moroccan independence. The country was shared until recently by France and Spain as two protectorates, former French Morocco comprising the bulk of the country. In 1956 and 1958 the French and Spanish protectorates and in 1960 Tangier (hitherto an international port) were handed over to Morocco.

Morocco can be divided into a number of regions, the fundamental distinction being between the Atlas mountains and the western low plateau and coastal plain. Political factors have, however, tended to separate Spanish from French Morocco and the Riff Mountains comprising the former *Spanish Morocco* are treated as a separate region.

SPANISH MOROCCO

This is a long narrow zone extending for some 200 miles from the Atlantic coast to near the Algerian border, with an area of less than 8,000 square miles and a population of just over 1 million. It is a rugged area of mountainous country consisting largely of the Riff mountains and a small strip of coastal plain to the west. Precipitous slopes tumble to the Mediterranean and Jurassic limestone forms a conspicuous feature in a bleak landscape. There are forests of juniper and cedar in the western areas which have a higher rainfall. The mountains are used largely for pastoralism, and settlement occurs in the valleys, some areas being densely populated, where vegetables, cereals and tobacco are grown, while on the western coastal plain cereals and vines are cultivated. The most important export is of iron ore worked near Mellila. The main towns are Tetuan, Ceuta and Mellila. *Tetuan*, situated some six miles from the coast and 20 miles south of Ceuta, was an important military and administrative centre and the headquarters of Spanish administration. Ex-Spanish Morocco is now being integrated more fully into the economy of Morocco and regaining its natural hinterland.

Tangier was an international port from 1923 to 1960, but is now incorporated in Morocco. At the head of a wide bay the picturesque town, with a population of over 140,000 and with white houses set on a hill crowned by a citadel, has lost some of its importance as a port with its loss of international status and the marked growth of Casablanca.

Ifni is a small enclave of Spanish territory in Southern Morocco about forty miles north of the Wadi Dra, the border between Morocco and Spanish West Africa. It has an area of less than a 1,000 square miles and a population of about 40,000.

FRENCH MOROCCO

The great bulk of Morocco consists of the former French Protectorate. The French who had been in Algeria since the 1830s only moved into Morocco in 1912. They were not able to control the whole area effectively until 1933, partly because of the rugged nature and high altitude of the Atlas ranges and partly the resolute resistance of the Berbers. There are two main regions, the western plateau and coastal plain and the Atlas ranges.

The western plateau and coastal plain

The western plateau and coastal plain is a triangular area bounded by the Riff mountains and High and Middle Atlas to the north, east and south, and by the Atlantic to the west. They consist of the Meseta, the coastal plain, the basin of the Sebou and the Taza gap. The Meseta with an average height of about 1,000 feet is composed of an Archaean platform, covered by sedimentary rocks, which acted as a stable block against which the Atlas mountains were pressed. North of the Meseta is the Sebou river which, by way of the Taza gap further to the east, affords the only practicable routeway to the coastal areas of Algeria. The northern areas receive sufficient rainfall for cereal cultivation, but farther south the rainfall drops below 12–15 inches and cultivation is impossible without irrigation.

Irrigation is essential for permanent agriculture in the long dry summer. Morocco is relatively well endowed, as the mountains have over 30 inches of rainfall and a snow cover which lasts for over six months above 8,000 feet. The limestone plateau of the Middle Atlas fulfils a valuable role as an aquifer and is the watershed for Morocco's major rivers. Little more than a third of the potential irrigable area of $2\frac{1}{2}$ million acres is now irrigated. There are major schemes on the Oued Beth and lower Sebou, where in the Rharb much potential land can be irrigated, and on the Oued N'Fis, Oued el Abid and Moulouya.

Only about one-sixth of Morocco is *cultivated* and most of this area is found in the western lowlands. Orchards, vineyards and gardens cover little more than 1% of the total area. Although there were far fewer European settlers in Morocco than in Algeria before independence their farms occupied nearly 10% of the cultivated area and they produced a considerable proportion of the exports of fruit, vegetables and cereals. Indigenous farmers tend to produce for the local market, and use backward farming methods with little mechanization and fertilizers. Barley and wheat are the main cereals, occupying 80% of the cultivated area, barley being produced in greater quantity and exported. Yields vary a great deal according to rainfall. There are over 10 million olive trees and their distribution is similar to that of the vine, along the coast and with concentrations at Fez, Meknes and Marrakech. Citrus and early vegetables are grown under irrigation for export and the local market. Nearly half the citrus production comes from the Rharb plain. Market garden products have the advantage of being several weeks earlier than in the rest of the Maghreb, but are handicapped by considerable distance to markets and high costs of transport. The production of citrus has increased nearly tenfold in the last twenty years. Fishing is important along the coast owing to the influence of the cold Canaries current, sardines being the main fish caught.

The production of *minerals* is significant. By far the most important are phosphates, $8\frac{1}{2}$ million tons of which were produced in 1963, accounting for nearly a quarter of exports by value. There are two main areas of production at Kourigba and Louis Gentil. The beds consist of a number of

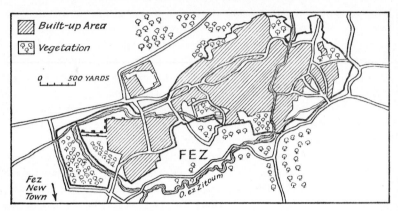

FIG. 19—The old town, Fez

horizontal layers of phosphate-bearing sand with a high content of lime phosphate. There are a number of lead and zinc mines, particularly in the Oujda area, although these deposits will be exhausted in 30 years, and anthracite is mined at Djerada. Iron ore is mined at Uixan and just over 2 million tons of iron ore was exported in 1962, mostly from Uixan. The production of cobalt is also important near Bou-Azzer.

The bulk of the *population* is situated in the western lowlands with densities of over 100 per square mile near Mazagan and 50–100 per square mile over most of the area. Villages are most numerous in the well watered piedmont zone below the Atlas mountains and in the central part of the Sebou basin. It is in this zone a hundred miles from the coast that the old walled towns of the interior, such as Marrakech, Fez and Meknes, have developed. *Fez*, a centre of Moorish culture with its ancient university, was founded in the ninth century (Fig. 19 and Plate II). It grew up at the crossroads of two important caravan routes, one from the west to Algeria, the other leading northwards from the Sahara to the Straits of Gibraltar. It has many local crafts of leather, pottery, carpets, copper and brass which are in great contrast to the modern industries of Casaablanca.

The walled city of *Marrakech*, with a population of ¼ million, is slightly larger than Fez. It is the main centre for the High Atlas and north-western Sahara. It extends over the irrigated plain of the Haouz and is surrounded by date-palm and olive groves, vineyards and orchards. The town commands two natural routeways over the High Atlas, to the south-east via the Rdat valley and to the south-west via the Nfis valley. There are also good communications to Mogador, Casablanca, Meknes and Ouarzazat.

The old walled town has many fine mosques, public buildings, palaces and specialized markets; carpets, leather, jewellery, weapons and pottery are manufactured locally.

The coastal towns, particularly Casablanca, have grown considerably. The main town and commercial capital of Morocco is *Casablanca* (the White House) which has expanded greatly in size under French influence, having quadrupled its population in little over 20 years to nearly 1 million. It has become the most important Moroccan port, handling most of the country's trade with nearly 10 million tons of cargo in 1961. Factors contributing to the town's rapid growth have been the construction of good breakwaters, the export of phosphates, central position on the Atlantic coast and good railway links to the rest of the country and Algeria.

Casablanca has developed in a semicircle round the harbour and old town or medina, at the southern tip of which is the Place de France, forming the focal point of the new town which has been planned by the French. The industrial area lies to the east and there are sugar-refining, fish-curing and canning, food-processing and engineering industries, while cement, soap, superphosphates and other products are manufactured.

Rabat is the capital of Morocco and is situated on the coast about fifty miles north-east of Casablanca at the mouth of the Bou Regreg river, on whose north bank lies the smaller town of Salé. Salé was a famous corsair centre during the Middle Ages and the most important commercial centre in West Morocco. It engaged in many struggles with Rabat, but now forms part of the twin settlement of Rabat-Salé. Rabat was founded by the Almohads in the twelfth century and prospered until the seventeenth century. It has declined in commercial importance with the growth of Casablanca and Port Lyautey.

The Atlas

The Atlas ranges of Morocco consist of the three separate areas of the High, Middle and Anti-Atlas. In the High Atlas the Hercynian rocks form some of the highest areas, above 12,000 feet, near Marrakech. This great range stretches for 200 miles in an almost unbroken line only pierced by occasional high passes. Snow lies for half the year above 8,000 feet, but the glacial relief was carved during the Ice Age as there are no permanent snowfields now. The Middle Atlas is more plateau-like in form, with bare limestone topography in the east which acts as an aquifer. Rainfall is considerable, with 30 inches and over, and the region is of much importance as it is the main watershed in Morocco. The Anti-Atlas lie to the south of the High Atlas and the downfaulted depression of the Sus, and consist of a high arid plateau at an average altitude of some 5,000 feet.

As in the Riff mountains *cultivation* is concentrated in the valleys of the High and Middle Atlas where cereals and fruit are grown. The southern part of the Atlas ranges are, however, too dry for cultivation. The Berber inhabitants practise both cultivation and stock rearing, but *pastoralism*, as in the meseta, is predominant. There are over 20 million small livestock, the majority being sheep, the rest goats. There are also about $2\frac{1}{2}$ million cattle, about 1 million donkeys and $\frac{1}{4}$ million horses and camels. Transhumance is common, particularly in the south, and flocks and herds leave the mountain and plateau tops for the deeply trenched valleys in winter. Overstocking is a serious problem with consequent destruction of pasture.

Population density on the Middle and High Atlas range falls to less than 10 persons per square mile, and on the Saharan fringe where the population is entirely nomadic to less than 2 per square mile. The High and Middle Atlas acted as a refuge zone for the Berber people when faced by invasion along the Mediterranean coastal plain or the desert fringe. Berber culture and language has survived, but is being rapidly modified by French influence and the impact of Moroccan independence. The traditional family and clan structure is tending to break down as in other parts of Africa. Population which formerly retreated into the mountains when forced back by invaders is now flooding back on to the western lowlands and to urban areas like Casablanca. Population density in the mountains is thus decreasing.

To the east of the folded Tertiary ranges of the Atlas is the stable block of the high plateaux which are continued eastwards into Algeria. This is a dry, sparsely populated region with a pastoral economy similar to that of Algeria.

ALGERIA

Algeria has a total area of about 850,000 square miles, the northern strip, excluding Algerian Sahara, with an area of 115,000 square miles, and a population in 1963 of 10·6 million. About 10% of the population was European until 1962, but with independence most of the Europeans have gone back to France. However, over 90% of the country is desert or semi-desert with a very small population. The northern strip comprising the Atlas ranges and the Mediterranean coast has less than 10% of the area, but over 90% of the people. Algeria has been occupied by France for more than a century and has long been governed as part of France, returning a large number of deputies to the French Assembly. The recent struggle between the nationalist Algerians, the F.L.N., the European settlers and the French administration ended in favour of the nationalists who now govern an independent Algeria. Nearly ¼ million people were killed in the war and Algeria suffered great damage.

WESTERN ALGERIA

A series of zones aligned approximately east and west for some 600 miles succeed each other rapidly from the well watered coastal strip to the desert edge, comprising coastal mountain chains and depressions, the high plateaux, Saharan Atlas and the Sahara. The *western coastal region* or the *Tell Atlas* is approximately 400 miles long and 100 miles wide and is situated between the Mediterranean coast and the high plateaux and blocked to the east by the Djurdjura massif and the Hodna chain. It consists of diverse relief and structure with a series of separate mountains and plateaux enclosing plains and river basins such as the Mitidja near Algiers and the Cheliff. These lowland hollows tucked in between the coastal and interior mountains are affected by a rain shadow and have a lower rainfall than the massifs of Tlemcen and Ouarsenis. Western Algeria has a lower rainfall generally than Eastern Algeria as the region is in the rain shadow of Morocco and the Iberian peninsula, and the coastal ranges are low.

There are striking differences in physical environment and human response between the well farmed coastal plains and the adjacent mountains; these coastal plains, with only 2% of the area of Algeria, have over 50% of the population. The plain of the Mitidja and the Cheliff river are good examples of the intensive *agriculture* developed by French and Italian settlers. Agriculture in Algeria supports about 70% of the population and supplies nearly a third of the national income. Of the total area of about 220 million hectares less than a quarter is classified as productive land, of which some 80% is pasture and rough grazing largely in the south. Europeans occupied nearly 37 million acres, approximately a third of the cultivable area and some of the best land in Algeria. Before 1963 nearly ¾ million peasants were landless, and about 600,000 peasants cultivated 64 million acres of land. The average European farm was ten times the size of the Arab holding of 30 acres. Most of the export crops such as wine, citrus fruit and early vegetables were produced by Europeans, while the Algerians produced cereals, olives, figs, dates and pastoral products largely for the local market.

The plain of the *Mitidja* near Algiers is about sixty miles long and ten miles wide, backed to the south by the Blida Atlas. It consists of a series of alluvial fans composed of clay and gravel, sloping away from the mountains.

Water seeping through the fans is used for irrigation, particularly in the reclaimed northern and eastern parts of the plain. Extensive vineyards and orchards and large farms have replaced what was once a marshy waste. On the hill slopes citrus and almond trees flourish, while tobacco, vegetables and flowers are widely grown. The main agricultural products of the coastal lowlands are wine, citrus fruit, early fruit and vegetables (tomatoes, beans and artichokes), tobacco and perfume flowers.

Wine production is by far the most important agricultural activity, most of the wine being vin ordinaire which is exported to France, making Algeria the main exporter of wine in the world. In 1955 vineyards occupied nearly 1 million acres and over 5% of the arable area. Vines are grown almost continuously along the coast from the Moroccan to the Tunisian border. There is considerable over-production of wine, a problem that is complicated by slight local consumption, as the local Moslem population does not drink wine and also because France has a wine surplus and is restricting imports of Algerian wine. It has been suggested that the area under vineyards should be cut by half. The Constantine Plan of 1958 has attempted to increase greatly the production of early vegetables, citrus and other fruits.

Fruit trees covered about 500,000 acres in 1955. Citrus is grown under irrigation in the coastal zone, the main area being near Algiers. There are over 11 million olive trees with a concentration in the area between Algiers and Bougie, while figs are important in the mountains between Algiers and Philippeville. Vegetable production is concentrated around Algiers.

As in Morocco *irrigation* is necessary for intensive crop cultivation. Rivers are more transient than in Morocco where melting snow of the High Atlas fills the rivers of early summer, but a larger area has been irrigated than in Morocco mainly because of the far larger numbers of European settlers. Irrigation schemes have been carried out on the Oued Makeira, Oued Hammam, Oued Tafna, and the river Cheliff and its tributaries. Widespread erosion and deforestation have caused siltation of dams and irrigation is facing serious problems.

Between the coastal ranges and the intensively farmed coastal plains and the high plateau rise a series of mountain chains such as the Ouarsenis massif. They form a barrier to *communications* and roads and railways have to make use of suitable river valleys and depressions; for example, the main coastal railway from Algiers to Oran passes down the Cheliff valley, then continues south and west to the Moroccan border round the eastern end of the Tessala mountains, along the Mékerra-Sig valley and Oued Tafna. Communications have been handicapped even more by the deep valleys and rugged relief of Eastern Algeria. There are over 3,000 miles of railway and, besides the main coastal line to Morocco and Tunisia with branches to the ports of Oran, Algiers, Bougie, Philippeville and Bône, there are also links to the southern oases, one to Ain Safra and Colomb-Béchar, the other to Biskra and Touggourt. The *Ouarsenis massif* consists largely of Cretaceous marls and clays, with brushwood and pine forests, from which rises the limestone massif of Kef Sidi Amar, over 6,500 feet in altitude.

South of the folded mountains of the Tell Atlas are a series of *depressions and plains* such as the Plain of Bel Abbes. This is an important centre of European settlement and one of Algeria's major cereal-producing regions. However, the bulk of the wheat and barley, which are the main cereals, is produced by the Berbers. Wheat is far more important than barley, in contrast to Morocco. Cereals are largely grown on the southern fringe of the coastal region and yields vary considerably. The area under cereals has remained static in recent years and there is little hope of further expansion. Attempts to grow cereals on the high plateaux by dry farming methods have not been successful.

Population density is highest along the coast, being closely associated with the zone of high rainfall, intensive agriculture and the marked degree of commercial activity and urbanization. This is the main centre of European settlement with the largest towns of Algiers and Oran where the French formed the majority of the population. Of the European population of about 1 million in 1962 a considerable proportion were originally from Spain or Italy, but became assimilated into the Franco-Algerian community which developed a strong local patriotism. However, suggestions that the French and Berber communities might eventually merge and form an Algerian nation have proved incorrect and the majority of the French have fled the country since independence.

The most important town in Algeria and the capital is *Algiers* with a population of nearly 870,000 in 1960. It is also the main commercial and industrial centre with the largest port. Founded in the tenth century the town became the main centre for the Barbary corsairs from the sixteenth to the nineteenth centuries. It has grown up on the western side of the bay of Algiers on the steep slopes of the Sahel and faces east and north overlooking the Mediterranean. A large part of the population were French, and there is the usual contrast between the native quarter, the Kasbah, with its steep dark alleyways and narrow streets lined closely by dimly lit booths and houses, and the modern French town. The harbour is artificial and the port handles between a third and a half of Algeria's trade, some 3 to 4 million tons per annum. Algiers has a wide range of industries producing consumer goods, cement, footwear, agricultural implements, paints and varnishes and processing the local agricultural products of wine, tobacco, flour, etc.

Oran is the second city of Algeria with a population of nearly 400,000, over half of whom were French. It stands at the head of the Gulf of Oran and is an important port exporting wines, cereals, vegetables and fruit. The industrial area near the port has textile, food-processing and other industries.

THE INTERIOR PLATEAU

The next region south of the Tell Atlas is that of the *high interior plateau*. Its high northern scarped edge composed of Jurassic limestone stretches for 200 miles and descends to the north by a series of steps. Limestones over 1,000 feet thick, creating a causse landscape, overlie sandstones and clays from which emerge copious springs which have attracted settlement and cultivation. The sandstone areas with their forests of evergreen, oak, juniper and thuya contrast with the bare stony limestone uplands. To the south the large area of the High Plateaux is nearly enclosed by the arms of the Tell and Saharan Atlas. It is a comparatively little disturbed part of the great African shield whose slight fractures and folds are hidden under a cover of debris. The high plateau, with an average height of nearly 4,000 feet, is a great contrast structurally and climatically to the Tell Atlas. A wide, monotonous surface with a rainfall of less than 10 inches per annum supports only a steppe vegetation. There are many hollows occupied by playa lakes, known as shotts, which fill with water during the winter, but dry out and become salt pans in the summer. The largest of these shotts is the Chott Chergui, about a hundred miles long; the other shotts are lower in altitude, the Chott el Hodna being at only 1,300 feet.

Extensive cereal cultivation under dry farming methods is only possible on the northern fringe but, if the marginal areas surrounding some of the shotts, for example the Chott el Hodna, were drained, reclaimed and irrigated, the intensive cultivation of a variety of crops would be possible. The predominant activity is pastoralism, although the number of $5\frac{1}{2}$ million sheep and 2 million goats is far less than in Morocco. There are two types of pastoralists, those who are almost entirely nomadic and those who are semi-nomadic. The northern Saharan tribes spend the winter with their animals in the desert and the summer on the highland pastures in the Saharan Atlas and high plateau. The semi-nomadic pastoralists grow cereals and rear livestock on an extensive basis. There are only about

640,000 cattle which are found in the better watered areas to the north and east. There are 200,000 camels used as beasts of burden on the plateau. As in Morocco, halfa or esparto grass is cut and exported for the production of paper.

THE SAHARAN ATLAS

The Saharan Atlas, a continuation of the Anti-Atlas, forms a dry mountainous fringe to the high plateau into which it merges. To the south stretches the Sahara which sends tongues of arid land into the mountains which, in some areas, rise steeply like islands from a sea of sand and gravel. For example, the Djebel Amour mountains, over 5,000 feet high, form a series of flat-topped sandstone massifs with great cliffs. But it is in the east that the contrast between the wide tectonic fosse of the Chott Melhrir, much of it below sea level, and the sharp bold relief of the Aures Mountains is most marked. This contains the highest peak in Algeria, the Djebel Chelia (7,638 feet) and others 6–7,000 feet high. High parallel ridges aligned south-west to north-east are separated by deep gorges. This area has long been a refuge for the Berber people, the other Berber stronghold being in the Djurdjura and Ouarsenis ranges. Villages are sited on hill tops and escarpments and the flat-topped houses cling in lines along the contour. A form of semi-nomadism is dictated by the variability of the rainfall and intrusion of semi-Saharan conditions. Villages are deserted periodically when the Berbers move off with their flocks in search of grazing to the south. A little cultivation is carried on in the valleys. This area was of considerable strategic importance in Roman times as the ruins of Roman towns now testify.

EASTERN ALGERIA

Eastern Algeria or the eastern block is the last region. It forms a great bastion between the Tunisian frontier, the coast, and the high plateau which narrows to the east and plunges beneath the mountains of the Tell and Saharan Atlas. The meeting of the coastal and desert chains of the Tell and Saharan Atlas and the bold relief so created in the Aures Mountains, Hodna and Biban chains and the Djurdjura massif is characteristic of this region of Eastern Algeria. The mountains of the Eastern Tell Atlas abut directly on to the coast and from Algiers to Bône there are no coastal plains or depressions as in Western Algeria. The mountains are old with a core of Archaean and pre-Cambrian rocks and consist predominantly of Pyrenean folds. Alpine orogeny lifted the old massifs and rejuvenated the rivers causing deep gorges, such as the Rummel near Constantine, a great barrier to communications. East of the mountains the Constantine plains are of considerable elevation and are divided up into a series of small basins by isolated ridges.

Rainfall is heavy along the coast and in the northern mountain chains, Bougie receiving over 40 inches a year. Snow fall can also be heavy on the Djurdjura and lies all the winter. A more luxuriant type of forest is found than in the south where rainfall is only 16 inches. Evergreen oaks and junipers grow up to about 4,000 feet and there are cedars at higher altitudes.

The *Djurdjura* massif, also known as Great Kabylia, has, like the Aures mountains, acted as an important refuge zone for the Berber people. Berber is still spoken by about a quarter of the Algerian population. There are comparatively few French in this area.

The massif is deeply dissected causing a series of narrow ridges and deep valleys. Villages cling to the sinuous line of the ridge top up to a height of 4,000 feet and avoid the valley floors. The high rainfall permits permanent cultivation and characteristic tree crops such as olives and figs are produced. Cereals and vegetables are planted under the trees and the ground is worked with plough or mattock. The small amount of cultivable land is insufficient, however, to support the population, which has to supplement its income by work outside.

Kabylia has 4% of the area of Algeria and 15% of the population. Population density is among the highest in Algeria and locally in the Djurdjura may reach 500 per square mile. There are considerable problems of overpopulation and pressure is relieved to some extent by migration to the towns and western coastal region and to France where some 300,000 Algerians are employed.

Mineral exploitation is of some importance in Algeria and increasingly so with the export of oil and natural gas from the Sahara. In 1963 about 2 million tons of high-grade iron ore was mined for export to Europe, the chief areas being at Ouenza and Bou Kadua near the Tunisian border. There are large deposits, including those recently discovered near Tindouf, in the western Sahara. Some 300,000–400,000 tons of poor quality phosphates a year are produced, the main deposits being at Le Kouif, Djebel Onk and Tocqueville. Zinc is also mined near Batna. Large deposits of oil and natural gas have been found in the Sahara and these should be a great economic fillip to Algeria. The problem of the lack of power for industry has been solved by the discovery of vast deposits of natural gas at Hassi R'Mel. Pipelines have been constructed linking Hassi R'Mel with the ports of Arzew and Algiers. Gas may be exported to Europe by tanker or a pipeline across the Mediterranean.

The long Algerian war has left much bitterness and caused great destruction of property and the concentration of people in temporary camps. After Algerian independence most of the European settlers went back to France. The Algerian economy is considerably affected by this as the European settlers produced most of the export crops. About $2\frac{1}{2}$ million acres of land vacated by the settlers is being reallocated to Algerians, and will be worked on a collective basis. In addition nearly $\frac{3}{4}$ of the small businesses and industries were abandoned causing widespread unemployment.

Algeria is also faced with problems of mounting population pressure in areas such as Kabylia. The population has increased fourfold in about a hundred years and is now surging ahead at the rate of $2\frac{1}{2}\%$ a year. There are divergent political currents pulling Algeria towards France and towards the rest of the Arab world. The great stretch of desert abutting on to the republic of Niger also gives Algeria a southward pull into Africa. A federation of Morocco, Algeria and Tunisia is at present somewhat remote in view of political differences between the leaders. The three states of the Maghreb have, however, many interests in common such as their Arab culture and religion, and a closer economic and political association may well come about in future.

TUNISIA

Tunisia is by far the smallest of the three Barbary states with an area of only about 48,000 square miles and a population estimated at $4\frac{1}{2}$ million in 1963. Of the original European population of 250,000 in Tunisia, over 200,000 have gone back to France. The southern half of the country is desert and virtually uninhabited, and the bulk of the population is to be found in the northern coastal area, with the highest density south of Sfax. About two-thirds of the people depend on agriculture for a living.

The French assumed control of Tunisia in 1881 when a protectorate was declared. Independence came in 1956 when the monarchy was abolished and President Bourguiba became head of the Republic. Facing east and north-east Tunisia tends to dissociate itself from Algeria. It is of great strategic significance and the former French naval base of Bizerta commands the Sicilian channel between the basins of the Eastern and Western Mediterranean.

There are two main regions of the northern Tell Atlas and the southern steppes and Sahel, divided approximately by the line of the Medjerda

Valley. South of the Chott Djerid the rainfall is below 8 inches and soon becomes less than 4 inches so that Southern Tunisia, comprising about half the country, is included in the Sahara.

NORTHERN TUNISIA

The northern Tell, or northern Tunisia, has the highest rainfall in Tunisia, a small area near the coast and the Algerian frontier having a rainfall of over 48 inches; two-thirds of this area, however, receives less than 24 inches. The folds of the northern Tell are continued from Algeria and reach the coast at right-angles, causing prominent headlands and sheltered bays. Tertiary sandstones overlie limestones and marls. The Kroumirie is well forested with various species of oak predominant and also with olive, pine, ash and elm. Lack of settlement and livestock has saved the forests.

To the south of the Tell are contrasted lowland areas comprising the plains behind Bizerta, the Medjerda Valley and the southern slopes of the Tell where underlying limestones are widely exposed. The northern Atlas folds die out in Cape Bizerta which shelters the bay of Bizerta and the former French naval base which is of considerable strategic importance. To the south-west is the lac de Bizerte which is connected to the sea by a channel. The alluvial plains of Bizerta are some of the richest in Tunisia and were colonized by the French. To the south these plains merge into the lower valley of the Medjerda and the plain of Tunis.

The lower plains and delta of the Medjerda are unhealthy and malarial, but large areas are fertile and with drainage and reclamation could support a large population. There are plans for irrigating and draining $\frac{1}{4}$ million acres. It is on the deep alluvial soils of the middle reaches of the Medjerda valley that agriculture is important, and with mechanized farming cereals give high yields. The 300-mile long Medjerda valley affords an important route from northern Tunisia into Algeria and there are numerous traces of former settlement showing the significance of this valley in former times.

South of the lower Medjerda and half encircled by the high Tell are the spade-shaped plains of Tunis grouped around the Gulf of Tunis, which is an area of submergence. The folds of the Tell reappear in the Djebel er Rahman and Cape Bon which forms the north-easterly arm of the Gulf of Tunis. The plains are open to the dry southerly winds and rainfall is only about 16 inches a year so that viticulture tends to replace cereals, and vineyards established by the French and Italians are found on the western edge of the plains. Tunisians grow a variety of crops—wheat, barley, olives, citrus fruit, tobacco and vegetables. Yields of cereals fluctuate widely. The European settlers, owning about 6 million acres of land in 1961, specialize in the production of wine and early vegetables.

The coastal area is closely settled and dominated by the *city of Tunis* which, with a population of ¾ million, has nearly one-fifth of the population of Tunisia. It is the capital and the major port. The city is situated on a shallow lagoon near the head of the Gulf of Tunis to which it is connected by an artificial cutting five miles long. As the result of its nodal situation the site has been used for settlement for a very long period, and Tunis is probably older than Carthage which lies some twelve miles to the north-west. During the ninth century it is said to have had a population of 100,000 and to have surpassed Cairo.

The city has separate native and European quarters, the latter on flat ground between the native quarter and the shores of the lac de Tunis. The native city has a large number of mosques and an important Moslem university founded in the eighth century. During the nineteenth century the port for Tunis was at la Goulette which has become an outport since the construction of the main harbour east of the city and the Canal de Tunis. The port handles much of the country's trade, exporting minerals, cereals, wine, olive oil, dates and sponges. Tunis is also the main industrial centre producing foodstuffs, linseed, olive oil, cork, textiles, carpets and cement. It is also the main communications centre in Tunisia controlling routes into the Tell, to Algeria and along the coastal plain.

SOUTHERN TUNISIA

The southern region is three times the size of the northern strip and consists of three areas of the high Tell, steppes and Sahel. Its southern limit is along the line of Chotts and salt pans to the west of Gâbes. The high Tell consists of a series of limestone and sandstone domes and basins typically aligned north-east to south-west. Drainage in the north is to the Medjerda, and in parts of the Mellégue and Siliana valleys cereals are grown. Here much of the population is sedentary and lives in a kind of windowless hut known as a gourbi with low walls and roof made of thatch or branches. The coming of the French has tended to stabilize the population, although in the south, where cultivation is difficult because of the low rainfall, the people live in large tents which may be forty feet wide.

To the south and east the plains become more continuous and drier and the steppes are sparsely populated by tent-dwelling nomads who own herds of camels and flocks of sheep. During Roman times the rivers that drain intermittently to the south and east were dammed and the water used for cultivation. There are numerous remains of water storage works and buildings showing that the area was well populated in Roman times. *Pastoralism* is now dominant in this area and the nomads move with their animals[1] from the desert and steppes to the Tell in the spring. The old pattern has been changing in recent years as much of the best land in the north was taken over by Europeans. As elsewhere in the Maghreb efforts have been made by the French to settle the nomadic pastoralists by encouraging permanent cultivation and olive growing. However, the prevailing aridity of much of the area will continue to dictate a nomadic way of life, as permanent agriculture is impossible except where a continuous water supply can be assured. Esparto grass is also an important export but the amount has declined considerably since 1949.

The low sandy plains or *Sahel* of Eastern Tunisia stretch for over a 100 miles from Enfidaville to south of Sfax and are 60 miles wide. This region is densely settled compared to the empty pastoral steppes to the west and is dependent on the growing of olives, which has been greatly extended in the last sixty years. Olive plantations stretch 50 miles inland near Sfax. The climate, although dry, is ideal for olive trees with abundant dew and a water-retentive soil. There are nearly 30 million trees, the main areas being near Sousse and Sfax. The olive trees are carefully tended and pruned

[1] In 1960 there were about 3 million sheep, $1\frac{1}{4}$ million goats, $\frac{1}{2}$ million cattle and $\frac{1}{4}$ million camels in Tunisia.

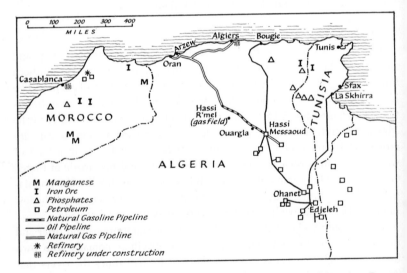

FIG. 20—The Maghreb: minerals, 1962 (from Petroleum Information Bureau *Newsletter*)

particularly in the south and olive oil of high quality is exported to Europe. *Minerals* are important and supply one-third of Tunisia's exports. The most important mineral is phosphates with a production of about 2 million tons per annum. The main deposits of low-grade phosphates are near Gafsa from where a narrow-gauge railway runs to Sfax, the main phosphates-exporting port. Iron ore, lead and zinc are mined in the north-west, Tunisia exporting about $\frac{3}{4}$ million tons of iron ore per annum (Fig. 20).

Gafsa on the southern fringe of the Sahel is a fertile oasis with a million palm trees, great olive groves, citrus orchards, apricot trees, fields of wheat and barley, and gardens with many varieties of vegetables. Cultivation is entirely dependent on irrigation from springs and artesian wells.

One of the most interesting cities in Barbary and one of the principal holy cities of Islam is Kairouan, nearly forty miles east of Sousse. There are five hundred mosques, most of which are small chapels dedicated to a holy man or sect. From the towers of the Great Mosque the traveller looks out over a sea of flat roofs towards the massive fortifications and the monotony of the steppe beyond.

Compared with Algeria and Morocco, Tunisia faces problems of small area and population with only a third to a quarter of the population of these countries. Much of the country is marginal for cultivation and farming methods are backward, handicapped by the usual problems of peasant conservatism, lack of education and outdated land laws. Considerable success has been obtained in certain aspects of agriculture, such as the growing of olives which could be extended on to the steppes in favourable areas. Over 6 million acres of tribal and common land have been subdivided or split up into individual holdings, and 1 million acres purchased from European farmers. Tunisia had a peaceful road to independence compared with Algeria and under a wise President is attaining a modest measure of prosperity. Politically, the country is affected by the divergent forces of Arab unity on the one hand and a pro-Western policy on the other. President Bourguiba is anxious to preserve friendly relations with France.

CHAPTER SIX

The Sahara

THE Sahara is the world's largest desert and with an area of over 3 million square miles covers more than a quarter of the African continent. It extends for 3,000 miles across Africa and is 1,000 miles wide. The name Sahara is derived from the Arabic word *sahra* meaning a wilderness. Everywhere the rainfall is very low, in some areas nil, and even on the northern and southern borders it is usually less than 10 inches and is highly irregular in periodicity and amount. The northern and southern limits of the desert are thus very transitional and fluctuate greatly according to the amount of rainfall. The southern limit is sometimes taken as the 15-inch isohyet, which runs east-west south of a line from the Senegal river mouth and Timbuktu to Agades. The northern edge is taken as the 10-inch isohyet, as rainfall has a winter maximum here and evaporation is lower than in summer.

The area is divided up politically into a number of different units, the majority formerly under French control, but now independent. The region has a distinct geographical unity, the political boundaries slicing across the desert in straight lines having little meaning until the recent discoveries of oil and natural gas. The basic regional distinction in the Sahara is between the oasis-dweller growing crops under irrigation and the wandering pastoralist with his herds of camels, sheep and goats. Oil and mineral discoveries have intruded a third element which is having revolutionary effects on the economy of the oases.

RELIEF AND STRUCTURE

The Sahara has an average altitude of 1,000 feet and from its low surface rise the broken remnants of mountains and plateaux which are being worn down to the level of the tawny, dusty wastes of rock, gravel and sand that surround them. The oldest rocks of Archaean and pre-Cambrian age constitute the fundamental complex which is overlain by extensive Palaeozoic deposits—limestone, shales and sandstones—and Cretaceous and Tertiary limestones and sandstones (Fig. 21). There was much warping and faulting in the late Tertiary with vulcanism and widespread lava flows causing spectacular scenery with extinct volcanoes in the central highlands of the Ahaggar and Tibesti, where some peaks of over 10,000 feet in height are found, for example Emi Koussi, 11,204 feet.

The Ahaggar is a rugged highland trenched by deep valleys and is composed of Archaean and Pre-Cambrian rocks elevated in a broad dome and studded with volcanic craters, necks and flows, with numerous peaks exceeding 6,500 feet. Surrounding the Ahaggar is a belt of Palaeozoic sandstone plateaux, the Tassili, cut by deep valleys, and covering a large area.

The sand or erg deserts are found mainly in a wide belt stretching from Mauretania to the Gulf of Gabes, in the great basins forming inland drainage areas for the ancient rivers, and also to the north-east of the Tibesti highlands. These sandy stretches are not as extensive as popularly imagined and cover only about one-ninth of the Sahara. The bulk of the desert is

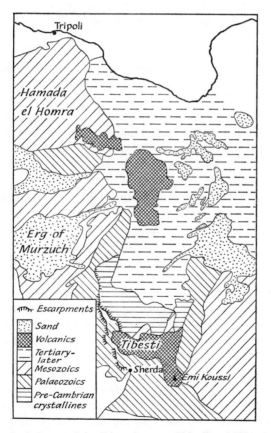

FIG. 21—Geological map of the Sahara between Tripoli and Tibesti (after Grove)

made up of sandstone plateaux, the hamada, which occur particularly round the highlands and the reg, or rocky deserts, wildernesses of stones and gravel, well developed in the Tanezrouft, west of the Ahaggar. This is an extremely flat, pebbly region, almost devoid of water, and therefore very difficult for caravans.

The Sahara has *no permanent rivers* except for the Nile and part of the Ahaggar. Large wadis radiate out from the Ahaggar and other highlands of the Central Sahara and are evidence of a former drainage system that developed in the Quaternary Ice Age, but has now fossilized because of aridity. The Oued Tamanrasset and other wadis flow from time to time as the Ahaggar receives a slight rainfall and the Oued Iherir in the Tassili des Ajjer flows from October to April. Water draining from the highlands soon disappears into the sandy wastes, but helps to increase the supply of underground water often tapped a considerable distance from the highlands.

CLIMATE

The interior of the Sahara apart from the central highlands receives *virtually no precipitation* as the north-east trades become increasingly warmer and drier as they blow across the desert. Much of the area receives no rain at all for a number of years, and what rain does fall tends to come in showers at very irregular intervals. Flash floods may occur, filling the wadis with water which soon evaporates. The highlands have some rain, perhaps 10 inches a year, the Ahaggar both in winter and summer. Snow is not uncommon on the highest peaks.

The air is very dry and relative humidity low, while the amount of sunshine is among the highest in the world. Dust and sandstorms are frequent and are often associated with a strong, southerly wind, known as the sirocco or Khamsin, which raises a wall of dust and sand that obscures visibility and halts travel. There are great temperature extremes and a marked diurnal range. Although the heat may be intense by day clear skies cause maximum radiation by night and temperatures drop over 50° F (28° C.) The annual range is much lower and seldom more than 25° F. (14° C.). The Sahara experiences some of the world's highest temperatures; Salah, Timimoun and Azizia have all had temperatures of over 130° F. (55° C.), and the mean July temperature for a large area north of the Ahaggar is over 100° F. (38° C.). In winter, however, minimum temperatures may be below freezing point and frosts are not uncommon. In fact, Touggourt and Biskra record more days with frost than stations on the Mediterranean coast which are four degrees further north. The Atlantic coastal strip has cooler temperatures with a lower diurnal range and cloudier skies than the interior. The cool Canaries current causes cooler and foggy conditions.

Timbuktu Altitude 820 ft.

J	F	M	A	M	J
71 (22)	74 (23)	83 (28)	92 (33)	95 (35)	94 (34)

J	A	S	O	N	D
89 (32)	86 (30)	89 (32)	89 (32)	87 (31)	71 (22)

Year 84° F. (29° C.)

Tamanrasset Altitude 4,429 ft.

J	F	M	A	M	J
54 (12)	58 (15)	64 (18)	72 (22)	78 (26)	83 (28)

J	A	S	O	N	D
84 (29)	83 (28)	79 (26)	74 (23)	65 (18)	58 (15)

Year 71° F. (22° C.)

The above figures for Timbuktu on the southern fringe of the Sahara and Tamanrasset on the Ahaggar highlands illustrate the high temperatures for Timbuktu, particularly in April, May and June, before the meagre summer rainfall of about 9 inches occurs. The effect of altitude at Tamanrasset reduces the mean temperature by 20° F. (11° C.) in April and the mean annual temperature by 13° F. (7° C.) compared to Timbuktu.

VEGETATION

Vegetation is very sparse, particularly in parts of the Tanezrouft and the erg desert, and has to be adapted to arid conditions (Fig. 22). There are two main types of vegetation, the permanent and the ephemeral. The permanent vegetation consists of a few perennial trees and shrubs—jujube and tamarisk—and a large number of herbs adapted to the harsh climate by means of small size of tree or shrub, with small leaves, hard foliage and the growth of hairs or spines. The ephemeral type of vegetation consists of herbs and tiny plants whose seeds lie dormant for long periods, then germinate and grow immediately after the occasional rainstorms. They flower briefly, covering the desert with carpets of colour, then die away as quickly to await the next shower of rain.

Where rainfall increases on the Ahaggar and other highland areas, or near the margins of the desert, vegetation becomes more varied. The

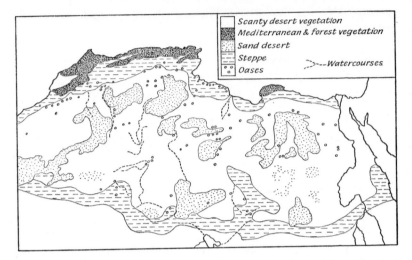

FIG. 22—North Africa and the Sahara: natural vegetation (from *Atlas of the Arab World and the Middle East*)

Ahaggar has a tropical type of vegetation of acacias and panicum grass on the lower slopes and a Mediterranean type with grass, shrubs and trees—olives and sumacs—higher up.

WATER RESOURCES

Water resources are of vital importance in the Sahara as the entire population rely absolutely on the meagre available sources. Water supplies were probably greater in former times and there is much evidence that parts of the Sahara, now desert, were inhabited some thousands of years ago. Animals such as elephants and hippopotami are portrayed in rock drawings and in the Egyptian desert the abundance of flint implements, the variety of wild animals shown on Egyptian monuments, dead tamarisks and trails left by grazing animals on hill slopes, now desert, are ample evidence that the region had a higher rainfall at times. It seems possible that rainfall in Egypt decreased greatly some three quarters of a million years ago at the close of the Plio-Pleistocene period, and that the subsequent, long period of drought was broken only by two rainy interludes in the middle Paleolithic and from about 8,000 to 4,000 B.C.

Except for a few permanent streams in the Ahaggar, or occasional springs in the oases, nearly all the Saharan water must be obtained in various ways from underground sources. Rainfall from the Anti-Atlas and

Saharan Atlas in the north and the central highlands and the southern savanna percolates through the wadis and depressions to replenish the fossil waters, forming a vast reserve which lies near the surface in the string of oases to the south of the Atlas and in scattered oases such as Kufra and Borkou, but elsewhere it may be over 2,000 feet deep. The commonest source of water is a well, the shaduf method being used to obtain water from a shallow well and the pit method from a deep well. Water at greater depth may be obtained by cutting long slightly sloping channels, or underground aqueducts, to tap the water table, so leading it into a nearby cultivable hollow. Such channels are known as foggaras and the system is widely employed in other desert regions such as Iran and Baluchistan (Fig. 23). From the air strings of miniature craters may be seen radiating out from the oasis, marking the lines of the foggaras and their shafts leading to the underground channel, and dug originally for the extraction of the spoil. The foggaras, constructed originally by slaves, are tending to choke up now for lack of labour. The Touat area south of Tademait hammada is well known for its foggaras. The French have increased considerably the available supply of water by sinking thousands of wells, particularly in the Albienne nappe south of the Atlas Mountains, a process helped considerably now by the use of petroleum for deep boring equipment.

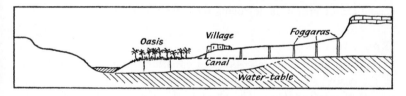

FIG. 23—Underground water from foggaras in the Sahara

POPULATION

The *inhabitants* of the desert are of diverse origin and consist of Negro, Berber and Arab stock; the Negroes are found mainly in the oases and the south, the Berbers and Arabs in the desert and the north.[1] The Tuareg, of Berber stock, live in the central highlands of the Ahaggar and Air massifs and are nomadic cattle-herders who have for long preyed on the Saharan routes. They are declining in importance, as the French control the caravan

[1] There are also the Haratin, who are ranked socially between the Negroes and the Arabs and Berbers.

traffic, and in purity of stock, through intermixture with Negroes. To the east are the Tibu of the Tibesti highlands and adjacent desert who represent a further stage of Negro admixture and who speak a Sudanic language. To the west and north are the Moors of Arab and Berber stock, far more influenced by Islam than the Tuareg or Tibu. They are partly nomadic wanderers, partly oasis-dwellers living on the northern and western fringes of the Sahara and near the hamada aquifers. The total population of this vast region is, however, very small, the majority living in the oases which are only a minute fraction of the whole Sahara.

The wide differences between oasis-dwellers and nomadic pastoralists are more fundamental than ethnic distinctions. *Oasis-dwellers* live in walled villages for defensive purposes. Winding alleyways thread between the houses, usually flat-roofed and mud-walled, a common type of architecture in the arid and semi-arid regions of Africa. As water is so precious it is owned rather than land, and careful regulations control its use, each oasis having a guardian of water supplies. The basis of oasis life is the date-palm which provides food, fodder, housing materials and the shade under which other trees and vegetables can grow. There is a three-tier cultivation with cereals and vegetables such as onions, beans and tomatoes, and also watermelons and tobacco, on the ground storey, apricots, peaches, almonds, figs and vines on the second storey and date palms at the top. Small rectangular plots of land are frequently watered by basin irrigation and two crops a season can be grown, barley being the chief crop (Fig. 24).

INDUSTRIES

However, life is circumscribed and opportunities of employment scant in the oases,[2] so that the population is on the decline, hastened by the impact of oil discoveries. The effect of the discovery of the *Saharan oil fields* has been profound both on the economy of the oases and, on a much broader scale, that of Algeria and Libya. The intensive search for oil was a post-war development, but it was only in 1956 that oil was first found at Edjeleh on the Libyan border and later in the year at Hassi Messaoud about four hundred miles east of Edjeleh, while near the end of 1956 a very large natural gas field was discovered at Hassi R'Mel two hundred miles south of Algiers. In 1957 production of crude oil was 21,300 tons, in 1959 1·2 million tons, in 1961 nearly 16 million tons and in 1964 26 million tons. The oil from Hassi Messaoud and Edjeleh comes from a

---

[2] It is estimated that there is only $\frac{1}{10}$ acre of cultivated land per person in the oases of the Algerian Sahara.

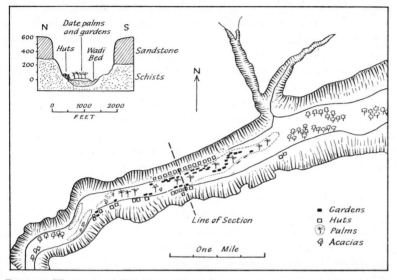

FIG. 24—The oasis of Debasser, twenty-five miles north-north-east of Sherda
(after Grove)

considerable depth, below 10,000 feet. The great bulk of it is pumped
approximately four hundred miles to the port of Bougie. Edjeleh produces
nearly as much oil as Hassi Messaoud and its advantage is that oil is found
at depths of only about 5,000 feet. A pipeline has been constructed from
Edjeleh to La Skhirra on the Tunisian coast. An adjacent field at Ohanet
sends its oil by pipeline to Hassi Messaoud and so to Bougie.

Natural gas is pumped from Hassi R'Mel to Arzew on the coast. It is
one of the greatest natural gasfields in the world with reserves estimated
at 2 billion cubic metres. From Arzew the pipeline bifurcates, one branch
going to Algiers, the other to Arzew and Oran. At the moment the gas is
used for local industry, but in future it will be shipped from Arzew to
European markets, the nearest being Sicily and Southern Italy. The
Libyan oil industry is described on p. 178. Problems for the Saharan oil
industry include the extreme climate, the future allocation of oil profits
and concessions between France and Algeria, and a world surplus of oil
at present.

COMMUNICATIONS

Another feature of general importance in the Sahara is communications.
Trans-Saharan camel routes have been of great importance for nearly

2,000 years since the camel was introduced into the Sahara from Arabia (Fig. 25). Camel caravan traffic was monopolized by the nomads, and commodities such as salt, ivory, gold, slaves and leather goods have long been carried from the Sudan belt and parts of the desert to markets in North Africa. Centres such as Timbuktu, Gao, Niamey, Zinder and Kano throve on this traffic. A number of these routes are still important despite the advent of motorized transport and aircraft. There are two main routes regularly used by motor transport, one following the line of the old caravan track from Colomb Bechar to Gao, and the line of the oft-proposed trans-Saharan railway, the other the Ahaggar. There are other routes from Southern Morocco to Senegal and in the eastern Sahara.

SPANISH WEST AFRICA

Spanish West Africa stretches for about 750 miles along the Atlantic coast from the mouth of the wadi Dra to Cape Blanco. The climate is arid, but cooler along the coast than in the interior. The topography is generally featureless. The area is about 125,000 square miles and the population less than 50,000, composed largely of nomadic pastoralists. The only important economic activity consists of fishing along the coast, which is carried on by Canary Island and Breton fishermen.

MAURITANIA

Mauritania is a large territory with an area of over 400,000 square miles lying to the south and bordering Spanish West Africa (Fig. 26). The name is a reminder of the Roman province of Mauretania which actually consisted of northern Morocco and Algeria. It is largely desert except for a small riverine strip along the Senegal in the south. In 1963 the population was estimated at just over $\frac{3}{4}$ million, largely concentrated in the south. The majority of the inhabitants are Moors, but there are some 130,000 Negroes in the south.

The country was occupied by the French in 1902; it was declared a protectorate in 1903 and a colony in 1920. It became a member state of the French Community in 1958 and an independent republic in 1960. Morocco claims Mauritania mainly on religious grounds as the Mauritanian Emirs owe allegiance to the King of Morocco.

There are *two main regions*, the crop and livestock region of the south and along the Senegal river, and extensive pastoralism, oasis cultivation and recently iron mining in the north. Crops are grown in the Chemama, the flood plain of the Senegal river by means of a primitive form of basin irrigation. Negro cultivators produce millet, sorghum, rice, sweet potatoes,

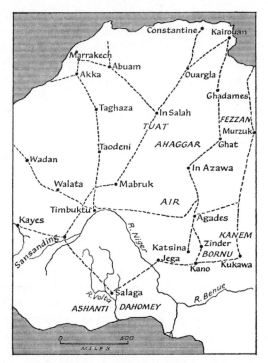

FIG. 25—The principal Saharan caravan routes of the nineteenth century (after Bovill)

water-melons, tobacco and a variety of vegetables. Locusts and the quelea bird are a problem. Livestock are more important than crops and the bulk of the 900,000 cattle and $5\frac{1}{2}$ million sheep and goats are located in the south. Their main value is for social rather than economic purposes, although about one-fifth of the sheep and goats and one-tenth of the cattle are sold annually to markets in the south.

In the oases of the north—Atar, Chinguetti and Ouedane—dates are the main crop, but a variety of fruit and vegetables is also produced. The main economic hope for the future lies in the rich deposits of *iron ore*, averaging 63% iron, near Fort Gouraud, which can be worked by opencast methods. Reserves of 40% iron ore are estimated at 8 billion tons. Mining started in 1961 and the railway, 400 miles to Port Etienne, is now moving stocks of iron; by 1967 an output of 6 million tons a year is planned. The shortest route would be about 200 miles across Spanish West Africa to Villa Cisneros, but boundary complications have necessitated building a longer line to Port Etienne which, until recently a small fishing port, will become a major ore loading port. It has deep water offshore and is able to cater for large ore-carriers without the expense of dredging, but there are problems of water supply and very high summer temperatures. There are deposits of copper at Akjoujt, with gold as a by-product, and more deposits of poorer quality iron ore near Atar.

The other towns of Mauritania, Kaédi and Rosso are very small and located in the south. Nouakchott is the new capital which is being developed from the start in desert country on the main road to the north and about two hundred miles north of St Louis the old capital, now in Senegal.

LIBYA

Libya consists of the former Tripolitania, Cyrenaica and Fezzan in the south. Impressive ruins at Cyrene, Sabratha, Homs, Leptis and elsewhere are a reminder of the importance of Libya in Greek and Roman times. The Greeks and Romans had a very efficient system of water conservation with catchments and diversion dams in the wadis, underground reservoirs in the hills and thousands of wells and cisterns, some of which have been cleaned out and serve as well today as they did 2,000 years ago.

The country was under Turkish control until 1912 when it was taken over by Italy. The country became independent in 1951 and the leader of the Senussi, a puritanical Islamic sect, became king. In the 1920s conquest of the interior was carried out and Italian settlers, of whom there were nearly 100,000 by 1939, were placed on the land. Now nearly 50,000 Italians remain, most of whom are urban. The total area is estimated at about 680,000 square miles and the population $1\frac{1}{2}$ million, nearly three-quarters of whom are in Tripolitania.

Although Libya consists largely of the Sahara there are a number of minor physical regions. From the Tunisian frontier to Misurata there are a string of coastal oases backed by a triangular semi-arid coastal plain, the *Jefara*, bounded southwards by the dissected plateau of the Jebel Nefusa. This region has a slight winter rainfall of less than 10 inches which permits cultivation, now supplemented by irrigation. At the edge of the Jebel Nefusa a spring-line occurs which supports a series of settlements. The main crops are cereals, olives, figs and groundnuts. Yields are highly variable and barley, which is the staple crop, has varied in quantity from 10,000 to 150,000 tons a year. Wheat is also grown, but requires irrigation and, therefore, the tonnage produced is considerably less than that of barley. In 1961 18,000 tons of barley and 8,000 tons of wheat were produced.

Groundnuts are now the most important cash crop, as they are best adapted to the uncertain rainfall. Esparto grass grows wild in the Jefara and the Jebel, but falling prices have restricted output. There are over 3 million olive trees in Libya, the bulk in Tripolitania, and production is increasing, exports depending on climatic and home demands. Almonds and citrus are also of some importance.

Tripoli is by far the most important town in this region and also the main port and biggest town in Libya. Its name is derived from the three towns or ancient Phoenician trading ports of Sabratha, Oea and Leptis Magna. Sabratha was a flourishing Roman town in the second century A.D. with a theatre seating 5,000 people and exports of olive oil, cereals and other products. Tripoli has a good natural harbour and can accommodate ships with a draught of 26 feet. From 1954 to 1960 the number of ships using the port doubled. Tripoli now handles the bulk of Libya's imports. The population has also increased rapidly from 130,000 in 1954 to nearly 200,000 in 1962, largely as a result of increasing economic opportunity in Tripoli and Benghazi afforded by the oil boom.

Over seven hundred miles along the coast and separated from the Jefara by a long stretch of desert is the second cultivable region of the *limestone plateau of the Barce and the Jebel el Akhdar* with a winter rainfall of 8–24 inches and terra rossa soils, some of considerable fertility. However, Tripolitania has a far greater area than Cyrenaica suitable for cultivation. As in Tripolitania barley is the main crop and it is grown widely by semi-nomads, with greatly fluctuating yields. Wheat is grown on the Barce (El Marj) plain, the most fertile area in Cyrenaica. About a thousand former Italian farms have been distributed among the local people. Wheat is grown continuously with little rotation or soil conservation. Progress has been disappointing and in parts soil erosion is serious. There is a great need for mixed farming and tree crops such as olives, almonds and citrus fruit, and vegetables. Another problem is to persuade the pastoral semi-nomadic population to settle down permanently and become sedentary farmers. Water resources have been improved, 3,000 cisterns dating back to classical times have been cleared out, many small reservoirs reconstructed and wells dug.

Benghazi is the main town and port for Cyrenaica. Its harbour was blocked for a long time by sunken ships, mines and other obstacles, but war damage has now been restored and when the reconstruction programme has been completed the harbour will be able to accommodate ships of up to 30-feet draught. As with Tripoli, Benghazi's population has increased greatly, from 70,000 in 1954 to over 120,000 in 1962. Benghazi now contains over half Cyrenaica's population. The city is acquiring a new skyline with many tall buildings in the central area, and even the medina is being reconstructed, while the ubiquitous shanty towns form a rash on the outskirts. The problem of a capital was only partially solved by moving it periodically from Benghazi to Tripoli and back. A new capital is being established at Beida in the Jebel Akhdar, but its great distance from Tripoli and the south makes it an inconvenient location.

Most of Libya consists of the *Sahara*, the desert even reaching the coast in the Gulf of Sirte, although here a slight rainfall of 2–8 inches permits extensive pastoralism. Temperatures can be extremely high; for example in June 1961 a temperature of 135° F. (57° C.) was recorded at Marsa Bregha on the coast. The Sahara consists of a number of low plateaux 1,500–2,000 feet in altitude, such as the Hamada el Hamra in the south and the east, partly enclosing wide sandy depressions and bordered by a number of oases such as Murzuq and Kufra. To the north of the north-west to south-east trending plateaux is the large expanse of the Libyan Sand Sea, much of it below 300 feet in height and even below sea level in the oasis of Giarabub.

The meagre rainfall and vegetation of the northern fringes of the Sahara only support *extensive pastoralism* and a nomadic or semi-nomadic existence is carried on by over a third of the population, particularly in Cyrenaica. Sheep and goats number some $2\frac{1}{2}$ million, and there are few camels and cattle. Sheep and goats are the main exports, largely to Egypt. Wool of poor quality and only suitable for rugs and carpets is sent mainly to Italy. The main oases, containing the bulk of the population, are widely separated and situated in such places as Ghadames, Murzuq, el Jofra, Kufra, Jalo and Giarabub. Date-palms are the mainstay of oasis life, as elsewhere in the desert. Under the impact of the oil boom the population of the oases is declining rapidly; the population of the Fezzan has dropped from 54,000 in 1954 to 35,000 in 1960.

The most promising economic activity in the desert is the exploitation of *petroleum* which was first exported in 1961 from the Bir Zeltan field nearly 250 miles south of Benghazi and about 100 miles from the gulf of Sirte. The two most important fields are the Zeltan and Dahra Fields. Production, which in 1961 was only about $\frac{1}{2}$ million tons, had reached nearly 8 million tons by 1962 and 40 million tons in 1964 with a potential production of 70 million tons by 1970. A pipeline has been constructed from the main field at Zeltan to Marsa Bregha and another oil terminal has been developed at Es Sider, 70 miles west of Marsa Bregha. The Libyan oilfield is better placed than the Algerian fields and there are now

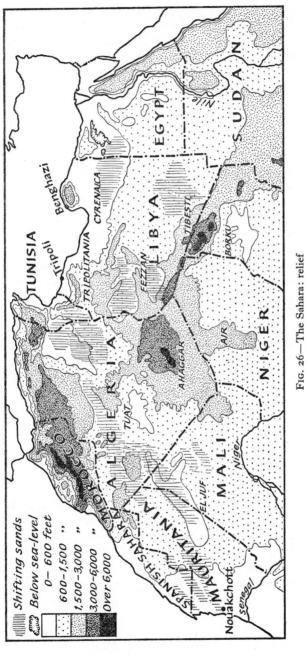

Fig. 26—The Sahara: relief

Shifting sands
Below sea-level
0– 600 feet
600–1,500 "
1,500–3,000 "
3000–6000 "
Over 6,000

two pipelines only 195 miles long, compared to the 1,300 miles (late 1961) of pipelines in Algeria which have to cross the Atlas mountains and the boundary of Tunisia. There are also considerable quantities of natural gas.

Expenditure by the oil companies has had a profound effect on the economy of Libya, but wealth, as in a number of Middle Eastern countries, has tended to become concentrated in the hands of a few rich men, emphasizing still further the great gulf between them and the poor. Foreign expenditure rose by 100% from 1950 to 1959 causing inflation and a spiralling cost of living. Little of the money obtained from the petroleum boom has been invested in agriculture and industry, but much in property and land speculation, causing the mushrooming of the population of Tripoli and Benghazi and the spread of hotels, office blocks and apartment buildings. Smaller towns such as Homs, Misurata, Derna and Tobruk are stagnating or declining in importance and there is a serious drift from the rural areas. The basis of Libya's economy should be strengthened and far more spent on agriculture, afforestation, tourism and other economic activities. However, the area that can be developed is extremely limited. Afforestation has been carried out with drought-resistant trees such as eucalyptus, acacia, olive, palm, cypress and carob. The Agricultural Department in Tripolitania has restored and extended the green belt which the Italians began to create twenty-five years ago around Tripoli.

Libya faces the problems of a harsh environment, aridity and great distances between centres of economic activity, unwieldy size and disproportionate distribution of population. Tripolitania, with only one-sixth of the total area, has three-quarters of the population; Cyrenaica has half the area and only one-quarter the population; while Fezzan with one-third of the area has less than 40,000 people. Considerable aid has been granted by the British and Americans, and oil revenues have boosted the economy.

The Romans and Greeks, with a rainfall which may well have been much the same as it is today, developed an intricate system of water storage in aqueducts and a thriving agricultural economy, with large exports of grain, olive oil and fruit to Rome. The Italians, two thousand years later, built good roads and placed thousands of Italian peasants on the land who, like the Greeks and Romans, began to develop a promising agriculture with crops of wheat, olive oil, almonds and other commodities. The ravages of war and Libyan independence caused the evacuation of most of the Italian farmers. The new Libyan farmers have much to learn from the past in the efficient use of their meagre natural resources of water and soil.

CHAPTER SEVEN

The Nile Valley

THE third region of the Nile Valley has a broad base in the Sudan and parts of Ethiopia and Uganda, and a long, narrow, winding strip extending northwards through Egypt to the Mediterranean (Fig. 27). The Nile snakes across the Sahara bringing abundance of soil and water from tropical regions and creating fruitful places in the desert. The Nile gives this region its unity, although there are great physical and human contrasts from the highland edge of Ethiopia to the great swamps of the Sudan, and between the nomadic pastoralists of the almost rainless wastes and the fellaheen of the populous Nile delta. It is a region that has little mineral wealth and relies greatly on agriculture, and that has in Egypt a very long and almost unique history and now a population problem of great magnitude.

Egypt, in particular, is absolutely dependent on the Nile, which gives a constant and assured supply of water and annual renewal of silt. These factors, combined with warm temperatures and man's early mastery of water control, encouraged the development of irrigation agriculture and a fruitful riverine civilization five to six thousand years ago. While natural obstacles of sea, desert and swamp enabled this civilization to prosper, it did not prevent outside influences from affecting Egypt, so that although there was protection it was not isolated.

DRAINAGE

A description of the physical background must largely be concerned with the Nile drainage system. The Nile is reputed to be the longest river in the world, being over 4,000 miles long. Its main sources are Lakes Tana and Victoria, which give rise to the Blue and White Niles respectively. The Blue Nile supplies over half the total volume of Nile water, the White Nile and Sobat about a quarter and the Atbara the rest. The ultimate source of the Nile is the headwaters of the Kagera river which discharges into Lake Victoria. From the northern end of the lake the river plunges over the Owen Falls, where it provides electricity for Uganda and Kenya, to flow through a short gorge section to Lake Kyoga, and thence westwards down the narrow cleft of the Murchison Falls into Lake Albert where it picks up additional water.

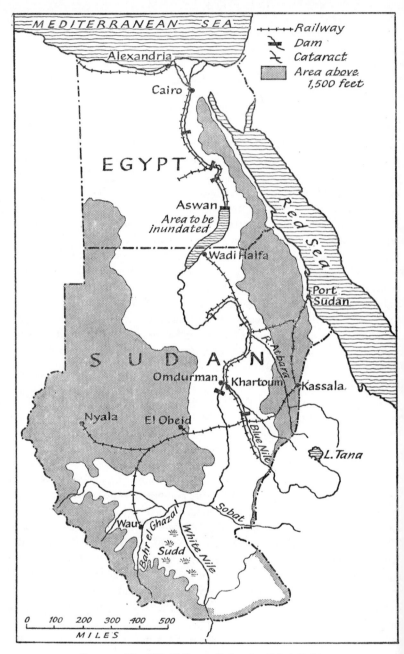

FIG. 27—The Nile Valley: relief and political division

From here the river known as the Bahr el Jebel, or 'river of the mountains', flows north into the Sudan through a wide basin at an altitude of less than 1,500 feet, having dropped nearly 3,000 feet from the African plateau. This is the region of the *sudd* which is extensively flooded each year, and where the river winds through numerous channels which are occasionally blocked by patches of floating vegetation forming green islands. Because of the high altitude of the sun and the wide areas covered by shallow floods evaporation is intense and carries off half the available water, a considerable loss to the cultivators of Egypt. Proposals have been put forward to offset this by cutting a straight channel, the Jonglei Cut, 300 miles long from Jonglei to south of Malakal. This would concentrate the waters of the Bahr el Jebel and prevent it from splaying out, so reducing by half the loss from evaporation. Other proposals to increase the effectiveness of the Nile waters are the construction of a number of dams on and near Lakes Victoria and Albert, on the Blue Nile at Roseires and the high dam at Aswan. The main purpose of dams on the White Nile is to hold back the flow of water at a time of year when the Blue Nile is at a high level (in late summer and autumn), and to release the stored waters down the White Nile when the Blue Nile is at a low level. This is partly achieved already by the Owen Falls dam on the Victoria Nile and the Jebel Aulia dam south of Khartoum.

The problem of increasing salinity of water will have to be faced if the Jonglei Canal is constructed. Lake Albert is far more saline than Lake Victoria and the sudd extracts or absorbs chemical substances from the water, so that water at Malakal is less saline than that flowing out of Lake Albert. The Jonglei Canal will reduce the absorptive effect of the sudd and, coupled with the effect of storage schemes along the Nile, there is a danger that soils in Egypt may become more saline, although this could be offset by good field drainage.

Near Malakal the Bahr el Jebel is joined by the Sobat river from the east and the Bahr el Ghazal, or 'river of gazelles', from the west. The latter tributary supplies less water than the Sobat as much water is lost through evaporation in the swamps (Fig. 28).

The high Abyssinian plateau, fed by heavy monsoon rain, is the major source of Nile water, the Blue Nile, Atbara and Sobat supplying three-quarters of the total Nile waters. The waters of the *Blue Nile* come from tributaries flowing into Lake Tana, which acts as a kind of natural reservoir in somewhat the same manner as Lake Victoria, although less than a twentieth of its size. The Blue Nile, or Abba, curves south-west, west and then north in a great gorge 5,000 feet below the plateau surface through wild and inaccessible country to join the White Nile at Khartoum. The Blue Nile and its tributaries descend swiftly from the Ethiopian highlands compared with the slow-moving and meandering waters of the Bahr el Ghazal and Bahr el Jebel.

From the sixth cataract just north of Khartoum to the fourth cataract the river flows in a narrow faulted trench and is unnavigable, the cataracts occurring where granitic rocks outcrop. The Atbara from northern Abyssinia joins the Nile 1,700 miles from the sea, and is the last permanent tributary. Navigation is again interrupted by the second and third cataracts between Wadi Halfa and Dongola. From Khartoum to Aswan there are in all *six cataracts*. From Aswan to the sea the amount of alluvium increases and green ribbons of cultivation follow both banks, widening out as the delta is reached. The boundary between the narrow strip of sown land and the great wastes of desert to east and west is very sharply defined. At Cairo the delta proper begins and the river splits into a number of distributaries which have been reduced to two main channels. Here the gradient is very slight and only a third of the water reaches the Mediterranean because of losses due to irrigation and evaporation.

Low water occurs in the Nile from February to July, but with the heavy summer rains on the Ethiopian plateau the river rises rapidly, over 20 feet at Khartoum, and from August to October there are floods, the maximum being reached in September at Khartoum and October at Cairo. The Blue Nile holds up the White Nile drainage and ponds it back, only letting it down when the Blue Nile peak has passed. In April and May over three-quarters of the Nile flow is supplied by the White Nile. Enormous quantities of fertile silt are brought down from the Abyssinian highlands and renewed annually, although with the spread of perennial irrgiation far less silt is deposited than with the former practice of basin irrigation. The total discharge of the Nile drainage system is about 85 billion cubic metres of which 21 billion leaves Lake Victoria, while the Blue Nile at Khartoum has an average flow of 50 billion cubic metres. A third of the Nile drainage enters the Mediterranean.

CLIMATE

There are great variations in *rainfall* ranging from desert conditions over most of Egypt and northern Sudan (for example, Cairo has 1 inch of rainfall a year) to the abundant rainfall of 50–80 inches in parts of the Ethiopian and East African highlands. Rainfall decreases rapidly northwards from Roseires, with less than 30 inches per annum, to Khartoum

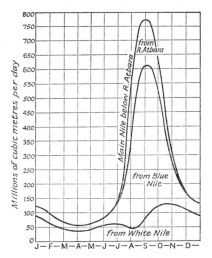

FIG. 28—Discharge of the main Nile at Khartoum and of the river Atbara (after Hurst and Phillips)

with 5 inches. From Khartoum to the delta the Nile Valley is practically rainless, only the Mediterranean coast receiving a slight winter rainfall of 7–8 inches. The régime is equatorial near Lake Victoria, but to the north it becomes increasingly tropical with a summer maximum in the Sudan. Strong south and south-east winds caused by depressions crossing Lower Egypt, particularly in spring and early summer, bring high temperatures and blow great dust storms from the desert into the cultivated area. The northern Sudan is also affected by hot dust storms, known as haboobs, which are most common in summer.

Mean annual *temperatures* vary from 67·5° F. (19·8° C.) and 68·2° F. (20° C.) at Alexandria and Cairo respectively, to nearly 83° F. (28° C.) at Atbara and Khartoum, the mean monthly temperatures from May to July being over 90° F. (32° C.) at Atbara. In the southern Sudan the highest temperatures occur from March to June before the rains come, while the coolest period is the rainy season, as at Mongalla from June to September. The mean annual range of temperature in the southern Sudan is far lower than in the northern Sudan and Egypt; Mongalla 5° N. has an annual range of 6·9° F. (3·7° C.) compared with Wadi Halfa 22° N. with a range of over 30° F. (17° C.). Winter nights are cool in the northern half of the Nile Valley, although the effect of winter warmth from the Mediterranean causes Alexandria to be a little warmer than Cairo.

VEGETATION

There is generally a close correlation between rainfall and *vegetation* but the effect of soil texture, site, man's influence and other factors are important as well. Rainfall decreases from over 70 inches in a few southern areas to nil, and vegetation reflects this in the change from small patches of tropical rain-forest to broad-leaved savanna, acacia woodland to scrub and desert types. In arid areas with a rainfall of less than 4 inches vegetation is confined to the Nile and wadis, and consists of coarse tussocky grasses, tamarisk and mimosa. In the semi-desert with a rainfall of up to 15 inches acacia is found on thin soils, but on deeper soils mixed grasses and herbs may provide a continuous cover. With a higher rainfall of up to 30 inches woodland and savanna is found. Areas of open grassland occur and trees tend to be larger, from 10 to 15 feet high, and often growing close together. In sandy areas with a rainfall of less than 20 inches the main type of tree is the Acacia senegal, the producer of gum arabic. With over 20 inches of rainfall thorny trees are replaced by broad-leaved trees such as Combretum, Dalbergia and Albizzia.

AGRICULTURE

The economy of Egypt and particularly the Sudan is greatly dependent on agriculture. Both countries are poor and with a low per capita national income, that of the Sudan being only £S27. In Egypt Suez Canal dues and secondary industry contribute a considerable proportion of the national income. Cotton is by far the most important crop in both countries, supplying over half the exports from the Sudan and a third to a half of Egypt's exports. The arable acreage is small in proportion to the area, only 3% in Egypt, and cultivation relies almost entirely on irrigation in Egypt and northern Sudan. Major development projects such as the Aswan High Dam are thus of vital importance. Secondary industry is of little concern in the Sudan, except for cotton ginneries, a cotton-spinning and weaving factory, cement factory, shoe factory and a few other factories.

POPULATION

The population of Egypt is basically Mediterranean, as shown by the fellaheen or peasants with their broad faces, high cheek bones, fuzzy hair and thickset figures, who form three-quarters of the population. They live in straggling villages where the mud houses crowd onto narrow, winding lanes. The townspeople are more varied in background, but with the great urban influx in recent years links with the fellaheen are strong amongst the lower and middle classes. Nubians are an important class in the large towns. There are over 1 million Copts, survivals of the pre-Moslem Christian era, who are mostly town-dwellers. In the eastern and western deserts are over 50,000 nomadic pastoralists who are declining in numbers. Over 90% of the population of Egypt is Moslem and Arabic is the official language, the colloquial form differing considerably from classical Arabic. An Islamic–Arabic culture is dominant in Egypt. Egypt faces one of the most serious problems in the world of *population pressure*, with an average density of 1,800 people per square mile in the Nile valley and a high rate of increase.

The Sudanese population is much smaller in number, more scattered and more varied in composition. Hamitic influence is strong in the north, where the Nubians inhabiting the Nile valley are the dominant element, although modified by negro blood. There are nomadic Arabs in the northern desert area, and to the east are the Red Sea tribes who have occupied their present area for over 5,000 years. The north is Moslem in religion and Arab in culture and contains the bulk of the population. The south has a smaller and more polyglot, pagan population. There are nomadic Arabs and sedentary cultivators of mixed Arab and Negro blood, Nilotes such as the Dinka, Shilluk and Nuer, and other Negro tribes such as the Azande.

HISTORY

The irrigable valleys of the Nile, Tigris and Euphrates, Indus and other rivers in the Near and Middle East were the mainsprings of *early civiliza-tion*.[1] The annual floods brought water and soil which when controlled could be made to bear fruit abundantly. But in some years, as related in the Old Testament, high or low water could lead to flood or famine when hunger and pestilence might strike the population. Strict control over the distribution of Nile waters was essential, and this was exercised by an increasingly centralized government which, besides being in charge of the rise and fall of Nile waters, controlled the use of waterways, the collective storing of harvests and the distribution of seed. There has been a remark-able continuity of life in the Nile valley since the unification of Upper and Lower Egypt which occurred over 6,000 years ago. Government has been marked by an almost continuous autocracy, from pharaohs to pashas and

---

[1] The lower Nile Valley has far less rain than the other river valleys so that Egypt's oasis-like nature is more pronounced; the annual Nile floods were, therefore, still more of a mystery to the peoples of the ancient world.

now a dictator. Egypt's deserts have afforded a measure of protection, and yet the Nile valley has long attracted invaders from the desert, the sea or along the Fertile Crescent. Control has often been imposed from outside by people such as the Hyksos, Persians, Greeks, Romans, Arabs, Turks and British.

The Arab invasion of the eighth century had a profound influence, the Coptic language was replaced by Arabic and Christianity by Islam. After many centuries of stagnation Mohammed Ali brought about an economic revival during the early nineteenth century. Cotton became the main cash crop and there was increasing commerce with Europe. The European powers, too, became more interested in Egypt's role as an isthmian state between the Mediterranean and the Red Sea and, therefore, the route to India and the Far East. The Suez Canal was built in 1869 and there was virtual British control of Egypt from 1882 to 1955 because the Canal was a vital link in the route to India. Egypt played an essential role in the Second World War, and the battle of El Alamein was one of the major turning points of the war for the Allies. With the suppression of the monarchy and control by President Nasser since 1954 British influence has disappeared, and the Suez Canal has been taken over by Egypt.

Egyptian influence has been significant in the Sudan for thousands of years. At times Sudanese culture has affected Egypt and the pre-dynastic Egyptian civilization may have derived from the Neolithic culture of the Sudan. The chief export to Egypt during the period of the New and Middle Kingdoms was gold with a wide range of other exports during the New Kingdom consisting of wood, gum, ivory, ostrich feathers and eggs, perfumes and oils, cattle, leopards and leopard skins, giraffes, dogs, baboons and slaves. These exports illustrate the sophistication of Egypt's culture, and the role of the Sudan as a major primary producer. From 725 to 660 B.C. the Sudanese kingdom of Cush actually conquered and held Egypt, only to be pushed out by the Assyrians using the new weapons of iron. This contact led to the exploitation and smelting of iron ore at Meroë, which became the capital of the kingdom in the sixth century B.C. Meroë was better sited than Napata in relation to the cattle-raising and crop-producing areas of the Central Sudan. But, as with the transfer of the capital from Napata, so problems of overgrazing and erosion at Meroë led, among other factors, to its downfall. Christian influence lasted for over a thousand years, but gradually succumbed under the effect of the Arab invasion and disappeared with the downfall of the Christian state of Dongola in the fourteenth century.

In 1821 the Sudan was invaded by Mohammed Ali, and a new capital set up at Khartoum. Egyptian control was upset by the Sudanese revolt of 1881, but was re-established with British help in the 1895-8 campaign. From 1898 to 1955 the country was known as the Anglo-Egyptian Sudan, although it was virtually ruled by the British. Since 1955 the Sudan has been independent.

UNITED ARAB REPUBLIC—EGYPT

Historically Egypt consists of the long narrow valley section of Upper Egypt stretching for about 500 miles from Aswan to Lower Egypt which consists of the delta. The great riverine oasis is now bounded by a far wider area, and the modern political boundaries extend south for another 200 miles to near Wadi Halfa and the Sudanese border, east to the Red Sea including the Sinai Peninsula, while to the west is the Western Desert comprising almost three-quarters of the total area of Egypt. The total area is 383,000 square miles, of which only 13,000 square miles is in the Nile Valley, Fayum and the delta. On this tiny area about *28 million people* (1963) are crowded with a population density of nearly 2,000 per square mile, and reaching 6,000 per square mile in places. The rate of increase is nearly 3% and the population may be 40 million by 1980.

THE NILE DELTA

The Nile delta or Lower Egypt is the most important region (Fig. 29). It is a large delta with an area of over 9,000 square miles, more than twice the area of Upper Egypt. Below Cairo the Nile splays out in two main branches, the Damietta distributary to the east and the Rosetta distributary to the west. The delta was cultivated in Pharaonic, Ptolemaic and Roman times, but after the Arab invasion it was increasingly neglected. Cultivation depended largely on basin irrigation until the early nineteenth century when a greater measure of perennial irrigation was introduced by Mohammed Ali. A network of canals was constructed and the delta reclaimed. Prior to this most irrigation was of the basin variety, whereby silt-laden water was led from the Nile in August and September into a series of basins, left for about a month, then the seed was sown in the moist silt and crops harvested in April and May. Only one crop a year could be grown under this system,

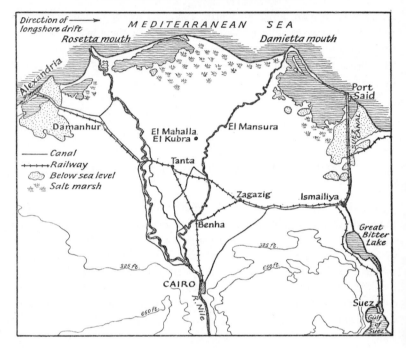

FIG. 29—The Nile delta and the Suez canal

whereas with perennial irrigation two to three crops may be produced in one or two years.

Settlement has spread rapidly all over the delta, except for the northern fringe where salt marshes and lagoons have only a slight population. Much of this area could be reclaimed, but costs are high, elaborate drainage works would have to be constructed and soils are very saline. The delta has become the major economic region with the largest population and the largest towns. The central and southern districts are the richest with the most fertile soil and more water available for irrigation.

Irrigation

The green plains of the delta are extremely level with a gradient of only 1 : 10,000, which facilitates greatly the work of irrigation. The system of canals and channels is now about 18,000 miles long. Canal water has to be raised to the level of the fields by a number of methods. In Upper Egypt the sakieh is used, which consists of a vertical wheel with cups that hold the water; it is turned by an ox walking endlessly in a circle. The shaduf, a long beam or pole weighted at one end, is still older and was used in dynastic Egypt. The most common instrument is the Archimedian screw, a wooden cylinder with a spiral which is turned by hand, so causing the water to rise in the tube. On state lands and large estates modern pumping machinery is used.

Agriculture

The peasant uses equally primitive implements, hoes with triangular blades and ploughs that have not changed in 4,000 years except for the tip or share which is now steel-shod. A sickle is used for cutting corn and rice, the stalks of which are heaped in the open and threshed by oxen drawing a heavy wooden sledge.

Of the 6 million acres of irrigated land some two-thirds is double-cropped each year; however, the usual practice on the large estates is to produce three crops in a two-year rotation, so as to maintain soil fertility. There are three main categories of crops, firstly the winter crops grown from October to April and consisting of wheat and barley, vegetables and birsim or Egyptian clover. Summer crops are grown from April to October and consist of cotton, sugar-cane and rice. The third type is the Nile harvest of maize from July to November.

Cotton is by far the most important crop, grown on about a fifth of the cultivable area, its yield and quality being the best in the world (Fig. 30). Production is 300,000 to 400,000 tons a year, most of which is exported. Its proportion of the total agricultural income fluctuates widely; in 1951 it was 48%, in 1953 only 30%, and in 1959 about three-quarters of all exports. Cotton-growing in Egypt is quite unlike the mechanized cultivation of a country like the United States of America and demands daily attention and a great deal of labour. Nearly all the work is done by hand, a careful watch being kept for pests which have to be removed by hand as few landowners are able to afford insecticides. The cotton flowers in June, but the bolls ripen at different times between August and the beginning of October. The whole village turns out to pick the bolls which are then placed in rough sheds for ginning and stripping. Children again provide the labour for sorting out the bolls, stripping the seed used for making the oil and detaching the long white fibres.

The main *cereals* are wheat, maize and rice, the latter being exported, although wheat has to be imported. Production of these cereals in recent years has been over a million tons each. Other important crops with high yields are sugar-cane, groundnuts, onions, tomatoes, beans and fruit. The following figures show exports of rice, onions, groundnuts and raw cotton from Egypt for the period 1958–60, in thousand metric tons.

	1958	*1959*	*1960*
Rice	360·1	23·2	280·0
Groundnuts	7·8	9·1	9·9
Onions	157·6	178·6	173·8
Cotton, raw	281·5	318·3	374·2

Population and distribution

Egypt suffers from a *population that is growing too rapidly* for the land to support it. The rate of increase is half a million people a year so that the population will reach 50 million by the 1990s unless this rapid annual rise is checked. The ratio of cultivated land to the total population has fallen from 0·2 hectare per head in 1907 to 0·1 in 1954, although yields have increased because of the change from basin to perennial irrigation. The amount of cultivable land has increased slowly compared with the population which has doubled in size since 1917. It is estimated that half the rural population is surplus to labour requirements, so that there is a great deal of under-employment and unemployment, rural poverty and indebtedness.

Another problem is that the *land is distributed unequally*; in 1950 2% of the landowners held half the land and 70% of the holdings were under one acre in size and, therefore, unable to support a family adequately. Recent surveys show that a minimum of five feddans[2] is needed for a family of eight. The agrarian reform law of 1952 limited holdings to a maximum of 200 feddans, and this was further reduced in 1961 to 100 feddans. Nearly 1 million acres were expropriated in the period 1952–62.

Although the technical reforms of the nineteenth century increased greatly the wealth of the few it did not help the fellaheen whose health and standard of living has declined. The perennial system of irrigation is partly to blame for the poor state of health of the fellaheen, 95% of whom are suffering from bilharziasis. Other serious diseases are anchylosetomiasis, malaria, eye diseases and malnutrition.

Nearly three-quarters of the population is *rural*. The fellaheen live in villages set back from the roads on mounds above flood level. Some of these

[2] One feddan equal 1·038 acres.

villages are very large in size, and those which are largely self-sufficient tend to remain isolated from one another. Houses are usually low and one-storeyed, windowless and built of sun-dried mud bricks. They crowd together onto narrow winding paths and alleys. The flat roofs are used for storing maize and cotton stalks, cakes of dung and straw for fuel. Dust rises from the fields and lanes, animals root in piles of rubbish, women and children work in the fields, men ride past with flapping gallabiyas (long, loose gowns) partly obscuring the tiny donkeys, and the plough may be drawn by an oddly assorted team of buffalo and camel. Through the fields twine irrigation canals and ditches that bring the life-giving water from the

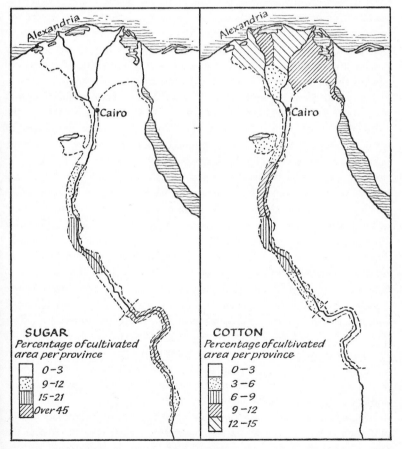

FIG. 30—The Nile Valley: sugar and cotton (from *Atlas of the Arab World and the Middle East*)

193

great river which dominates the scene, providing water, soil and the mud for the houses. It is a busy scene that has been enacted for thousands of years, only the slaves of Pharaonic Egypt depicted on the walls of rock tombs are now replaced by the serfs of the twentieth century dominated by the demands of an irrigation civilization.

There is little or no *local rule*. All authority is vested in Cairo and wielded through the umda or village head, who is all powerful, and must own at least ten feddans of land. He directs the crop rotation, administers justice and is responsible with the auxiliary police for the maintenance of law and order.

Nearly a fifth of the Egyptian population live in Cairo and Alexandria. *Cairo* with a population of about 3 million is the largest city in Africa and by far the largest in Egypt (Fig. 31). It was first founded by the Arabs in the tenth century A.D., but has only become of major importance since the nineteenth century with the economic growth of the delta. Cairo is well sited at the junction of the delta and Upper Egypt, and was built initially at the lowest crossing point of the Nile on the east bank of the river and on a spur of the Mokattam hills (Plate III). To the south-west are the ruins of Memphis, the ancient capital of Upper and Lower Egypt, and of great importance until the Arab conquest. Cairo now extends along the base of the hills to Heliopolis and across the plain to the Nile and Giza on the west bank. The city is dominated by Saladin's citadel, and there are hundreds of mosques, prominent among them being the Mosque of El Azhar, which houses the largest Moslem university in the world, founded in the tenth century. There is a mixture of ancient and modern, of glitter and squalor, with Cities of the Dead, Pyramids, skyscrapers and slums. There are ten- to fifteen-storey buildings, wide tree-lined avenues and spacious apartments lining the river banks next to slums and shanties as bad as any in Africa. The modern commercial quarter of Ismailiya contrasts greatly with the booths and bazaars of the medieval Musky. Cairo has good communications by rail and canal with Upper Egypt, Alexandria, Port Said and Suez and there is a busy Nile trade. Cairo also has an important international airport. The city is the most important industrial centre in Egypt with nearly a quarter of the country's industrial establishments, consisting of textiles, clothing, shoes, perfume, cigarettes, soap, furniture and agricultural processing factories.

Alexandria, with a population of about $1\frac{1}{2}$ million, is the summer capital and main port of Egypt and a leading seaside resort. It is situated at the western edge of the delta on a strip of land between Lake Maryut and the Mediterranean some 150 miles from Cairo, with which it has good canal

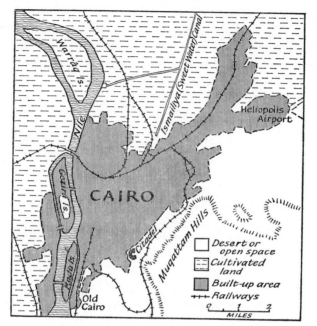

FIG. 31—Cairo

and rail connections. There is a ship canal to Rosetta. The city was founded by Alexander the Great in the fourth century B.C. and became the largest city in the west before the rise of Rome. It was a great centre of Hellenistic and Jewish cultures, but declined after the Arab Invasion. The construction of the Mahmudiya Canal in the early nineteenth century caused Alexandria to regain its former importance. It is Egypt's other main industrial centre with cotton, leather, footwear, paper, metal and other industries.

The rest of the delta towns are far smaller, Tanta being the largest and a commercial and industrial centre of some importance.

Industrial development

Attempts to solve the problem of population pressure in Egypt by developing industries have not been very successful. Despite state control of the economy and determined efforts to industrialize, the purchasing power of most of the people remains as low as ever, thereby restricting the size of the local market.

195

Cairo and Alexandria are the most important centres, containing nearly three-quarters of the factories and employing over half the factory workers. Most of the factories employ less than ten workers, but a small number of large concerns employ the majority of the workers and produce the bulk of the output. The main industries are textiles, food-processing, tobacco, cigarettes, and constructional materials.

A feature of the post-revolutionary industrial pattern is the growth of heavy industry, such as an iron and steel plant at Helwan, fertilizer factories near Aswan and at Suez, and an oil refinery at Suez. Much of this industrial growth has been badly planned and costs miscalculated, thereby increasing the already high cost structure of Egyptian industry. Egypt has no coal deposits, but there are important iron ore deposits near Aswan, and the country produces limestone and manganese, which are required for an iron and steel industry. There is a small but growing oil industry, and output is now over 3 million tons of petroleum a year, mostly from Western Sinai.

THE UPPER NILE

The Upper Nile is the second region, with an area of 4,000 square miles and a population of about 8 million. The contrast between the green strip of the sown in the Nile Valley and the tawny wastes to east and west is vivid and dramatic. The river is thronged with craft, and a succession of mud villages perch on the desert edge, or on banks above the reach of the Nile floods. From the Sudan border to Aswan the river flows through a gorge cut in Nubian sandstone. This part of the Nile valley has a very slight population.[3] Near Aswan granites cause a series of rapids and a break in navigation. North of Aswan softer clay has been eroded away to form a series of basins, while north of Idfu there is a continuous alluvial plain to the delta, usually wider to the west than to the east. The valley is bordered by limestone scarps which reach a height of 2,000 feet near Luxor. At Qena a great bend of the river encloses much fertile land which is densely populated. This was the *core of ancient Egypt* with its capital at Thebes for a period of 1,500 years until the seventh century B.C. The Qena bend is the nearest point to the Red Sea and caravans travelled from here to the Red Sea ports. At Nag Hammadi a barrage permits intensive cultivation with cash crops such as cotton and sugar-cane and food crops grown on a 10–12 mile wide alluvial plain. Asyut is the main commercial centre of the Nile valley and lies at the northern end of the Nile gorge. From here to the delta the valley opens out and the scarp to the east is nowhere more than

[3] It will soon be flooded so that population is being evacuated.

700 feet high. From the Asyut Barrage the Ibrahimiya Canal, nearly 200 miles long, waters the most intensively cultivated section of the valley where cotton, wheat and maize are the main crops. Near Cairo market gardening becomes increasingly important.

Between Asyut and Aswan basin *irrigation* is now being converted to perennial irrigation. Well water is also important, the water coming from the beds of sand and gravel. Besides the large dam at Aswan the Nile is regulated by barrages at Esna, Nag Hammadi, Asyut and Zifta. These are, however, insufficient to regulate water for the pressing needs of a rapidly increasing population, which is the major reason for building the High Dam at Aswan (Plate IV), 5 miles south of the present dam. The dam will store the water from year to year to even out the flow of the Nile. Its siting at Aswan will leave control of the dam in Egyptian hands. Besides increasing the irrigable area of Egypt by almost a quarter the scheme should provide large quantities of hydro-electric power and greatly improve Nile navigability. The completion of the upstream coffer dam and the diversion canal for drawing off the water of the Nile was completed in May 1964, when work on the main dam started. Here the river is about half a mile wide and hemmed in by granite walls. When the work is completed the dam will be nearly 350 feet high, 2·6 miles long and will create a lake 300 miles long, with the largest storage capacity in the world. The lake will drown many villages and the town of Wadi Halfa, and cause the displacement of 60,000 Egyptians[4] and 70,000 Sudanese; the latter are being rehoused 600 miles away in the new town of Khasm el Girba and 26 surrounding villages in eastern Sudan. The dam will also drown a number of ancient monuments such as those at Abu Simbel.

Agreement between Egypt and the Sudan was reached in 1959 as to the use of Nile waters in their respective countries, with an allocation of about 22% to the Sudan and 65% to Egypt, the rest being lost by evaporation from Aswan lake and discharge to the Mediterranean.

THE FAYUM

The Fayum is a third region 40–50 miles south of Cairo with an area of nearly 900 square miles and a population of three-quarters of a million. It is a very large oasis below sea level in the Western Desert, and almost surrounded by a prominent escarpment. It is, however, unlike other Saharan oases as it is connected by the Ibrahimiya Canal to the Nile from

---

[4] The Egyptians are being rehoused in about 75 government townships such as New Daboud north of Aswan. Here landowners are being given 5 feddans of irrigated land, and landless peasants 2 to 5 feddans each.

which it draws its water for irrigation. Nile mud is also brought down the distribution canals and high yielding crops are grown. Three *crops* are produced in two years, and they include cereals, cotton, olives, figs, grapes and many other types of fruit and vegetables. Sheep of good quality are pastured on the waste land. A brackish lake, Birket el Qarun, occupies the northern end of the depression and fishing is of considerable importance. There are many ruins of ancient towns and villages, showing that the Fayum has long been a favoured area for settlement. After a long decline the Fayum is coming into its own again, as in Roman times, and its agricultural products are of excellent quality.

THE WESTERN DESERT

The Western Desert covers nearly three-quarters of Egypt and is an area of great uniformity compared to the rugged nature of the Eastern Desert. It consists of a low plateau usually below 1,500 feet in altitude with sandstones and limestones of Nubian to Miocene age. To the north and centre are a number of oases and depressions, the Qattara depression being the largest with an area of nearly 7,000 square miles. The oases Kharga, Dakhla, Farafra, Bahariya and Siwa have a total population of about 50,000. The New Valley Scheme plans to make the five oases into a continuous populated area, which with modern irrigation methods could support a far greater number of people. Artesian, and not Nile water, will be used. The Wadi Natrun near the delta is the source of an important soda and salt industry. The coastal plain, the scene of many a swiftly moving battle in the Second World War, was once a region of prosperous farms and gardens, as evidenced by the remains of Roman and Greek occupation with abandoned wells, cisterns and irrigation channels, and ruined buildings. Cereals and grapes were then the main crops, but the present meagre winter rainfall of 5–6 inches is only sufficient for pastoralism and uncertain crops of barley and wheat. There are *settlements* at Salum, Sidi Barrani, El Daba, El Hammam, Burg el Arab and Mersa Matruh, the last the terminus of the coastal railway and a growing summer resort. The coastal plain has the bulk of the settled population which numbered over 100,000 in 1947, the desert having little more than 50,000 nomadic pastoralists.

THE SUEZ CANAL

Besides the Nile valley leading into Africa Egypt has an internationally important routeway, the Suez Canal, linking the Mediterranean to the Gulf of Suez and the Red Sea. The 107 miles of canal cuts across the narrow waste of desert connecting Africa to Asia. The Suez Canal, unlike the Panama Canal, has no locks and is open to the sea at both ends. It is a vital link in world communications, shortening the voyage from Europe to the Far East by 4,000 miles and saving the long voyage round the Cape. The first canal was dug in 800 B.C. from the Nile via the Bitter Lakes to the Red Sea. The Ptolemies, Romans and Arabs restored and used this waterway, but it fell into disrepair after the sixteenth century with the use of the Cape route by the nations of Western Europe. It was not until the 1860s that Ferdinand de Lesseps was able to promote the idea of another canal, this time cutting right through the narrow neck between the two seas. The Suez Canal was opened in 1869 with a depth of 26 feet. The canal has been successively widened and deepened, and by 1957 the maximum permissible draught for a ship using the canal was 34·5 feet, which was extended to 37 feet by 1961 and in future will be 40 feet. The canal is now 650 feet wide and may be widened in future to allow the simultaneous passage of ships both north- and south-bound. In 1875 the British acquired nearly half the Suez Canal shares and for over seventy years from 1882 to 1955 British forces were stationed in Egypt to safeguard the vital link to India. Exploitation of the great oilfields of the Persian Gulf, particularly since 1945, has further reinforced the significance of the Suez Canal, affording the shortest route from the Middle East to markets in Western Europe. Nearly three-quarters of all cargo is oil from the Middle East. Despite Egyptian nationalization and the Anglo-French landing of 1956 the Canal has prospered. In 1913 about 5,000 ships of 20 million tons passed through the canal, compared to nearly 18,000 ships with a tonnage of 154 million in 1958.[5] British ships are still the largest users with over 21% of the vessels going through the canal, compared to Liberia with 15·8% and Norway 12·9%. The value of the canal to Egypt is shown by the fact that in 1961 gross revenue from canal tolls was about £50 million compared with £32 million in 1955.

The two main ports at the northern and southern ends of the canal are *Port Said and Suez* which handle a large entrepôt trade. Port Said is an important fuelling station, and Suez has an oil refinery, engineering and repair facilities.

[5] In 1963 the tonnage of goods moving north–south was about 34 million, but moving south–north it was nearly 160 million.

THE SUDAN

POPULATION

The name Sudan is derived from the Arabic word meaning black, as it was here that the Arabs met the negroid races of the south. The word Sudan is also used geographically to describe the great savanna belt stretching east to west across Africa. The total area is nearly 1 million square miles and the population is estimated at about 13 million. The population may have been nearly as great before the Mahdist revolt which greatly reduced the numbers. There has been a rapid increase in the last sixty years. The population pattern is irregular and distribution unequal, reflecting the great contrasts between the north and the south. The north is Moslem and Arabic-speaking, and the people are of mixed origin with a far greater proportion of Arab and Hamite than in the negroid and Nilotic areas of the south. Here the people are pagan and speak a number of distinct languages and there is little unity among the different ethnic groups.

HISTORY

The history of Egypt and the Sudan has been interconnected on many occasions over a very long period. During the early years of the nineteenth century Mohammed Ali occupied the country. At the end of the century the Mahdist revolt was crushed and the Sudan was virtually ruled by Britain until 1956, despite the Condominium with Egypt. Since 1956 the Sudan has been independent. There is much talk of Nile unity and closer political association between the Sudan and Egypt. The economy of the cultivated strip along the Nile in the northern Sudan is somewhat similar to that of Upper Egypt, but the southern tropical base is completely different from Egypt. Good relations are essential between the two countries as Egypt is so dependent on the water of the Nile.

ECONOMY

The recent economic development of the Sudan has been based almost entirely on *cotton*, grown on the irrigated clay plains. Although a number of alternative crops have been tried in order to widen the Sudan's economic base no other crop has been found to be as profitable as cotton. As in Egypt only 3–4% of the land is cultivated, although the potential area is far greater.

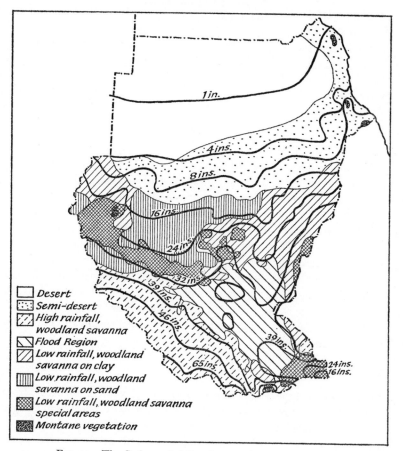

FIG. 32—The Sudan: rainfall and vegetation (after Barbour)

RELIEF AND CLIMATE

Vast clay plains and depressions extend on either side of the Nile, and are half surrounded by the uplands comprising the Nile watershed. The climatic belts stretch approximately east and west across the Sudan and grade off from hot desert in the north to a tropical rainy climate in the south (Fig. 32). The various regions reflect the conflicting trends of the climatic belts and the flow of the Nile from south to north.

THE SOUTH

The south is a large flat triangular area of basin and plateau. It has adequate rainfall and high temperatures, but soils are leached and the region is isolated from the rest of the Sudan by poor communications. The basin region of the Nile sudd is a swampy area watered by the meandering tributaries of the Nile, the Bahr el Jebel and Bahr el Ghazal. The basin is floored with clay, and this, combined with the high rainfall of the southern Sudan, causes widespread seasonal flooding which determines the use of the region (Fig. 33).

Population and way of life

The population consists of Nilotic tribes, the Dinka, Shilluk and Nuer, who are very tall, of slight build, with frizzy negroid hair and black to dark brown skin. They are predominantly pastoralists whose main interest lies in their humped, long-horned cattle which are very important for ritual and social reasons. They have a great love for their cattle and there are many songs about the colours, shape and movement of the oxen whose horns can be curved according to whim when the animals are young. Despite the apparent superfluity of water in the south it is difficult to find sufficient water and grazing in the dry season. There are problems of pests and insects and dung fires are burnt during the wet season to ward off insects. Grasses growing on the clays are only usable during the wet season and in the dry season there is no grass.

Migration

With the rise of the flood waters in April and May migration occurs from the river banks to the ridges of higher ground where the permanent villages are located. These consist of a group of homesteads, each homestead comprising a number of round thatched huts and a cattle kraal. With the lowering of the flood waters in October the move back to the river bank and lower ground is possible. The river bank settlements consist of temporary cattle posts containing the young men in charge of the cattle. Milk is the main product, although the yield is very low and not more than a few pints a day. Meat is only eaten when animals are killed for sacrificial purposes.

Agriculture and fishing

Except for the Dinka and Nuer, who are almost entirely pastoral, the other Nilotic peoples cultivate a variety of crops, while fishing is also important. The staple crop is millet of which there are many varieties. Maize, beans, groundnuts, sesame, pumpkins and marrows are grown. Various methods of fishing are used, spear, net, trap and hook and line, and

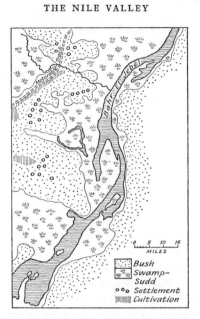

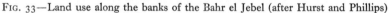

FIG. 33—Land use along the banks of the Bahr el Jebel (after Hurst and Phillips)

a variety of fish are caught, such as Nile perch and Tilapia. The cutting of the Jonglei Canal will affect directly over ½ million Nilotic people through loss of riverine pastures and fisheries, and will modify greatly traditional methods of land use.

THE PLATEAU

The plateau region to the west of Equatoria Province is physically similar to adjacent areas in Uganda and the Congo Republic. Rainfall is higher and up to 6o inches in places and duration is longer. The widespread incidence of tsetse fly largely precludes the keeping of cattle. There are a variety of tribes, the largest being the Dinka and the *Zande*. The administration has been able to persuade the people to settle in compact areas through the necessity to control sleeping sickness. In Zande territory 25–40 acre holdings subdivided into fields have been laid out in a grid-iron pattern. The Zande are good farmers and their system of agriculture is well adapted to the local environment. The main crops are maize, eleusine, cassava and groundnuts, and minor crops are pumpkins, sweet potatoes, okra, chillies, rice, tobacco and hemp. A variety of fruit is also grown—mangoes, pineapples, pawpaws, etc. Cotton is the only important crop, but yields are below those of the country as a whole, and production is small. The Zande resettlement scheme carried out from 1946 to 1951 settled nearly 200,000 people in a thousand villages. The new settlements consist of individual holdings fronting on roads and extending for a thousand yards into the bush.

IRRIGATION SCHEMES

The economic core of the Sudan is the small irrigated area south of Khartoum between the Blue and White Niles, consisting of the Gezira and the Managil Extension, which produce the bulk of the Sudan's main cash crop, cotton; exports of cotton and cotton-seed in 1961 were worth nearly £35 million.

The *Gezira*, which means 'island' in Arabic, was selected as the most suitable area for a large-scale agricultural and irrigation project (Fig. 34 and Plate V). Work was begun on the scheme in 1914, but the Sennar dam was only completed in 1925. However, by 1926 over $\frac{1}{4}$ million acres were under irrigation. With the recent completion of the Managil Extension to the south-west the total cultivable area of the whole scheme is nearly 2 million acres, cotton being grown on about $\frac{1}{2}$ million acres. A quarter of the area in the main scheme is sown with cotton, a quarter with food and fodder crops and the rest under fallow, while in the Managil area the relevant proportions are each one-third.

The long dry season allows effective control of pests and diseases, while the soil has a high mineral content and impermeability, so reducing loss by seepage. The site, consisting of level clay plains with a gradient of 1:5,000 to 1:10,000, is very suitable for irrigation, and it was comparatively easy to lay out the minor distribution canals. The Blue Nile floods occur in August and September, and irrigation is carried on from July to the following April. Villages are situated at distances of one to two miles apart near the main canals. The problem of bilharziasis has been partially overcome by treating the infected water with copper sulphate. Initially each tenant was given a holding of about 40 acres which was too large, but now holdings are half this size. They are divided into four separate fields which produce cotton, dura (millet), lubia (a useful food and fodder crop) and other crops. Cotton is only grown one year in four compared to one year in two or three in Egypt. It is sown in August and picked from January to April. A large labour force is needed, particularly during the picking season, and there is a considerable movement of migrant workers from other parts of the Sudan to the Gezira. The cotton produced is of good quality and of a very long staple, although the yield is much below that of Egypt.

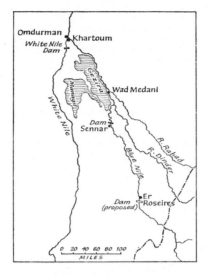

Fig. 34—The Gezira

The scheme was originally started as a partnership between private enterprise, the Sudan government and the tenants. It was taken over by the Sudan government in 1950 and is now operated by the Gezira Board. Tenants are carefully supervised by the Board and advised what crop rotation to follow, the types of cotton and other crops to plant and the proper use of fertilizers and sprays. Recent legislation has revised the distribution of profits, so that government and tenants receive 44% and 42% respectively and the Gezira Board only 10% compared to a previous 20%. The Gezira Scheme has been a great success both socially and economically and should have considerable significance for the regional planning of natural resources elsewhere in Africa. Production of cotton has increased sevenfold since 1925-6, and the number of tenancies from 8,000 to 90,000. However, there are problems of the lack of participation by tenants in managing the scheme, and of an increasing dependence on hired labour, half the work being done by immigrants from West Africa.

Other areas in the Sudan where irrigation schemes are planned are at Kenana to the south of the Gezira where a further $1\frac{1}{4}$ million acres could be irrigated. This will be dependent on the building of another dam at Roseires on the Blue Nile. Another scheme involving the irrigation of $\frac{1}{2}$ million acres may be developed on the west bank of the Atbara river, and a dam built at Kashm el Girba. There are also over 2,000 pump schemes in the Sudan, nearly all privately owned, and producing almost one-third of the Sudan's cotton. Millet, lubia and vegetables for the Khartoum market are grown as well.

KHARTOUM, KHARTOUM NORTH AND OMDURMAN

North of the Gezira and at the junction of the Blue and White Niles is the triple settlement of Khartoum,[6] Khartoum North and Omdurman, separately administered, yet forming one built-up area with a total population of about ¼ million. Khartoum, in the fork between the branches of the Nile, is the capital of the Sudan and the centre of the main commercial organizations and banks. It is spread out more horizontally than vertically and covers an area of eight square miles, the hot climate making a widespread distribution of buildings desirable. Growth has been annular, away from the waterfront where the main administrative area is situated. To the south of this is the commercial zone fronting on the main square which contains the central mosque, to north and south of which are separate European and Arab markets. Between the mosque and the railway station is a middle zone of ethnic minorities and foreign elements such as Italians, Levantines and Ethiopian Copts.

Khartoum North across the Blue Nile and connected to Khartoum by a bridge is more industrial, with a dockyard and light industries along the railway line. The combined population of Khartoum and Khartoum North is about 125,000. Omdurman with a population of nearly 120,000 has for long been an important caravan centre. It is situated across the river and to the north of Khartoum. The town has a regular layout with the characteristic mud houses of a dry region. Local craftsmen produce pots, glassware, domestic utensils, furniture and other articles. There is also an important animal market. Khartoum is a major communications centre by river and rail with railway links southwards to Sennar from where branches go to El Obeid, Roseires and Kassala, and northwards to Atbara and Port Sudan.

The area away from the irrigated cotton belt is one of nomadic pastoralism with a settled population only occurring where there is a permanent water supply. The area extending east from the Blue Nile to Gedaref has a more reliable rainfall and a richer soil, and it has become the main grain-growing area of the Sudan, producing mostly millet.

---

[6] Khartoum means 'elephant trunk'.

NORTH-WESTERN SUDAN

North of the sudd and west of the White Nile is the north-western Sudan, a very large region occupying about one-third of the area of the Sudan flanked by the plateau of western Darfur and Jebel Marra to the west and the Nuba Mountains to the south-east. The volcanic range of the Jebel Marra, rising to over 10,000 feet, has a number of terraces. The Nuba Mountains consist of isolated peaks of Basement Complex rocks with a height of 5,000 feet in places. The southern half of this great region is known as the Qoz and consists of fossil sand dunes now stabilized by rainfall and vegetation and forming gently undulating ground covered with grasses and trees increasing in density to the south with higher rainfall. The northern area, inhabited by nomadic pastoralists, has a lower rainfall and is a semi-desert area with sparse vegetation. The total *population* is 3 million, but density in the northern semi-desert zone is less than 3 per square mile.

THE CENTRAL ZONE

The central zone is occupied by *cultivators*, while the population to north and south is nomadic. The somewhat anomalous location of the cultivating zone is connected with the distribution of the main belt of Acacia senegal, the source of *gum arabic*, at the border of savanna and steppe, with the availability of permanent water, the railway line from Kosti to El Obeid and the road from El Obeid to El Fasher. The location of villages is dictated by the availability of well water. The main crops are various kinds of millet, maize and groundnuts. Yields are very variable and crops often fail completely. The main economic crop is gum arabic, exports being worth nearly £7 million in 1960. The distinctive feature of the Sudan as a supplier of gum arabic lies in the fact that large areas have only one instead of several varieties of trees producing gum of a high quality. The small lumps of gum exuded from the Acacia verek trees are collected in the dry period from about November to April and are left exposed for some weeks before being sold. The Sudan supplies over three-quarters of the world's output of gum arabic, used in printing and the manufacture of confectionery and textiles.

North and south of the cultivated zone are areas of *nomadic pastoralism*. In the north rainfall is below 12 inches and the Arab pastoralists are mainly camel owners whose movements are related to the rainbelts and watering places. At the beginning of the rainy season animals are taken south, while in the dry season water supplies are sought in the clay depressions, wadi beds and granitic outcrops. The lack of well dispersed watering points restricts the range of grazing, so that in many areas close to existing water-holes the vegetation has been eaten out. The problem of water supply is serious as in other semi-arid parts of Africa. Several hafirs or rectangular basins, some 15 feet deep, have been dug in the Central Sudan since 1950 to store surplus water. In the savanna area of the south cattle, sheep and goats are more important than camels. There has been a considerable increase in the sale of cattle in recent years. The Sudan is estimated to have nearly 7 million cattle, 7 million sheep, nearly 6 million goats and 2 million camels.

THE NORTH

Northwards beyond the zone of semi-desert is the wide expanse of the Sahara where there is practically no population and the only activity is pastoralism on a very extensive basis. To the west of the Nile the *desert* merges into the Western desert, and there are nomadic Arab tribes. To the east of the Nile are nomadic Hamites, the Beja who, because of their great mops of hair, are also known as the Fuzzy-wuzzies. The strip of desert is bordered by the Red Sea hills which receive about 5 inches of rain in the south. Erkowit was developed as a hill station during the war, but is now neglected. The Sudan's main port is Port Sudan on the *Red Sea*, which was developed in 1906 to replace Suakin. Port Sudan is a hot dusty place, enervating in summer, but with a good water supply from the Red Sea hills. It has a population of 50,000 and is connected by rail with Khartoum. The port has good berthing and bunkering facilities, and will be able to take 14 ships alongside when harbour extensions are completed. In 1961 the port handled nearly 2 million tons of cargo with exports of cotton, cotton-seed, gum arabic, hides and skins, cattle and sheep. There are no oases in the Nubian desert, but the few oases in the Libyan desert, such as Salima and Bir en Natrun, provided a route in the past for the export of slaves, gold, ivory and other commodities to Egypt.

The Nile describes a great S-bend through the desolate northern part of the Sudan providing a thin cultivable strip in a series of terraces and basins where the valley has widened, particularly in the Merowe–Dongola section. A total area of only 500 square miles supports about 600,000 people. The stretch from Shendi, nearly 100 miles north of Khartoum, to Berber is the most densely populated. Agriculture has been transformed with the introduction of diesel pumps which have replaced traditional methods. As in Egypt there are three cultivating seasons of summer, winter and flood, most crops being grown in the latter two seasons. Wheat, barley, millet, maize, birsim, lubia and pulses are the main crops. Promising crops for the

[*Office National Marocain du Tourisme*

I. A small village in the Rheraia valley with the High Atlas in the background.

II. A view of Fez, one of the old cities of Morocco.

[*Office National Marocain du Tourisme*

[Egyptian State Tourist and Information Burea]

III. The Nile at Cairo.

IV. The High Dam under construction near Aswan.

[Egyptian State Tourist and Information Burea]

future are vegetables such as onions, and fruit such as citrus. Where the amount of cultivable land is slight nomadism may be important as at modern Merowe, and herds of camels and flocks of goats are moved from the river bank to the desert and back. Some of the towns along the upper Nile have developed as market and route centres, for example Shendi, Atbara and Berber in the south near Khartoum. *Atbara* with a population of 36,000 is an important route centre, situated at the junction of the Atbara river and the Nile and with railway links to Wadi Halfa and Port Sudan.

COMMUNICATIONS

The Nile is an important artery of communications in the Sudan. South of Khartoum there are long stretches of navigable water from 850 to 1,100 miles long to Wau, Juba and Gambeila on the Nile and its main tributaries. North of Khartoum Nile navigation is interrupted by rapids, except for short navigable stretches such as the 210-mile reach from Shellal to Wadi Halfa and the 208-mile reach from Karima to Kerma. Navigation in certain areas has declined, for example from Dongola to Khartoum, because of the competition of motor transport. Railways have increased in mileage with the construction of a new line from the Sennar–Nyala line to Wau. The total railway mileage is now over 4,000, compared to about 2,000 miles of river navigation, and the most important links are from Khartoum to Port Sudan, Wadi Halfa and Sennar. Owing to the great distances, the aridity of the north and lack of economic development over much of the country roads are of little importance and are often no more than rough tracks. There are at present little more than 30,000 licensed vehicles in the Sudan.

FUTURE PROBLEMS

The Nile is not only a symbol of unity between Egypt and the Sudan, but is of vital concern to Egypt which is therefore anxious to secure equitable distribution of the Nile waters. The basic problem of Nile control is to even out fluctuations between high and low water. To achieve a more regular flow of water throughout the year more dams and conservation works are necessary, although the completion of the High Dam at Aswan should be of major benefit to Egypt. Envious eyes are cast from Egypt's overcrowded delta to the comparatively empty areas of the Sudan. The United Arab Republic's rulers are, however, pursuing the more popular appeal of Arab unity and nationalism, so that a union with a country like Syria or Jordan is a more attractive proposition than a political link with the

Sudan. However, an economic link with the Sudan might be attempted. The Sudan also faces its own problems of exploiting the under-developed yet potentially rich areas of the south and linking it more effectively with the more powerful and dominant area of the Moslem north. The South has resisted strongly attempts by the North to enforce their rule, and sporadic fighting has occurred.

CHAPTER EIGHT

West Africa

WEST AFRICA extends from about 5° to 15° N. with an area of about $1\frac{1}{4}$ million square miles and a population of about 70 million in 1961[1] (Fig. 35). There is a sharp distinction between geographical as opposed to political West Africa. Using criteria of relief, climate and vegetation *geographical* West Africa is a fairly well defined region stretching from near the mouth of the Senegal river, across the great bend of the Niger, for over 2,000 miles east to Lake Chad. The northern boundary selected here is based mainly on rainfall, using the 10-inch isohyet, and is thus a fluctuating and transitional zone as rainfall is very variable in the Sahel near the southern edge of the Sahara. The western and southern limits are the Atlantic. The eastern boundary extends to the Cameroon and Bamenda–Adamawa highlands, which have for long been an important physical and human divide.

If *political* instead of geographical factors form the main criterion for the delimitation of West Africa the Saharan areas of the ex-French territories must be included. The Niger republic, Sudan and Mauritania have a combined area of about $1\frac{1}{2}$ million square miles and political West Africa would thus have an area of nearly 3 million square miles. Excluding the desert areas of these countries one is left with an area of about $1\frac{1}{4}$ million square miles for geographical West Africa.

For the most part the climatic and vegetation zones are arranged in east–west trending belts which are cut across by the political boundaries along the Guinea coast from Sierra Leone to Nigeria. This has had a profound influence on the human geography of West Africa and a country like Nigeria or Dahomey includes a number of sharply contrasting regions, varying from luxuriant rain-forest to poor savanna, within its boundaries. The economic, social and political strands of life in West Africa therefore represent a complex interaction with these basic factors of east–west trending physical belts and north–south political boundaries. Some of the elements of structure and drainage are, however, aligned more north–south

---

[1] If the 1963 Nigerian census is accepted, the population of West Africa would be over 90 million.

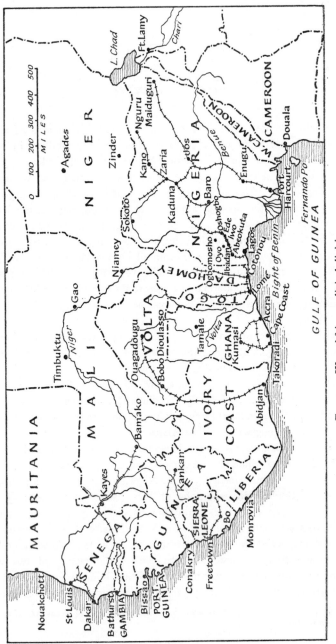

Fig. 35—West Africa: political divisions

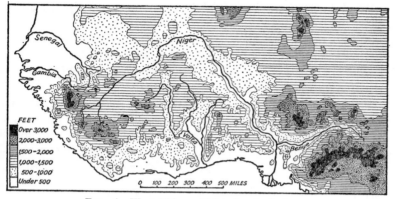

FIG. 36—West Africa: relief and main drainage

than east–west; for example, the Akwapim–Togo hills, Bamenda Highlands and most of the rivers which drain south to the Gulf of Guinea.

STRUCTURE, RELIEF AND DRAINAGE

Structurally West Africa consists of a low dissected plateau composed of ancient crystalline rocks which have been worn down by long exposure to erosion. Compared with the high plateau of East Africa, which is largely above 4,000 feet in altitude with considerable areas above 8,000 feet, West Africa is rarely more than 2,000 feet in altitude. Pre-Cambrian rocks, with folds often trending north-east to south-west and affecting subsequent relief, are exposed over two-thirds of the area from about 6° to 12° N., forming part of Africa's ancient fundamental complex. North of this are large areas covered with Quaternary sand, particularly to the north of Nigeria and the Voltaic Republic, which add greater monotony to the widespread peneplaned surfaces (Fig. 36). Here is found one of the fundamental features of West Africa, the great central depression through which flow the Senegal river to the west and the Niger at first to the north-east, then south-east and south.

South of the central depression the plateau reaches its highest elevation in the Guinea Highlands, Jos plateau and Adamawa Highlands with altitudes of over 6,000 feet. Much of the higher plateau levels have been caused by intrusive volcanic rocks, as in East Africa. The effect of various periods of uplift, faulting and vulcanism has been to raise the south-western part of the plateau to create a major watershed from which flow the Niger, Senegal and other rivers. The *Niger* is nearly 3,000 miles long and rising in the Guinea Highlands follows this peculiar semi-circular route to the sea, providing one of Africa's greatest geographical puzzles which was only solved during the nineteenth century. The key to this puzzle lies in the evidence supplied by the Quaternary pluvial cycles. A wetter phase caused the river to break through the Inland Delta near Timbuktu to join a great river, now fossilized, which flowed from the Air Massif to the existing south-east flowing Niger.

The Niger and its great tributary the Benue afford useful navigable waterways for considerable distances. The Niger is navigable for much of its length during the wet season, with navigable reaches from Kouroussa to Bamako, Koulikoro to Ansongo, Niamey to Gaya and from Jebba to the sea. The Benue is navigable right up to Garoua in August and September.

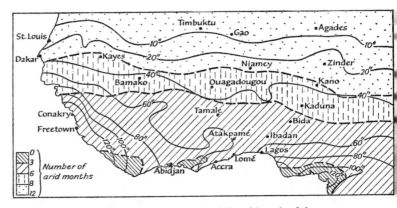

FIG. 37—West Africa: annual rainfall and length of dry season

South of the ancient rocks of the plateau are younger rocks comprising the coastal plain in Sierra Leone, Ivory Coast, Dahomey and Nigeria. In Nigeria the Lower Niger and Benue valleys are composed of Cretaceous and Eocene marine beds. Much of the coastal zone consists of lagoons fringed by sand bars, long curving sand-spits, mud-flats, mangrove forests, or distributaries and channels whose mouths are obstructed by sand bars. Many of West Africa's rivers, particularly those that rise on the plateau and have short courses obstructed by rapids, are thus of limited navigable value. The sea is shallow for a considerable distance offshore causing heavy surf that is dangerous for shipping. The inhospitable coastline and lack of harbours has been a serious economic drawback and the occasional natural harbours such as Freetown and Lagos are of great importance.

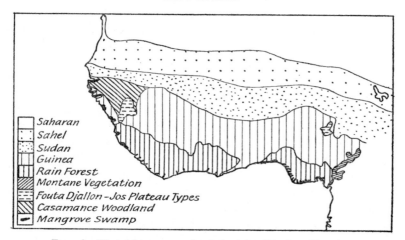

Saharan
Sahel
Sudan
Guinea
Rain Forest
Montane Vegetation
Fouta Djallon - Jos Plateau Types
Casamance Woodland
Mangrove Swamp

FIG. 38—West Africa: vegetation belts (after Harrison Church)

CLIMATE AND VEGETATION

The climatic and vegetational zones, trending east–west, differ somewhat from the structural features aligned more in a north–south direction. Climates vary from a monsoonal and semi-equatorial type along the coast, with well distributed rainfall for most of the year (although there is a marked dry period for part of the coastal zone),[2] to increasingly seasonal tropical savanna types of climate in the interior (Fig. 37). The temperature range seasonally and diurnally tends to increase as well away from the coast. Vegetational belts reflect climatic influence to a considerable extent, but also man's influence (Fig. 38). Mangroves and rain-forest are found in the equatorial type of climate along the coast. The northern edge of the rain-forest has been retreating rapidly before man's advance, and a widening belt of derived savanna bears testimony to the wasting effect of axe and fire. In the Guinea savanna zone trees are of a deciduous variety to withstand the effect of a dry season,[3] and gnarled and thick-barked as a result of their annual firing. With the increasing seasonality and decreasing amount of rainfall in the Sudan savanna, trees are more scattered and the grass shorter, creating a brown park-like effect in the dry season. On the Saharan fringe the vegetation degrades to scrub and thorntrees.

[2] From Cape Three Points to the Dahomey border the rainfall is less than 35 inches and there are little more than 100 rain days. This contrasts with the high rainfall of the Niger delta and Bight of Benin.
[3] There are, however, many trees that have leaf in the dry season and shed their leaves in the rain, the common *Acacia albida* for example.

West African soils generally lack plant nutrients and are acidic because of the rapid decomposition of vegetation and the high degree of leaching to be found, particularly in the hot and humid climates of the coastal areas. The forest areas are deceptively fertile, luxuriant tree growth occurring in areas of poor soil, the forest helping to provide its own humus. When the forest cover is removed soil deterioration occurs rapidly.

Man's increased cultivation of the coastal zone may have caused retreat of the forest in a region which produces most of West Africa's valuable tree crops for export, such as cocoa, oil palm products, coffee, rubber and also timber.[4] Laterite occurs freely in West Africa, particularly in areas of marked seasonal rainfall, the alternation of wet and dry seasons probably causing the formation of the hard ferruginous crust, which when exposed is more like a rock than a soil. The Sudan belt has large areas of sandy soils, with more fertile lacustrine soils in the Delta region of the Middle Niger. Grey and black clayey soils are found in waterlogged areas such as the shores of Lake Chad.

HISTORY

Historical and cultural influences in West Africa have tended to come either from the Negro areas to the east and south-east, from the Mediterranean world southwards across the Sahara, or down the Nile Valley and westwards along the Sudan belt. Pastoralists have invaded the savanna belt over a long period, from the east came a knowledge of iron working and perhaps kingship, while later from the north came the religious influence of Islam. An extensive trans-Saharan caravan trade developed with gold, ivory, slaves, leatherwork and other commodities being exported to the North African coast. Powerful and extensive Sudanic empires such as Gao, Songhai and Mali grew up on the grasslands of the Western Sudan, their power based on control of a wide tributary area by means of cavalry, a varied trade, and a complex administrative system. A number of important trading centres developed in two main zones, the first with towns such as Agades and Timbuktu at about 17° N., the second with towns such as Kano and Katsina which were established further south at about 12° N. The latter is an important line of contact marking the northern limit of the tsetse fly, the southern limit of the camel, and a zone in which horses and donkeys can be used for transport. Here too the grasslands aided east–west movement and it was a zone where Islam encroached on pagan religions and pastoralists on cultivators.

[4] Although the examination of air photographs taken in 1953 and 1963 in Southern Nigeria suggests that there has been little change in the forest–savanna boundary. (Information from Dr W. B. Morgan.)

In contrast to the land-based contacts of the Sudan zone is the later sea-borne trade of the coastal zone. West Africa was the first tropical region to have direct contact with Western Europe in the fifteenth century and, being only about 2,000 miles by sea from southern Portugal, is also the nearest to Europe. The West African coastal trade was based initially on gold, tropical products and slaves, but soon became concentrated almost entirely on slaves, for whom there was a constant demand in the rapidly growing plantation areas of tropical America and the Caribbean islands. Partly as a result of the political pressures and economic possibilities afforded by the slave trade there arose strong forest states such as Ashanti, Dahomey, Oyo and the Yoruba states which grew in importance. Thus an opposing cultural and economic current set in, causing a decline of the northward-flowing stream of Saharan trade, and a growth of the maritime trade from the Guinea Coast to Europe and the Americas. This economic shift was consolidated by European political and colonial penetration in the late nineteenth century and early twentieth century.

In the scramble for African territory during the latter half of the nineteenth century the French and British gained the lion's share. The French operated mainly along the Sudan savanna belt and eventually linked up with their Saharan interests in Algeria, and the British pushed into the interior from their coastal footholds. The ex-French territories are now hampered in general by problems of aridity and continentality with long lines of communications and high proportion of land to coast, whereas the ex-British territories have a more favourable environment, greater population and wealth, shorter lines of communications and a higher proportion of coast to land.

In contrast to East and South Africa white settlement has never been an important feature in West Africa for a variety of reasons such as the effect of tropical diseases, the inhospitability of the coast with its barrier of a fringe of mangroves in places and dense rain-forest, the hostility of coastal tribes, the high density of population in some coastal areas, and the reluctance of the British government to allow white settlement in their West African colonies. Economic and political development has been based largely on *indigenous African initiative* aided by the former colonial powers. This is in marked contrast to East and South Africa where white and Asian settlement and enterprise have played a major part in the development of these regions. The great bulk of the export crops are grown by Africans on small plots rather than by white settlers on large farms and plantations.

Economic growth has been most rapid in the coastal zone with the export of tree crops such as cocoa, oil-palm products and rubber, and forest products such as sawn logs and planks, veneers and plywood, and minerals such as gold, diamonds, manganese, bauxite and iron ore (Fig. 39). Along the Guinea coast cassava, yams and maize are the main food crops, while further west where rainfall is more monsoonal rice becomes important. In the interior with a drier and more seasonal climate guinea corn and groundnuts are important as food crops while cotton and groundnuts are major cash crops. Nomadic pastoralism is significant north of 14° N. and, although management of livestock is of a low standard, there are large exports of cattle to the protein-deficient areas of the south.

POPULATION

There has been a steady stream of population moving into the coastal zone with the expansion of cash crops, mining and commerce and the great increase in urbanization. Many of the coastal and forest towns have quadrupled their population over the last twenty to thirty years. With the advent of independence and the choice of ports such as Dakar, Abidjan, Freetown, Accra and Lagos as capitals, these coastal towns have grown even more rapidly.

Over the last twenty years roads have become more important than railways, especially in the rapidly developing coastal areas of Ghana and Nigeria. Whereas railways were built by colonial administrations for strategic and economic reasons and are not government controlled, road transport is predominantly a private concern. Lorry, bus and taxi services have been developed with great vigour in the rural areas despite the disadvantages of a tropical climate and lack of capital. Railways, although slow and often badly aligned, are still essential for long-distance haulage of minerals and crops such as groundnuts from Northern Nigeria.

The *population map* of West Africa shows two main zones of dense population, particularly in Nigeria, in the Sudan belt of the north and the coastal and forest area of the south. These two areas are separated by a far more thinly populated Middle Belt which owes its lack of population partly to the fact that it has long been a shatter zone between the states of the northern grasslands and the southern forests.

West Africa with a population of 70 million has over $\frac{1}{4}$ of Africa's people on only about $\frac{1}{10}$ of the area, which contrasts with other less favoured areas of Africa. Within the broad framework of the region there are local concentrations. Nigeria with less than $\frac{1}{3}$ the area of West Africa has approximately $\frac{2}{3}$ of the population; and in Nigeria a further regional breakdown reveals that an area such as the Eastern Region has densities of over 1,000 per square mile in Owerri and Onitsha Provinces. In Ghana

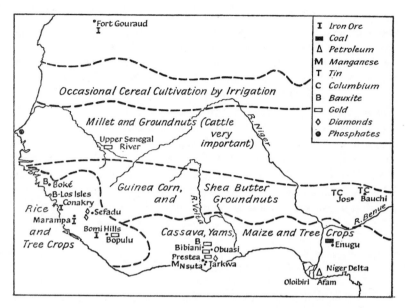

FIG. 39—West Africa: minerals and crop production zones (after Harrison Church)

too the bulk of the population is concentrated in the southern $\frac{1}{3}$ of the country. Thus from the human geographical point of view the main physical zones have been emphasized by cultural and historical trends, with a northern savanna Islamic region practising extensive pastoralism and growing seasonal crops, separated by the shatter zone of the Middle Belt containing many small broken tribes with a variety of faiths and customs from a coastal forest semi-Christian region of Negro cultivators producing perennial crops.[5]

By comparison in East Africa the line of contact between Hamitic and Negro Africa has given rise to local states and pastoral communities with little outside contact before the nineteenth century. The coastal region, not unlike the Sudan belt of West Africa, has long had widespread commercial contacts which reached their zenith before the Portuguese arrived, subsequently declined, but have revived again in the twentieth century. In East Africa it is the productive tropical highlands of Kenya and northern Tanzania and the Lake Victoria basin, i.e. the interior, and not the coastal belt, that has asserted its economic predominance.

[5] Although some forest tribes, such as the Yoruba, are strongly Islamic.

Using physical criteria such as climate and vegetation, West Africa has been divided into three major east–west trending regions. These major regions have been further subdivided for convenience into minor regions based on political units. The first major region consists of the northern Sudan zone with Senegal, Gambia, Mali, Upper Volta, Niger, northern Ghana, northern Dahomey and northern Nigeria as minor regions. The second major region to the south consists of the Middle Belt with Guinea, Portuguese Guinea, northern Sierra Leone, and parts of Ghana, Dahomey, Ivory Coast and the Middle Belt of Nigeria. The third region consists of the coastal and forest zone with southern Sierra Leone, Liberia, southern Ivory Coast, southern Ghana, Togo, southern Dahomey and southern Nigeria. Within this three-tiered classification of major regions and minor political units, there are a large number of smaller regions.

REGION 1—THE SUDAN BELT

This zone is situated between the southern edge of the Sahara and the Middle Belt to the south. Much of it consists of the central depression through which the Niger and Senegal rivers flow; to the south is the dissected plateau reaching altitudes of 1,500 to 2,000 feet. The basement complex consists of ancient granites, gneisses and metamorphic rocks of pre-Cambrian age worn down to form extensive level surfaces. In the central depression sedimentary deposits have been laid down over a very long period from Permian to lower Cretaceous times culminating in the great inland basins of the Middle Niger and Chad overlain by deposits of Quaternary age. The effect of long exposure to erosion and Quaternary deposition has been to produce a landscape enlivened by occasional monadnocks and sandstone plateaux, as for example the Ordovician and Silurian sandstones of Mali forming the Manding Mountains, and the long scarp from Bamfora in the Upper Volta to the Bandiagara scarp south of the Niger bend.

The *tropical climates* of the Sudan savanna are characterized by a greater seasonality and temperature range and less rainfall than those in the Middle Belt. Rainfall varies from +40 inches on the southern margin to 10–15 inches on the northern edge, the rainy season being from 3 to 6 or 7 months' duration. Rainfall is very variable, especially in the Sahel zone. There is intense evaporation and a high diurnal and annual temperature range, the latter being nearly 24° F. (13° C.) at Timbuktu compared to less than 6° F.

(3° C.) for most stations in the coastal zone. Maximum temperatures can be very high when the sun is overhead; for example, Kayes has average temperatures of over 90° F. (32° C.) for the three months of April to June. Highest temperatures tend to occur before the rains come, and decline considerably during the wet season, Kayes being 10° F. (6° C.) cooler from July to September.

Vegetation is affected greatly by the amount and distribution of rainfall and the relative humidity of the air during the dry season, which may drop to less than 10%. An area with low humidity for a long period cannot sustain forest and therefore the prevailing vegetation is Sudan savanna in the better watered areas and a kind of thorny scrub in the semi-arid areas.[6] The Sudan savanna has a park-like appearance most marked in the wet season when trees and grass are greenest. Acacias and economically important trees such as the dum palm, baobab, locust bean, shea-butter and kapok (*Bombax costatum*) are scattered about on the level plains carpeted with grasses up to five feet high, which are more useful for grazing than in the Guinea zone as they are less coarse and tussocky.

Further north in the Sahel the dry season is up to nine months long and the number of drought-resisting acacias with thin leaves and thorns increases greatly. Some acacias such as the *A. senegal* and *A. lacta* are economically useful and yield gum. There are also smaller thorny shrubs. Grasses form a discontinuous cover and are only 2 to 3 feet high. *Soils* are brown in colour, of medium fertility but shallow and poor in organic content. In the southern cultivable areas friable sandy soils are found, and alluvial soils suitable for rice production are widespread.

In the northern part of the Sudan–Sahel zone *livestock rearing* is most important. Transhumance rather than nomadism is characteristic of much of the area, animals moving south in the long dry season and towards the Niger Valley and Lake Chad; the degree of movement is controlled by availability of grazing and water. Improvements in the standard of cattle keeping are difficult to achieve because of the low carrying capacity of the grass, high mortality, and lack of watering points and supplementary feed.

With the increase of rainfall above 15–20 inches per annum *crops* such as pearl millet and sorghum can be grown and further south cultivation becomes important with bush fallowing the rule. Here millets and sorghum are the basic crops, but maize is important and near the edge of the Guinea zone root crops such as cassava begin to appear. Rice is grown extensively in the river valleys. The main export crops are cotton and groundnuts. In

---

[6] There is acacia in abundance in Mauritania and Bornu.

the western part of the Sudan zone continuous cultivation often replaces bush fallowing and a high standard of farming is achieved with the aid of manures, and household waste to fertilize the fields.

SENEGAL

Senegal was the starting point for French penetration into West Africa, the Senegal river providing a historic route into the interior, now replaced by the railway from Dakar to Kayes and Bamako. St Louis was the main centre for French activity in West Africa for two hundred and fifty years, and was only comparatively recently supplanted by Dakar. With an area of 77,000 square miles and a *population* of about 3⅓ million, Senegal has a number of different tribes, the most important being the Wolofs who number about ¾ million, the Peuls and the Toucouleurs.

Most of the country is composed of *Post-Eocene Continental beds* which form light sandy soils used extensively for groundnuts. The northern part of Senegal comes into the Sahel zone of thorn trees and meagre grass, while the southern part bordering on Portuguese Guinea has a higher rainfall and more continuous grass cover. Dakar has an average annual *rainfall* of about 25 inches, over 85% of which comes in the three-month period from July to September. Tambacounda, 300 miles east of Dakar, has 40 inches of rain a year and higher temperatures.

The small territory of *Gambia* is embedded in Senegal which is thus denied the use of a valuable waterway. The Senegal river suffers from sand bars and low water for most of the year, but is navigable for a distance of about 600 miles to Kayes from July to October. The Saloum river is navigable to Kaolack and the Casamance to Ziguinchor.

Groundnuts or peanuts, are by far the most important economic crop in Senegal (Plate VI), production varying from about ½ million to 800,000 tons, the bulk of which is exported, forming about 80% of total exports, mostly in the form of groundnut oil. The Wolof and Serer farmers grow large quantities of groundnuts as cash crops. Planting occurs from May to July, harvesting in October and November and marketing from December to March. The average yield is about ½ ton an acre, the highest yields occurring in the Casamance, which is somewhat isolated from the rest of Senegal by the Gambia river. There is now a trans-Gambia highway which links up the two parts of Senegal more closely. Centres of production have gradually moved away from the coast as the crop exhausts the soil and the Kaolack district now produces half the Senegalese crop. The government is concerned about the problems of soil exhaustion and undue reliance on one crop and is trying to stimulate interest in other crops and encourage the use of fertilizers and higher-yielding groundnuts. Seasonal labour is required for cultivation and harvesting of groundnuts and large numbers of casual workers known as *navétane*, are engaged each year. There is an important local refining industry, the main centre being Dakar. Rice and palm kernels are produced in the wetter area south of the Gambia. The Serer combine the growing of food crops and the keeping of cattle, the dung being used to maintain soil fertility. Food crops consist of millet, beans, groundnuts, rice, maize, cassava; millet (suited to the dry climate and grown in rotation with groundnuts) is the main food crop.

The valley of the river Senegal fulfils a dual role in the local economy and society; sedentary farmers such as the Toucouleur use the moist soil for the cultivation of millet and maize after the floods have gone down and nomadic pastoralists such as the Fulani use the river water for their flocks and herds. Irrigation is still at an elementary stage of development, the main scheme being on the delta near Richard-Toll with rice the main crop. About 20,000 tons of rice are produced a year by mechanized methods and at high cost. There are also plans for the regional development of the Senegal river, such as the valley and delta sections, although the delta would need expensive drainage and desalinization programmes before cultivation. Senegal has very large deposits of phosphates which are now being worked by opencast methods in two areas near Dakar, from where the phosphates are exported.

Dakar is the main port and capital of Senegal. Until 1959 it was the administrative capital of French West Africa. Of its population of nearly 400,000 about 30,000 are Europeans, forming by far the largest number of Europeans in any West African city. The city is of great locational and strategic importance and is situated on the southern end of Cape Verde, Africa's most westerly point. It is the first large settlement south of the Sahara, commanding entry to the Sudan zone and a wide hinterland. There is a good, well protected and largely artificial harbour which has attracted a great amount of maritime traffic, and nearly 4,000 ships now use the harbour. The large volume of traffic, much of it bunker fuel, gives a misleading impression as the hinterland of Senegal is not very productive. There is also an international airport at Yoff, ten miles to the north-west, strategically placed for trans-Atlantic (only 2,000 miles to Natal in Brazil) and trans-African flights. In addition Dakar is a naval base, and an industrial centre of some importance.

The city is not unlike a French provincial town. The European quarter with a radial and quadrangular street pattern and palm-lined avenues is south of the harbour, the African quarter with a more rectangular street pattern lies to the north-west of the European area. Beyond this is a European residential suburb overlooking the sea. An important factor is the cool Canaries current which modifies temperatures and humidity despite Dakar's location only 14° 40′ N. The average minimum temperature for the year at Dakar is 70·6° F. (21·4° C.) compared with 74·4° F. (23·5° C.) at Conakry which is 9½° N. Dakar thus tends to have a more pleasant climate than most West African cities with the tempering effect of sea breezes.

GAMBIA

Gambia really forms part of Senegal but through an accident of history has developed British rather than French and Senegalese links. It is a long sliver of territory, some 250 miles long and on an average 12 miles wide. The boundaries follow the approximate line of the *river* and are most unsatisfactory as they cut off the river from its hinterland and divide a number of tribes such as the Wolofs and Mandingos. The Gambia river like the Senegal afforded a natural routeway to the interior of West Africa, much used by nineteenth-century explorers. It is now navigable for small boats right up to the eastern boundary and for small ocean-going ships to Kuntau-ur, some 150 miles upstream.

The total area is about 4,000 square miles and the *population* 300,000. Most of Gambia consists of the former Protectorate, but the nucleus is the former Colony and particularly the town of Bathurst. The original settlement was founded in 1816 on St Mary's Island, a sandbank near the mouth of the river, although Fort James was established two centuries before. *Bathurst* is the main town and capital with a population of about 20,000. The site is difficult, being low-lying and insanitary, although the town is well placed. The port has a deep-water wharf and handles groundnut exports. *Groundnuts* are the predominant cash crop and constitute over 90% of the exports. They are grown on the sandstone terraces and in the light, sandy soil near the river. About 50,000 tons are exported each year, approximately one-tenth that of Senegal, and mostly in the unprocessed state. As in Senegal, migrant labourers, known as 'strange farmers', come from outside the Gambia to assist with cultivation and harvesting. Attempts have been made to develop other economic activities to lessen Gambia's undue reliance on groundnuts, but various schemes such as the Gambia Poultry Scheme, established after the last war, have proved abortive. Rice is perhaps the most promising alternative crop and is grown along the alluvial banks of the middle section of the river. Millet is the main food crop and as in Senegal is grown in rotation with groundnuts.

V. An air photograph of the Main Canal at Gezira.

VI. Groundnuts in Senegal.

[*Aut*

VII. Grain storage in northern Ghana.

VIII. A town in northern Nigeria with closely packed mud-walled houses.

[*Nigeria Ho*

Gambia's *political future* is uncertain. Increasing links with Senegal, although a logical development, are not liked by the chiefs who fear loss of power and by some of the people who prefer to pursue an independent line. A federation with an ex-British territory such as Sierra Leone is impracticable, although the inclusion of Gambia in a wider federation including Guinea, Senegal and other territories of the Western Sudan region might be a solution to the problem. Senegal has artificially high prices for groundnuts and a link with the European Economic Community on special terms, so that Gambia might be better off economically if united to Senegal.

MALI

Mali (formerly French Sudan), with a broad base consisting of the Upper and Middle Niger basins, extends northward into the Sahara for over 800 miles, her Saharan boundaries being shared with Mauritania, Algeria and the Niger Republic. It is only the southern part of Mali's great area of nearly ½ million square miles, consisting of the Sudan savanna and Sahel zones, that is included regionally in West Africa. Nearly all the *population* of about 4·4 million is situated in the southern half of the country.

Most of the country is flat with *pre-Cambrian basements* exposed in the east and extreme south, dipping under Ordovician and Silurian sandstones and shales and giving rise to well marked scarps. Much of the central area is covered by Quaternary sands. *Rainfall* varies from 50 inches to less than 10 inches, the amount decreasing rapidly north of Bamako.

The *Niger* is of great importance to Mali and attracts a ribbon of population along its banks. It is navigable from Kouroussa to Bamako but, between Kouroussa and Koulikoro, rapids restrict navigation. Cement and petroleum products move downstream and rice and groundnuts upstream. The railway from Koulikoro via Bamako to Kayes, where it connects with the Dakar–Kayes line, is of vital importance to landlocked Mali. The political crisis between Mali and Senegal caused a disruption of traffic, as the railway was blocked and goods had to be sent via Abidjan. The difficulty was resolved in the middle of 1963 and normal traffic has been resumed on the Dakar–Niger railway. The riverine region was more densely populated a few hundred years ago, and this area was the core of the Mandingan and Songhai empires, the town of Djenne, protected by its marshy defences, withstanding many a siege.

The potential value of the Niger lies in the possibilities of extensive irrigation in the former great inland delta or lake that extends from Segou to Timbuktu. An area of approximately 30,000 square miles may be inundated. Over much of the area lacustrine deposits provide fertile alluvial soils from which high yields could be obtained. The French have long been interested in the development of an irrigation scheme and during the 1930s a great deal of research was carried out as to the suitability of various crops. The first step was the building of a small dam across the Niger at Sotuba which was finished in 1939. The major scheme is at *Sansanding* where a dam over a mile long has been built across the river, and an extensive system of irrigation canals laid out. About 130,000 acres are irrigated at present in the Niger Scheme. It was hoped eventually to irrigate 2 million acres and support 1 million people, but results have been disappointing.

An attempt has been made by the French to blend features of African communal life with the needs of modern technology. Each farmer is given a few acres of irrigable land and basic agricultural implements and seed. Farmers belong to cooperatives known as Native Agricultural Associations and, after paying rent for the land and other dues, are allowed to dispose of the balance of the crop as they choose. The Sansanding scheme, which was very slow to begin with and also expensive, has made steady progress. Rice is the main crop, but millet, cotton and other crops are also produced. A major economic problem is to market rice at prices competitive with imported rice. In 1962 it was decided to concentrate more on cotton, and also to introduce sugar-cane. Other problems are the poor quality of much of the soil and the attacks of queleas.

South of the Niger in the better watered savanna area *groundnuts* are the major economic crop providing, as in Senegal, the bulk of the exports. In the Sahel zone livestock are important and there are exports of livestock to Guinea, Ghana, Senegal, Ivory Coast and Nigeria, some 60,000 cattle and 200,000 sheep and goats being sent out each year. Timbuktu, which was of great importance as a centre for trans-Saharan trade until the sixteenth century, has declined greatly and now has a population of less than 10,000. Bamako, the capital of Mali, is an important market and route centre with a population of about 120,000.

UPPER VOLTA

Upper Volta is only about one-fifth the size of Mali but, situated farther south, has higher rainfall and lies partly in the Sudan savanna and partly in the Middle Belt; the northern tip being Sahel. The western part forms the basin of the Upper Volta which is the main river. Perched on the northern boundary of Ghana this region was historically controlled by the *Mossi* for a long period. The *population* of 4½ million, of whom the Mossi form about half, is fairly dense by French West African standards, particularly near Ouagadougou the old capital of the Mossi Empire. Much of the area consists of thin lateritic *soils* with infertile sandstones in the southwest. *Rainfall* declines to the north from about 40 inches in the south to less than 10 inches in the extreme north-east. Despite the rainfall of over 30 inches in places water supply is meagre and the water table deep.

Over 90% of the population are engaged in agriculture. *Groundnuts* are the main export crop grown in the south and south-west and exports are about 20,000 tons per annum. These are exported by rail to Abidjan. Other crops are guinea corn, millet, beans and cotton, the latter accounting for about 10% of total exports. Increased quantities of rice are being grown in swampy regions in the west. *Livestock* are important, particularly in the tsetse-free zone of the north and, as in Mali, many thousands of cattle, representing nearly one-third of exports in 1960, are exported each year, chiefly to Ghana and the Ivory Coast.

Because of the *pressure of population* and poverty of environment there is a considerable seasonal migration of workers from the Ouagadougou area to the more prosperous regions of southern Ghana and the Ivory Coast to work on cocoa and coffee plantations and farms; a Labour Office, set up in 1956, helps to control the former somewhat haphazard movement of labour. Increasing pressure on the overworked soils of the Mossi area will make it necessary to resettle some of the Mossi in the better watered and more fertile but under-populated Lobi region to the south-east of Bobo-Dioulasso.

Upper Volta, 500 miles from the Gulf of Guinea, is also landlocked and suffers from isolation and lack of coastal outlet. There are no navigable waterways, but there is a railway from Ougadougou via Bobo-Dioulasso to Abidjan, much of which is in the Ivory Coast. In 1961 the Upper Volta section of the line supplied one-fifth of the traffic handled by the Abidjan railway.

NIGER

Most of the country is *Saharan in climate and vegetation* and only a narrow belt in the south comes within the Sahel zone with about 20 inches of highly variable rainfall along the Nigerian border. Nearly half the country has less than 4 inches of rainfall a year. The pre-Cambrian rocks of the Air Massif rise from the surrounding cover of Quaternary sands and form a minor sub-Saharan region.

The total area is about $\frac{1}{2}$ million square miles, of which less than one-third lies within the Sahel zone. The total *population* of about 3 million, with a density of only 6 per square mile, is concentrated largely in the south; the Hausa, akin to those of Nigeria, are the largest group with just over 1 million, while the nomadic Fulani, numbering about 400,000 range the thin grasslands near the Nigerian border. The Tuareg, who for hundreds of years preyed on trans-Saharan caravan traffic, inhabit the Air Massif. Agades was an important centre for Saharan trade.

In the south some guinea corn, millet and beans are cultivated while groundnuts are the main export crop; rice also is produced in the Niger Valley. Groundnuts provide two-thirds of the total exports and are found in the south and south-east. Production has increased markedly since 1950, partly because of protected prices on the French market. *Livestock* are of some importance and there is a considerable trade in cattle, sheep and goats and hides and skins with northern Nigeria. The main problems affecting the economic development of Niger are the isolation, aridity and the prevailing lack of water, which is only obtainable near the Niger and Lake Chad and south of the Air Massif. Situated to the north of Nigeria the nearest ocean outlets are at Lagos, which is nearly a thousand miles away, or through Cotonou, an expensive route that involves a number of breaks of bulk from river to lorry, and rail transport. The Niger river is navigable between Niamey and Gaya from October to March. There are two main roads, one linking Niamey with N'Guimy through Zinder, and the other linking Zinder northward with Tamanrasset in Mali. There are also road connections to the railhead in Nigeria, which provides the most economic route. The completion of the Kainji dam on the Niger in 1968/9 will provide an all-water route from the Niger Republic to the sea, the rapids at Bussa being bypassed by a canal.

NORTHERN GHANA

Northern Ghana is densely populated and in parts of the Wa and Mamprusi district densities approach 1,000 people per square mile (Fig. 40). The reasons for this high density are to be found in the granite soils, impermeability of the rocks and fairly high rainfall. There is considerable *overpopulation* and large numbers of migrant workers go south to the cocoa and mining areas to seek work each year. This area, in contrast to northern Nigeria, has no towns of any size and the population is dispersed in a large number of farm compounds with conical and flat-topped huts, grain bins (Plate VII) and peaked cattle byres grouped together and surrounded by a wall. In the densely populated areas of the north-east compounds are circular, compared to the square compounds in the less densely

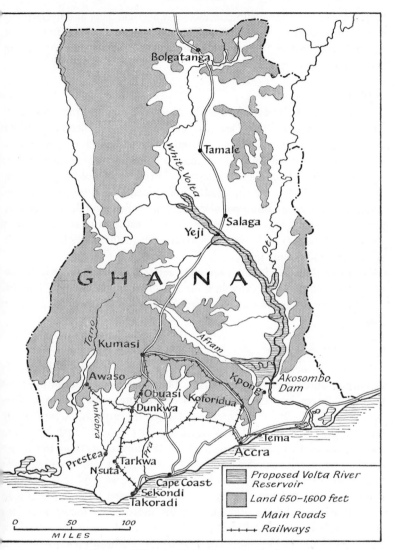

Fig. 40—Ghana: relief and communications

opulated north-west. Round the compound are small permanently cultivated plots growing tobacco, gourds, hibiscus, melon, okra, sweet potatoes,
pepper and paw paw, with grain crops such as guinea corn, millet, and maize
on the compound farm, beyond which is unmanured rotational crop land.

Cattle are of considerable importance, but over-stocking and over-cultiva
tion causes declining soil fertility. There is as yet no railway to the south
and this area suffers from isolation and the lack of a major export crop
although livestock are important.

NORTHERN DAHOMEY

In northern Dahomey, north of the Atacora mountains, rainfall is below
40 inches per annum. There are large expanses of poor Voltaian sandstone
with a degraded Sudan savanna vegetation. Characteristic of this region i
the shea tree, from the nuts of which shea butter is extracted and used fo
the preparation of food and the manufacture of vegetable fats and soap
Both nuts and butter are exported. Groundnuts are also grown on a smal
scale. Large areas are underpopulated and nomadic pastoralism is the mai
activity. There are only about 300,000 head of cattle, with small exports to
the south. Attempts are being made to raise the quality of the livestock with
pilot farms and mobile units.

Northern Dahomey is better served by rail than northern Ghana. Parakou
is the railhead of a line extending for 272 miles from Cotonou. This line
may be advanced northwards another 200 miles to Niamey, the capital o
Niger.

NORTHERN NIGERIA

Perhaps northern Nigeria is the region that best sums up the characteris
tics of the Sudan belt in West Africa. The airborne traveller of the 1960s
arrives at Kano airport; the green and brown patchwork landscape of the
Kano region affords a welcome relief after the arid immensity of the
Sahara. The scene, unusual in West Africa, has touches of both North and
East Africa about it, with small fields, continuous cultivation and home
steads of round huts like those of Kenya and Tanzania, and square huts and
houses reminiscent of the Mediterranean lands. In December the stubble
stooks of millet and brown grasses of the dry season bespeak a markedly
seasonal climate. The scattered, graceful shea-butter, locust bean and
other trees with stands of feathery grass are reminiscent of parkland in
Europe, although in December and January a faded brown park rather than
the lush greens of trees and grass in an English park in July. It is the
seemingly protesting baobabs and, near the Niger Republic border, the
acacias that bring one back to Africa.

From an aircraft or some high point the *landscape* stretches interminably
to a flat horizon; these are the high plains of Hausaland, over 2,000 feet
high near Zaria, formed of pre-Cambrian rocks on an ancient erosion

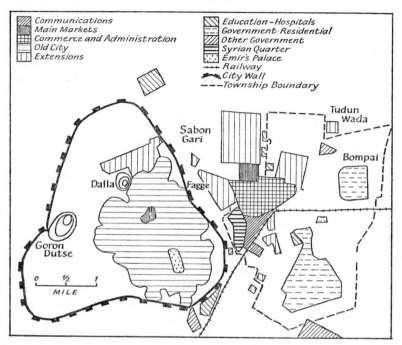

FIG. 41—Kano city (after Buchanan and Pugh)

surface and masked by recent sands, from which occasional smooth granite
inselbergs rise steeply and increase in number to the south. The mature
rivers flow slowly in shallow valleys across the wide plains to the Niger and
Lake Chad. The river courses are dry for over half the year, further testimony to the seasonality of the climate.

An important feature of the Sudan zone of West Africa is the number of
old trading centres and, although many have decayed, those of northern
Nigeria (Plate VIII) like Kano, have not only survived but increased in
size with the coming of the railway and the southward orientation of overseas trade (Fig. 41). *Kano* consists of many different sections, the old city
being dwarfed by the recent commercial, administrative and residential
accretions that have developed over an area of about ten square miles to the
east of the old city. There is a large modern administrative and commercial
area, commercial activities being controlled to a considerable degree by
Lebanese and Indians. Separate residential quarters, the Sabon Gari and
Fagge to the north and west of the commercial area, accommodate stranger
Hausa and Nigerians from the south. Government officials, European and

Nigerian, live in Nassarawa and Bompai to the south-east and east of the commercial area. The new town is laid out in a definite pattern with wide tree-lined streets and is a marked contrast to the congested Hausa city with its meandering lanes and haphazard layout, in which lies its fascination to the visitor. The old city was founded over a thousand years ago near sources of iron ore on Dalla and Goron Dutse Hills which also acted as defensive posts on the flat pediplained surface of northern Nigeria. The modern city owes its predominance to its nodality as reflected by its international airport, the well developed Hausa-Fulani culture and favourable physical environment.

The best view of Kano is from one of the towers of the main mosque. A jumble of flat mud roofs, borrow pits (from which mud has been taken to build the city) courtyards, palm and acacia trees, twisting alleys and paths, occasional straight wider roads, the great market place and the Emir's palace constitute the foreground. In the distance can be seen the bare hills of Dalla and Goron Dutse, the crumbling mud walls and gateways, and the large area occupied by gardens and waste ground within the walls; the gardens and cultivation are a reminder of the necessity to keep Kano's population fed in time of war. On descending the tower and walking through the city the houses and the main markets are the major attraction. The houses, some three to four storeys high, are built of mud and face directly on to the street. The walls built of small conically shaped mud bricks are very thick and reflect the climatic extremes of temperatures of over 100° F. (38° C.) and heavy rain storms in the wet season, and in the dry season contrastingly low temperatures of below 50° F. (10° C.), low humidity and desiccating Harmattan winds.

Pinnacles often surround the roof which, with the massive doorways and small windows, gives an embattled appearance offset to some extent by the practice of colour-washing the walls and decorating them with bas-relief. Timber and stone are practically unobtainable locally so that massive vaulted mud houses are built.

The *market* is the social centre of most West African towns. Kunya, for example, has a small roadside market selling anything from kola nuts, cakes and other foodstuffs to bicycles still wrapped in brown paper.

The greatest market-place in the north is the main market at Kano, a bustling, noisy hive of activity, a place of many races and pushing crowds in colourful costumes. A great variety of goods is offered—potash, salt, kohl, antimony, sugar, dates, sweetmeats, medicines and herbalistic remedies of all kinds, charms and jujus, saddles, bags, sandals, cotton cloth and clothing, jewellery, spades, hoes and other implements, etc. Many trades and crafts are practised in separate sections, in small booths or out in the open, and there are gold, silver, tin and iron smiths, dyers, weavers, sandal-makers, potters, herbalists and many others.

Local industries are still important in Kano and, as in medieval Europe, are still organized on a guild or handicraft basis.

The dyeing of cotton cloth with indigo is one of Hausaland's most typical industries and its products, including the blue Tuareg gowns, were exported widely. It is now declining in importance owing to the competition of cheaper mass-produced cloth. Indigo is a local product and the dye is produced in tanks, a familiar feature in Kano and many other centres.

Blacksmiths turn out a wide variety of agricultural implements, knives, scissors and domestic ware, using imported and local scrap rather than local iron ore. From this crude, although ingenious, craft has sprung an important machine repair industry as the bicycle and now the motor age has invaded West Africa. Leather workers are important in the north. A considerable proportion of the skins are tanned with the fruit of the tamarind and used locally. In addition to the traditional crafts a number of modern industries have also developed in Kano and there are soap factories, groundnut, oil and flour mills, cotton-weaving mills, tanneries, a cannery and factories producing canvas shoes, tarpaulins, floor tiles, tubular furniture, soft drinks, sweets, scent and powders.

However, for part of the nineteenth century *Katsina* eclipsed Kano, but the effect of the Fulani Jihad on Katsina and later the lack of railway communications meant that Kano soon recovered its primacy of position in northern Nigeria. Kano now has a population of over 130,000[7] compared with Katsina which has about 55,000 people. Other important towns in the north are Sokoto and Zaria, the former the religious centre of a wide area beyond the confines of Nigeria, the latter the main educational centre of the north with schools, training colleges, and a university. Zaria is also an important regional and administrative centre. *Kaduna* is the northern capital.

The northern cities and towns reflect strongly their rural and agricultural background. Constant caravans of donkeys laden with vegetables and fruit come into Kano early each morning and leave later in the day with panniers piled with refuse to be used as manure on vegetable plots and small farms near the city. At times the population of Kano is doubled as peasants pour in from the surrounding countryside. The zone of *intensive cultivation* stretches on an average for some 30 miles from the city, with an approximate area of 1,000 square miles, and population of over 1 million. The pattern of agri-

[7] A 1962 estimate of Kano's population was 250,000, great growth having occurred from 1958 to 1962.

culture has changed greatly over the past fifteen years. The subsistence family holding of 3 to 4 acres is tending to be replaced by a cash economy based on groundnuts, vegetables and other produce, paid labour and the hiring of land. In some areas land has been acquired by investors for the mechanized production of groundnuts.

Compared to the extensive farming methods of much of Africa the standard of cultivation in the Kano area is high and the hard-working and resourceful Hausa have been cropping the land continuously for hundreds of years, using manure and compost and a primitive irrigation system with a kind of shaduf (a weighted pole and container for water) in the dry season.

A variety of crops such as maize, sugar-cane, tomatoes and other vegetables are grown throughout the year on irrigated plots near the towns and on the banks of rivers. Millet and guinea corn are staple food crops usually planted in April or May with the onset of the rains and harvested from August onward. There is a useful arable-pastoral link between the *Hausa and the Cow Fulani* who are nomadic pastoralists. Their cattle are encouraged to graze on the stover after the crops have been cleared away and, although the manure dries up, the phosphate content which it contains is beneficial to the soil. The link has not become permanent, however, and the amount of mixed farming is negligible with less than 10,000 mixed farmers in northern Nigeria.

A number of *economically useful trees*, such as the locust bean and baobab, have been left to add to the park-like appearance of the savanna and provide useful sources of food, fodder and household needs.

The *main cash crops* of the north are groundnuts and cotton. Groundnuts epitomize the agriculture of the Kano region which produces half the Nigerian output. They are by far the most important export crop, on which some 5 million people depend. Groundnuts grow well in the Kano region with its light sandy soils and 30–35 inches of rainfall per annum, and occupy about one-third of the acreage on the small farms of the region. Intensive production dates from 1912 when the railway reached Kano. Exports are now up to $\frac{3}{4}$ million tons, worth £22 million in 1960.[8] From the agricultural officer's point of view groundnuts are a useful crop as the bacteria nodules that form on the plant increase the nitrogenous content of the soil, although soils may become exhausted, as in coastal Senegal, if cropped continuously under groundnuts. Groundnuts are also useful as food and fodder. Nearly all the crop is exported in the decorticated state, although groundnut oil is produced in Kano on a small scale. Great pyramids of groundnuts, looking

[8] In 1964 exports of groundnuts were worth £34 million.

strangely Egyptian with frieze-like figures balancing sacks on their heads, have been a familiar feature since the war owing to the problem of moving each year's harvest south by rail before the gathering of the next season's crop. Further extension of the groundnut and other cash cropping area depends partly on the extension of the railway net, for example the 400-mile railway which is being built from Kuru to Maiduguri.

Cotton becomes more important further south in Zaria Province where soils are heavier. The area of land under cotton has increased greatly with the introduction of a Uganda-American long-staple variety. The cotton belt of about ¼ million acres stretches north-west for about 200 miles from south of Zaria to Kaura Namoda, very little being grown in the Kano region. Yields are still low but, with improved methods of cultivation and increased use of fertilizer, could be trebled in some areas. The quality of cotton is high as distribution of seed is controlled by the government. Compared with groundnuts, Nigeria's contribution to the world trade in cotton is small.

Northern Nigeria is an important *cattle-rearing* region and there are large herds of cattle, estimated at from 6–9 million in number, although the official figure is only 4 million. The commonest breeds are the white and red Fulani, long-horned and humped, other varieties being the Sokoto, Chad, Bornu and Shuwa. The Fulani cattle-herders follow a nomadic existence in search of grass and water and avoiding tsetse fly infestation in the valleys during the wet season. A type of transhumance has thus evolved, with cattle being driven on to the cooler drier areas in the wet season and returning to the valleys and Hausa farmlands from October to April during the dry season. Although the Cattle Fulani are skilful herdsmen the standard of cattle keeping is generally low. Cattle are still valued more for social and ritual than economic reasons, so that only about one-tenth of the cattle are slaughtered annually compared to other countries where the rate is four or five times as great. There is a great demand for meat from the south and some cattle are driven to southern markets along well used cattle routes losing much weight on the long journey.[9] Sheep and goats are also widely kept and there is an important trade in hides and skins. So-called Morocco leather used to come from northern Nigeria. Over ½ million hides and 5–7 million goatskins are sold annually, exports in 1960 being worth about £4 million.

[9] Most cattle are now railed or sent by motor transport to the south.

The Middle Belt stretches from 8° to 13° N between the Sudan zone and the coastal forest zone. *Structural zones* are aligned more north and south compared to the east-west trending climatic and vegetation belts. Much of the area is underlain by pre-Cambrian granites and undifferentiated rocks with patches of Lower pre-Cambrian and Ordovician and Silurian sandstones and shales, particularly in Ghana and Guinea. The trend lines of the pre-Cambrian folds run north-north-east to south-south-west in Guinea, but change to north-west, south-east, farther to the east. Later Hercynian movements reinforced the pre-Cambrian trend, as in coastal Ghana. Miocene uplift and volcanic outpourings occurring at various periods were of great importance in moulding the landscape of the highland regions of Futa Djallon, Guinea and the Jos Plateau of Nigeria. The Futa Djallon plateau and the Guinea Highlands are West Africa's major watershed and the source of rivers such as the Niger and Senegal. The sandstones of Devonian and Cambrian age in the Futa Djallon cause flat plateau surfaces scored by deep gorges, and afford a great contrast to the rounded hills composed of ancient granites, gneisses, schists and quartzites in the Guinea Highlands. The broad structure of the Guinea zone was determined by the end of Palaeozoic times.

Rainfall is from 40 to 60 inches with a 4–5 month dry period during which low humidity prevails. Temperatures tend to be lower than in the Sudan zone, the diurnal range being only about 13° F. (7° C.) in the wet season, but increasing to 30° F. (17° C.) in the dry season.

In the southern part of the zone grasses, such as *Hyparrhenea*, tend to be taller, up to 10 feet high, and coarser than in the Sudan zone, and therefore less useful for livestock. Trees are deciduous, and because of the ravages caused by annual grass-burning are gnarled and fire-resistant. There is a narrow belt of derived savanna where rain-forest suddenly thins out and woodland savanna appears, the boundary being sharp, well defined and caused by man.[10] In the northern part of the Guinea zone the *vegetation* is not unlike that found in the miombo woodland of East Africa, although *Brachystegia*, common in East Africa, is not found here. The western end of the Guinea zone is occupied by Casamance woodland where a longer dry season hampers the growth of rain-forest. As in the Sudan zone economically useful trees such as the shea-butter and locust bean are protected.

[10] The mosaic of change between forest and savanna in southern Nigeria is over 50 miles wide, and strips of gallery forest remain interspersed among the patches of savanna.

Bush fallowing may be of long duration in the under-populated areas of the Middle Belt. Maize and cassava are the most important *crops*, yams are also grown and in the west rice cultivation, both upland and swamp, is of major importance. The main export crops are cotton and groundnuts and small quantities of coffee, cocoa and oil palm products come from suitable areas on the fringe of the zone. There is a considerable potential for the production of food crops for sale in the southern forest areas which concentrate on the more lucrative export crops. The amount of small grains grown has decreased because of bird and rodent attacks and the lack of children to scare off these predators. Maize and cassava are tending to replace the small grains.

Because of widespread tsetse infestation the Middle Belt is not important for the rearing of *livestock*, but certain breeds tolerant to trypanosomiasis, such as the N'dama, are found in scattered localities and goats and sheep are widespread.

The Middle Belt is transitional between the well distributed rainfall areas of the coastal forest belt with a wealth of tree crops and the seasonal rainfall areas of the northern grasslands with a varied economy based on seasonal crops and livestock. It lacks the resources of either zone and has suffered historically from slave raiding carried out by groups of peoples from the savanna and coastal zones. The *population* pattern reflects the poor physical environment and troubled past, and is very patchy, with local concentrations in refuge zones such as the Jos plateau but with large areas of slight density over much of the region.

PORTUGUESE GUINEA

Portuguese Guinea is the most western of the regions of the Middle Belt. It has an area of about 14,000 square miles and a *population* of about ½ million. It is bordered by Senegal and Guinea and is an enclave of Portuguese colonial territory set admidst independent African states. The Portuguese claimed the territory in 1446, but for the last 500 years there appears to have been little change in the life of the rural African. Most of the country consists of lowland with a swampy indented coastline with numerous inlets and islands such as the Bissagos archipelago. The vegetation of the interior is a thin forest and savanna woodland.

Compared with Senegal, economic development has been very slight. The contrast is most striking if Portuguese Guinea is entered by road across the northern frontier with Senegal. There are very few vehicles there, although there are a few roads, and most Africans walk, while villages may be cut off for months during the wet season. The main *cash crops* are groundnuts, palm oil and rice. *Bissao* is the capital and main port and has been a free port since 1869, but has a population of only about 6,000. Strict control is exercised by the Portuguese government. There have been attempts recently to break down some of Portuguese Guinea's colonial isolation and the number of assimilados has increased greatly. The future of Portuguese Guinea is in doubt because of rebel incursions from Guinea and strong attempts to overthrow Portuguese colonial rule.

GUINEA

That part of Guinea which is situated within the Middle Belt comprises more than half the country and excludes the coastal plain and a small strip of the wetter south-east near the Sierra Leone and Liberian borders. It includes the Futa Djallon plateau and the basin of the upper Niger river. Miocene uplift has caused rejuvenation of the rivers; these have cut deeply into the Futa Djallon plateau which constitutes West Africa's major watershed, much of which is above 3,000 feet in altitude. The *Fulani cattle-keepers* live on the flat plateau surface while negro cultivators, formerly slaves of the Fulani, raise crops such as bananas in the deep valleys that thread through the pleateau. The Fulani, who occupied the plateau in the eighteenth century, adapted their economy to a tsetse-infested environment by exchanging their Zebu cattle for a type that was more resistant to trypanosomiasis. Despite the *poor infertile* soils developed on the ancient sandstones the population is dense and erosion a serious problem. Although there is a heavy rainfall there are few trees because of wide lateritic exposures and bush firing. On the Niger plains to the north laterite is also extensive. This is more typically savanna country and N'dama cattle are kept, while crops of poor quality such as millet, cassava and groundnuts are grown.

The Niger is navigable from Kouroussa downstream into Mali, but the Milo tributary is more easily navigable. This had led to the increase in size of Kankan as compared to Kouroussa. Kankan has a population of nearly 20,000 and is a route and regional centre of some importance with a well developed lorry trade.

NORTHERN SIERRA LEONE

A small strip of northern Sierra Leone also comes within the Middle Belt, although most of the country is treated separately in the coastal forest region. This is a poor plateau area forming part of the Guinea Highlands and consisting of orchard bush, poor soils and steep slopes, with a population density of under 25 per square mile.

CENTRAL AND NORTHERN IVORY COAST

The Middle Belt of the Central and Northern Ivory Coast covers a much larger area than in Sierra Leone, occupying about half the Ivory Coast. The rainfall, although seasonal, is higher, with 50–60 inches average per year, these totals occurring on about 100 days a year. The vegetation is thus Guinea and derived savanna in type. The terrain consists of low undulating plateaux and plains forming a monotonous landscape underlain by pre-Cambrian rocks and with lateritic and sandy soils as in northern Guinea. Because of the infestation by tsetse-fly cattle-breeding is limited except near the northern border. Livestock are being bred selectively on model farms at Korhogo and Bingerville. *Cotton* is the main cash crop, particularly between Korhogo and Bouake, and nearly 8,000 tons of cotton seed were produced in 1963. Sisal is a minor cash crop, while groundnuts, millet and other crops are also produced. Bouake is an important route-centre and has a hosiery factory, cotton-spinning mill and sisal cordage factory using locally grown cotton and sisal.

CENTRAL GHANA

South of the heavily populated narrow savanna belt near the northern border of Ghana's Northern Territories is the larger Middle Belt. Much of this comprises the *Voltaian Basin*, bisected by the middle Volta and form-ing the least developed and populated area of Ghana. The southern border of this region is the line of rain forest along the Wenchi and Mampong scarps, a significant and even dramatic boundary between the backward, sparsely populated and tsetse-infested Afram Plains to the north, and the rich, forested country of Ashanti to the south. The northern edge of the forest is protected as far as possible, although it is slowly retreating under man's influence and being invaded by the tall grass and fire-resistant deciduous woodland of the Guinea zone.

The basin of the middle Volta occupies about half of Ghana, but only has about one-tenth of the population. There is a high but seasonal rainfall with floods in the wet season and lack of water in the dry season. In addition soils developed on the Voltaian sandstones are infertile, disease is prevalent and slave raiding was rife during the pre-colonial period, so that the *population* density is very slight. Small clusters of population in the form of incipient trading centres are found on the main road to Tamale. Firewood, oranges and yams are the most common commodities for sale, but at larger centres such as Yeji and Salaga a variety of stalls and small 'chop bars' cater for the passing lorry trade. Compounds consist, as in the north, of a number of round huts surrounded by a wall and are very different architecturally from the square or rectangular houses of the forest belt.

Cultivation is slight and patchy, only a small percentage of Ghana's Middle Belt being under cultivation, and yields are very low. Yams, grown in mounds (Plate IX), groundnuts, maize, cassava, guinea corn, millet and oranges are produced, while tobacco is increasing in importance as a cash crop. Yams are sent by lorry to markets at Kumasi and in the coastal belt. Guinea corn, millet and maize are stored in grain bins of a number of different shapes and materials, many bulbous or of a sagging oval shape and protected by a small conical cap. Crops are grown under the characteristic system of shifting cultivation, but north of Tamale with the increase of population the fallow is shortened.

Tobacco is a promising crop and the number of curing barns has increased near Ejura. Sweet oranges of large size could be grown more extensively. Experiments are being carried out with *urena lobata*, a fibre somewhat like sisal, for the production of cocoa sacks.

The incidence of *cattle keeping* increases north of Tamale and near the savanna border. The main types of cattle are the West African Shorthorn and the N'dama, the former being a small non-humped animal with a very poor milk yield. The main interest is in beef rather than milk, and there is a large market in the south. The cattle are bought in the north by itinerant dealers and sold in the main centres of Prang, Kintampo and Wenchi. Experiments are being carried out at the main Government Veterinary Station at Pong Tamale with N'dama, Sanga and white Fulani cattle in an attempt to increase the carrying capacity of the local grasses.

Mechanization has been attempted at *Damongo* as a possible solution to the problem of undeveloped land in proximity to overcrowded areas. The Gonja Development Company was established in 1949 with this in view. It cleared and contour-ploughed the bush, each settler being allocated 12 acres, cows and seed were given on loan, social facilities and tools provided and advice given on the correct methods for growing crops such as groundnuts, tobacco, rice and sorghum. People were taken from overpopulated areas in Mamprusi District where densities are up to 1,000 per square mile.

The Damongo Scheme has had a very limited success and by 1960 less than 1,000 people had settled there. Perhaps the major reason for the lack of success has been opposition by local priests and chiefs in the congested areas to a large-scale movement of people, which would mean a loss of prestige and income. Other reasons are the disruption of traditional customs and the strangeness of the new environment. Social rather than physical reasons have been the main stumbling block at Damongo.

The Volta River Scheme should prove of considerable economic benefit to the area, providing a cheap waterway for cash crops grown under irrigation and, if the tsetse fly can be eliminated, ample sources of water for livestock.

Tamale is the main town in northern Ghana and is of some commercial and administrative importance.[11]

CENTRAL DAHOMEY

That part of Central Dahomey within the Middle Belt is from about $9°$ to $10\frac{1}{2}°$ N. and consists of a granite and gneiss plateau, crossed by the north-east to south-west trending Atacora mountains, up to 3,000 feet in altitude. This part of Dahomey has poor soils and is thinly populated. *Cotton* is the main cash crop, the Allen variety being grown, as in Nigeria, with fairly high yields. 11,000 tons of cotton seed were produced in 1963. Other crops are yams, cassava, groundnuts, millet, beans and peas. As elsewhere in the Middle Belt shea-butter trees are important to the local economy. The main towns are *Djougou* and *Parakou*, the latter being the terminus of the railway from the coast. From here a motor road runs north to the Niger river at Bodjecali.

THE MIDDLE BELT OF NIGERIA

The Middle Belt of Nigeria, stretching from about $8\frac{1}{2}$ to $11°$ N. is difficult to define geographically and owes its regional distinctiveness more to negative than to positive factors. With thin skeletal soils over much of the area, a seasonal, albeit high, rainfall and infestation by tsetse fly the Middle Belt lacks a major commercial crop.[12] It has little of the ecological advantage of the rain-forest belt with its tree crops to the south, or the savanna zone to the north with its annual cash crops and livestock. It has also acted as a kind of labour reservoir for slave-raiding parties from both north and south, and the many Pagan tribes inhabiting the area have tended to concentrate in the dissected country and rocky outcrops near Abuja and the Jos Plateau which afforded refuge from slave raiders.

The *society* of the Middle Belt is very fragmented and there are a hundred different tribes with less than 10,000 members each. A remarkable form of *agriculture* has developed on the rocky slopes with terraced and fenced fields and manuring of crops. There has been a recent movement of some of these people on to the western plains where the prevailing backward methods of agriculture have been adopted. Near the Niger benniseed is

[11] Northern Togo is described on p. 262 and is included in southern Togoland.
[12] Cotton, groundnuts and other cash crops can be grown in the Middle Belt.

grown, some cotton is produced in Kabba Province and rice is grown on alluvial soils. As in Ghana resettlement schemes, such as that at Anchau, have only been partially successful or have failed altogether. Perhaps the most ambitious scheme is to dam the Niger and Kaduna rivers at three places and use their waters for hydro-electricity and irrigation. The Kainji scheme costing over £70 million will provide over 1 million kilowatts of electricity, and make the River Niger navigable from the Niger Republic to the sea.

The most productive part of the Middle Belt is the *Jos Plateau* which has an area of over 3,000 square miles with an average altitude of 4,000 feet. Almost flat grassy plains curve gently to the horizon, punctuated periodically by flocks of volcanic cones and by granitic outcrops which rise 1,500 to 2,000 feet above the plateau surface. The plateau has impressive scarps to the south and east which rise 2–3,000 feet above the surrounding country. In many ways the Jos plateau is like the high plateau country of east and southern Africa, a resemblance which is heightened by the tin-mining operations.

The town of *Jos* is the main centre on the plateau and, with its altitude of over 4,000 feet and healthy climate, fulfils the function of a hill station to a limited extent. The town spreads out in a number of belts between granitic ridges, the railway and two small rivers. It is a great contrast to the sprawling towns of the north with its gridiron pattern and zonal arrangement of commercial quarter, railway sector and higher class residential area, separated from each other by a golf course. On the outskirts of Jos and tucked away amid granitic outcrops on the plateau are Pagan villages with small square fields mounded for yams and cassava and bordered by cactus hedges, tiny mud huts and great storage bins. In the grounds of the Zoo at Jos an open-air folk museum is being developed illustrating the various styles of huts and storage bins to be found on the plateau and its environs. With its zoo, museum, rest houses and hotels, clubs and attractive shops Jos has attracted a considerable tourist trade.

The economic mainstay of the plateau is *tin mining*. There are over 70 companies and many thousands of African tributors who work for individual operators and the companies. The total labour force is over 30,000 Africans and about 250 Europeans. There are great contrasts between the somewhat primitive methods used by the Africans with pan, pick and shovel and nearby stream sluices and the most modern gigantic draglines which eat away the overburden. The modern mines are great opencast pits with powerful pumps sucking the tin-bearing gravel out of the pit in a slurry. It is then pumped to a separation plant at the side of the pit where specks of black cassiterite are precipitated on to a moving jig and the muddy residue washed away. In 1961 the first smelter was opened near Jos, and a second is planned which should be able to cope with present production.

Columbite is a valuable by-product, but production has recently declined. The annual quantity of tin produced has fluctuated considerably, but in 1960 was worth about £6 million.[13] Tin deposits are alluvial and at present are obtained from former stream-beds by opencast methods. By the 1970s, however, it will be necessary to bore through the basalt overburden and use expensive deep mining methods to exploit the tin deposited in older stream-beds. This expense will only be justified if tin prices remain high. If tin prices continue to fall and mining ceases the economy of the Jos plateau would suffer considerably and have to depend far more on livestock products and agricultural development. The Jos plateau is tsetse-free and could specialize in dairying in contrast to the beef production of the north.

REGION 3—THE COASTAL BELT

The last major region in West Africa is the coastal and forest belt stretching in a south-easterly direction for nearly 700 miles to Cape Palmas, then in an approximately east-west direction for over 1,200 miles to Mount Cameron and the Bamenda Highlands. It is generally a *lowland* area with relief usually below 650 feet. From Gambia to Sierra Leone there is a highly indented coastline which lacks the sand bars and spits of the coast further east. There is a high tidal range and strong south-west and north-east winds which prevent sandspit formation. From Liberia to Eastern Nigeria the effects of longshore drift are more prominent and the coasts have been smoothed by silting and the building of sand bars across the mouths of most rivers. There are short stretches of rocky coast near Cape Palmas and Cape Three Points, but for the most part the coastline is fringed by mud flats and mangroves and by sandspits and bars enclosing lagoons, which form a useful network of waterways in the Ivory Coast, Dahomey and Nigeria. A prominent feature of the coastal belt is the Niger delta which has a radial pattern of distributaries and mud banks.

The coastline consists of narrow strips of alluvium and recent marine sediments, which, however, are wider and older in Nigeria where Creta-

---

[13] Year ending 31 March 1965—£16 million.

ceous marine beds extend for considerable distances up the Niger and Benue valleys. Behind the coastline is the dissected edge of the low interior plateau, possibly downwarped towards the coast in Nigeria. Pre-Cambrian and Archaean rocks are exposed over wide areas. The landscape, developed under humid conditions, is surprisingly rounded considering the general lack of altitude, and steep forested slopes are common. Inselbergs and bare rocky domes rear up from the floor of the forest and, if climbable, provide far-reaching views.

The *climate* of the forest belt is characterized by continuous heat and humidity and well distributed heavy rainfall. The sun is almost constantly overhead and temperatures thus vary only slightly both seasonally and annually, the annual range being little more than 5° F. (3° C.). No month has a mean temperature below 75° F. (24° C.), but maximum temperatures are rarely above 90° F. (32° C.) and thus not comparable to the high temperatures of the savanna belt. Wet-bulb temperatures are of greater significance in West Africa and stations with wet-bulb temperatures of 80° F. (27° C.) and over have an uncomfortable climate, as, for example, Lagos, which has wet-bulb temperatures of 80° F. (27° C.) at 3 p.m. on 100 days in the year. A comparison of annual average wet-bulb temperatures for Dakar, Accra and Lagos shows that Dakar has the most pleasant climate of the three stations. These figures should be compared with London's 63° F. (17° C.) in summer and only 34° F. (1° C.) in winter.

	A.M.	P.M.
Dakar	68° F. (20° C.)	72° F. (22° C.)
Accra	69° F. (21° C.)	74° F. (23° C.)
Lagos	76° F. (25° C.)	78° F. (26° C.)

Rainfall is unlike that of the Congo basin and is more monsoonal in character with a relatively dry period from November to January. One of the major reasons for this is its location from 6° to 9° N. compared to the more equatorial latitude of the Congo basin. During the short dry season the north-easterly winds known as the Harmattan bring their dust and low humidity into the coastal region, which is usually dominated by the moist south-westerlies. Freetown, with about 140 inches average annual rainfall, as only 3 inches from December to February. Heavy rainfall occurs where relief is pronounced near the coast and at right-angles to the south-west winds, as in the Futa Djallon plateau and Mount Cameroon. There is a remarkable area of low rainfall east of Cape Three Points in Ghana, Accra having only 27 inches; possible reasons are the upwelling of cold water close inshore and change in direction of the coastline. A feature of local physiological importance is the occurrence of land and sea breezes along the coast which cause air movements and a sensation of coolness.

The climate of the West African coast, and particularly that of Sierra Leone, has gained an unmeritedly bad reputation brought about by the heavy mortality among white administrators, traders and missionaries during the nineteenth century which was caused by malaria, yellow fever and other diseases. For example, in the first 25 years of operations in Sierra Leone, the C.M.S. lost over 100 missionaries. Now, however, Freetown can be as healthy as any other tropical city.

The *vegetation* of the coastal area varies from mangrove forest to high rain-forest and degraded woodland and scrub in places. The strip of man-grove forest is at its widest in the Niger Delta where evergreen trees up to 40 feet high perch on the slimy mud with stilt-like branches and roots. On patches of higher ground adjoining the creeks swamp forest is found. Backing the mangrove swamps on the margin of the coastal plain and the southern edge of the lowland plateau the tall trees of the tropical rain-forest tower up to 200 feet high. There is a lower tier up to 100 feet high and an under canopy of lower trees 50 to 60 feet high. In the densest areas of rain-forest there is little or no ground vegetation as little light penetrates the gloom. Lianas and epiphytes, however, drape the trees in sinuous embrace in the general struggle to reach the light. There are very many species of trees, hardwoods being characteristic, but less than twenty species are commercially valuable; the commonest are mahogany, sapele, wawa, iroko and obeche. Trunks are tall and straight with few branches and are thus commercially valuable, although often buttressed at the base. One of the major problems of commercial forestry in this area is the fact that exploitable trees are widely scattered.

Tropical forest has probably been reduced in area over the last hundred years by slash-and-burn cultivation. With the increase of population in the forest region and demands made on the soil for cash and food crops, the period of fallow becomes reduced and the secondary growth of trees is not given sufficient time to restore the fertility of the soil before it is used again. Fertility levels remain reasonable with a two-year cropping period and a twelve-year fallow, but with a reduced fallow the humus equilibrium level may drop below 50% and become dangerously low. In Ghana determined efforts have been made by the Forestry Department to maintain forest reserves along the south-east trending Mampong-Kwahu scarp and prevent invasion of the area by grass and deciduous woodland. In heavily populated areas such as Iboland in eastern Nigeria the original rain-forest has been largely removed and replaced by oil palms.

In contrast to the large farms of the savanna zone where clearing the ground is far easier, the *farms of the forest belt*, where destruction of secondary bush is difficult, tend to be much smaller. Polyculture with interplanting of food and cash crops in a 2–5 year cycle is characteristic. Bush fallowing is probably a more accurate term for the type of cultivation carried out in this zone as settlements tend to remain permanent and cultivation to rotate round the villages.

Root crops and rice, particularly west of the Bandama river, are the most common subsistence crops. The length of the cropping period and type of crop grown depends on many factors such as physical environment, social customs and the density of population and markets, both local and external. The crops for export, cocoa, coffee, oil palm products, bananas, rubber and coconuts, are the most valuable products of the coastal zone. In addition to their commercial value tree crops are important ecologically as they cover the soil continuously, protecting it from the ravages of erosion, from the heat of the sun, and the force of rain and wind.

The southern forest region is the most highly developed economically of the West African regions with a variety of tree crops, timber and minerals for export and reasonable rail, road and port facilities; there is a dense population in areas such as southern Nigeria and Ghana and high standard of living in the cocoa areas. This coastal region is also the most advanced politically and has had long contact with Europe. The coastal region consists of the following minor regions: Guinea, southern Sierra Leone, Liberia, southern Ivory Coast, southern Ghana, southern Dahomey and Togo and southern Nigeria.

GUINEA

The total area of Guinea is about 97,000 square miles and the population $2\frac{1}{2}$ million. Because of the difficulties of navigation and the climate this part of the West African coast was not exploited as actively as areas further north and east, but during the early nineteenth century both the British and the French became interested in this area. From 1820 onwards several expeditions made their way into the Futa Djallon and Upper Guinea. It came under French influence in the second half of the nineteenth century and in 1889 became a separate colony. The Los Islands remained British until 1904.

Rainfall is heavy along the coast with 169 inches at Conakry, but there is a dry period similar to that of Sierra Leone from December to March. Behind the mangrove-fringed coastline is the coastal plain approximately 40 miles wide and stretching to the Futa Djallon foothills. This area is still well forested and is well populated. *Agricultural products* are coffee, palm kernels, palm oil, rice, bananas, and pineapples. Exports of coffee in 1959 were about one-third of the total value of exports.[14] Palm-oil products and bananas are cash crops of considerable importance, the latter being largely grown on European plantations although an increasing share is being grown by Africans. The banana acreage in 1962 was nearly 13,000 and the centre of production has shifted from the valleys of the Futa Djallon to the coastal area where transport is easier and cheaper. Production has declined considerably since independence and the departure of the French planters. Rice, as in Sierra Leone, is grown in the mangrove swamps, production being about ¼ million tons annually.

The *mineral resources* of Guinea, mainly bauxite and iron ore, are of considerable importance. Annual production of bauxite on the Los Islands was about ½ million tons until 1962, when nationalization of the plant caused a drop in production. There are very large deposits, estimated at 3 billion tons, in the coastal area and in the foothills of the Futa Djallon. Deposits at Fria are also being worked and the bulk of the production here, over a million tons in 1963, is converted into alumina at the Kimbo plant, providing nearly half of Guinea's exports in 1961. An ambitious scheme for the development of hydro-electricity and local refining of aluminium on the Konkouré river is handicapped politically at present. Iron ore is worked in the Kaloum Peninsula five miles north-east of Conakry to which it is connected by a light railway. Probable reserves are very great and estimated at 2,500 million tons with an average iron content of nearly 50%. Exports are mainly to the United Kingdom.

The capital and main town of Guinea is *Conakry* (Fig. 42). It has developed on a small island near the end of the Kaloum peninsula to which it is attached by means of a causeway. The harbour is on the north-western side of the island and has the advantage of deep water alongside. The port handled over 1 million tons of freight in 1958. The town quadrupled in size from 1936 to 1951 and now has a population of over 100,000. Exports of minerals during the 1950s contributed to the growth and the railway to Kankan gives it access to a wide hinterland. The Niger, which is navigable

---

[14] Production of coffee declined to about 105,000 tons in 1963.

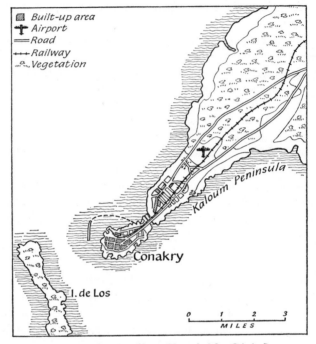

Built-up area
Airport
Road
Railway
Vegetation

Kaloum Peninsula

Conakry

I. de Los

0   1   2   3
MILES

FIG. 42—Conakry (from *Nouvel Atlas Général*)

between Kouroussa and Bamako (Mali) from July to December, provides a further link in the communications system.

In 1958 Guinea was the only one of the French West African Territories to break away from France's tutelage and refuse to join General de Gaulle's new Community. But aid from the Communist block has proved insufficient and ineffective and Guinea is once more seeking French economic aid. Work may be resumed on the Pechiney project for damming the Konkouré river and Guinea's considerable economic potential be realized more quickly in other sectors.

SOUTHERN SIERRA LEONE

The bulk of Sierra Leone comes within the southern forest zone. The origin of the country may be sought in the fine harbour afforded by the Freetown peninsula and Rokel Estuary which attracted European ships long before the colony of freed slaves was established here from 1787 to 1792. During the first half of the nineteenth century Freetown became the main naval base for the suppression of the slave trade and freed slaves were

248

settled in the peninsula to join the original nucleus. The descendants of these freed slaves are known as Creoles today. British interests for most of the nineteenth century were confined to the mountainous peninsula with an area of only 270 square miles, but by 1898 the Protectorate was established over an area of nearly 28,000 square miles. Today the total area of Sierra Leone is about 28,000 square miles and the *population* about 2½ million.

Sierra Leone consists of a *number of zones* parallel with the coast with a narrow belt of coastal swamps backed by the coastal plain usually below 500 feet in altitude, beyond which is the interior plateau 1–2,000 feet high, but reaching nearly 6,000 feet in the granitic Futa Djallon to the north-east. A prominent feature of the coastal zone is the Freetown peninsula of intrusive igneous rocks which rises sharply to nearly 3,000 feet from the drab border of mangrove swamps. This miniature range of green mountains gave the name 'Sierra Leone' to the whole country. Sir James Alexander, who visited Freetown early in the nineteenth century, said that the shape of the land, high and covered in part with forests, reminded him of a West Indian island.

Of the *rivers* that drain away from the plateau the most important is the Rokel which provides Freetown's fine harbour. The coast lies athwart the tropical maritime air masses so that there is *heavy rainfall*, with over 150 inches per annum in places. Being situated 7° to 9° N. the climate is more monsoonal than in southern Nigeria, and most rain falls between May and November, the three months from January to April being very dry; for example, Freetown with about 140 inches average annual rainfall has only 5 inches in the five-month period from December to May, and nearly 100 inches in the three months July, August and September. The concentration of heavy rainfall into such a short period causes serious soil erosion as the forest cover has been greatly affected and reduced by man. The *soils* are generally poor and leached and there are large exposures of laterite on the plateau. Less than 5% of the forest cover now remains and the bulk of the country consists of secondary forest and savanna.

The coastal swamps consisting of Pleistocene clays and sands have increased in area through soil wash from the plateau. They are flooded with fresh water for much of the year and have some good alluvial soils, so that they are suitable for the growing of swamp rice. *Rice* is a staple crop in Sierra Leone and much of the upland variety is grown inland with resultant damage to the soil. An increased use is being made of the swamp area for rice cultivation, some of it mechanized, but only about 10% of the potential land area has been used so far. It is essential for the future of upland soils to concentrate on tree crops instead of annual crops. The swampy regions which are so suitable for rice should play an increasingly important part in the agricultural economy of Sierra Leone.

The *second minor region* consists of the lower part of the interior plateau which is underlain by pre-Cambrian schists, granites, gneisses, and sediments. The most important crop is palm-oil products which account for nearly half of Sierra Leone's exports. Palm kernels provide the bulk of the exports, most of the palm oil being used locally. The quality of the palm oil is poor and attempts are being made to improve quality by using better types of trees and more oil mills. Cocoa is grown in the south-east near the Liberian border and the area under cocoa is increasing despite the handicap of a dry season. Coffee is also grown in the south-east. Other cash crops are ginger and kola nuts, the latter being most numerous in the south and of a high quality. Bananas and citrus fruit are produced in the Freetown peninsula, but production could be expanded in other areas. There are *valuable iron ore deposits* consisting of high grade haematites, now nearly exhausted, and large reserves of poorer grade ore in the Marampa area which is connected by a 55-mile long railway line to Pepel, a port on the northern side of the Rokel estuary. Exports were over 2 million tons in 1965, and there are further deposits to the north-east in the Sula Mountains awaiting exploitation.

The *third minor region* is the interior plateau about 1,500 feet in altitude but with residual domes such as Bintimani rising to nearly 6,400 feet. Soils over much of the plateau are very poor in quality and are usually farmed only one or two years in nine. Rice is the main crop with subsidiary maize, beans, yams and cassava. The main economic value of this region lies in its *mineral production*, particularly of diamonds and chrome. Diamonds—production for the period June 1964 to July 1965 being worth £6·7 million—are found in a number of scattered areas round Sefadu, Pendembu, Kenema and Bo and are easily mined after the shallow overburden is removed. A diamond rush occurred after 1952 and illicit digging became widespread, disrupting tribal life, causing the neglect of agriculture and serious loss of revenue to the government, and ruining much diamondiferous ground. The situation is more under control now and exploitation of diamonds has been rationalized. About 9,000 square miles have been declared alluvial mining areas, and up to 30,000 people are employed on the diamond diggings. Chromite is mined in the Kanbui hills and exports go mainly to the United Kingdom.

Freetown is the only town of any size in Sierra Leone and is the capital and main port. Its population is now about 130,000 and it has more than doubled in size since 1945. One of the main reasons for the importance of Freetown lies in its fine natural harbour, the deepwater quays of which accommodate ships with a draught of 32 feet. The traveller usually approaches by sea, as Lungi airport is on the northern side of the Rokel river which must be crossed by launch, and lands at Government Wharf. The Creole metropolis spreads along the northern slopes of the peninsula, the site being restricted by the steep slopes. Houses vary greatly in size and shape, but many are distinctive in style with a brick, concrete or laterite first floor topped by a wooden second storey and sometimes a latticed verandah. The town also clings to the hills with the administrative and educational sections on Hill Station and Mount Aureol, looking down on the sea of roofs and the majestic sweep of the harbour.

The main railway to the interior is 2 feet 6 inches in gauge and is single track from Freetown to Pendembu. Communications are still inadequate, although a network of roads is being developed. A mixed future awaits Sierra Leone; much of the country has worn-out soils and a low standard of agriculture, heavy downpours of rain increase erosion and the population is increasing fast, but on the other hand the coastal swamp region has considerable potential for increased rice production and tree crops could be grown more extensively in upland areas and mineral exploitation, already a valuable source of revenue, could be increased.[15]

LIBERIA

To the south-east of Sierra Leone is the republic of Liberia which has had a somewhat similar historical relationship with the United States of America as Sierra Leone has had with the United Kingdom. Liberia also owes its origin to colonies of freed slaves established by the American Colonization Society from 1822 onwards. As in Sierra Leone, these settlements faced great difficulties initially, both physical and human, with heavy rainfall, dense vegetation, many tropical diseases and tribal hostility. The area of the state is about 43,000 square miles and the *population* is variously estimated at less than 1 million and over 2 million. Nevertheless, the total number influenced by Western culture is probably less than 70,000, of whom some 15,000 are Americo-Liberians, who control Liberia politically.

The *structural zones*, as in Sierra Leone, trend parallel to the coast, but the area of coastal swamp is far smaller and is lacking entirely in places, so that the coastal plain is the first region, backed by a low hilly region up to 50 miles wide and rising to a more highly dissected foreland zone of scarps

[15] The political influence of the Creoles has declined greatly since independence.

and hills before the usual plateau is encountered. The plateau is 1,500 to 2,000 feet high and culminates in the north in the Guinea Highlands. Liberia's rainfall régime is not unlike that of Sierra Leone, amounts varying from 100 to 150 inches and lasting from May to November. Man's destructive influence on the vegetation cover is far less apparent in Liberia and the incidence of forest is far greater than in Sierra Leone. The *rain-forest* is impressive with the tall *Funtumia elastica* reaching a height of 200 feet and many other species of trees and palms such as the cotton tree, dracaenas, figs, Borassus palms, oil palms and the coconut along the coast.

The *main commercial crop* is rubber which accounted for over 95% of the exports in 1950 and was worth about £8 million in 1962. The main area of rubber production is on the Firestone Rubber Company plantation of over 100,000 acres, situated about 30 miles east of Monrovia. The Firestone Company has had a great influence on the economic development of Liberia and over 20,000 workers are employed on the plantation, where yields are very high, and in associated factories. Nearly one-sixth of the rubber is now produced by independent small farmers, who increased in number from about 300 in 1951 to nearly 3,000 in 1961. Other commerical crops produced in small quantities are coffee, cocoa, bananas, palm products and sugar. In the interior agriculture is almost entirely subsistence and the highly leached soils can only be worked one or two years in ten; rice and cassava are the main crops, Liberia being at the junction of these two major crop zones. The second most important export is iron ore, initially from the Bomi Hills and now from three other areas, two near Monrovia and the third in the Nimba Mountains. The Nimba Mountains area has the largest production and this is planned to reach over 10 million tons in a few years' time. Much of the ore is 66% grade, and lower grade ores are treated at the mine before export. Exports of iron ore are expected to reach a total of over 20 million tons by 1967. There are also large deposits of alluvial diamonds which are now being worked in the north.

Liberia, in common with Panama, has allowed owners of foreign ships, mostly American, to register under her flag under less stringent conditions than apply with many major seafearing nations. She now has a registered fleet of over 2 million tons of shipping, the fourth largest in the world.

Monrovia with a population of about 80,000 is the major town and capital of Liberia. It was the original centre where freed slaves settled. Growth has been rapid owing to recent economic development and many new buildings are being constructed in a town whose colonial style of architecture is reminiscent of Freetown.

Before 1945 the country had no paved roads, but in the period 1952–61

£10 million was spent on road construction from Monrovia to the mineral and plantation areas.

In contrast to many African countries Liberia is faced with the problem of a small population and lack of labour. The legacy of past stagnation is being gradually overcome, but too much reliance is still placed on American aid, as the major exports of rubber and iron ore are controlled by American companies.

SOUTHERN IVORY COAST

To the east of Liberia and broadly based on a great curving sweep of coastline is the Ivory Coast, a former French territory and now an independent country forming part of the French African Community. The Ivory Coast is considerably larger than Sierra Leone and Liberia and has an area of about 123,000 square miles but a *population* of only 3·3 million, little larger than that of Sierra Leone. The coastal forest region is nearly half the country and in contrast to Liberia and Sierra Leone has a more pronounced equatorial régime of *rainfall* with two maxima of rainfall, as exemplified by the figures for Abidjan, showing monthly totals of rainfall in inches.

J	F	M	A	M	J	J	A	S	O	N	D
1·6	2·1	3·9	4·9	14·2	19·5	8·4	2·2	2·8	6·6	7·9	3·1

Total 77·1

No month is really dry and there is a constant high humidity, ideal conditions for the growth of rain-forest which thins out to semi-deciduous forest and savanna at 8° N.

The shoreline is very regular and lagoon-bordered in the eastern half. From the alluvial coastal plain some 40 miles wide the characteristic low plateau rises to over 1,000 feet, reaching the highest point in the Man Mountains of the north-east, an isolated offshoot of the Guinea Highlands.

The *coastal forest* belt is again the most important economic region of the Ivory Coast and, as it is here some 300 miles wide, is of greater value than in most West African countries. In contrast to Ghana, coffee is the major *export crop* of the Ivory Coast, accounting for nearly half the total value of exports. The Ivory Coast is the chief producer in French West Africa. Over 1 million acres are under coffee, most of the crop being produced by Africans. The coffee is of a high quality, *robusta* being the main variety grown, and most of it is exported to protected markets in France. However, yields are low and there is need for new plantings. Over-production by

African countries has caused a reduction in coffee prices and in 1960 the Inter African Coffee Organization was formed to restrict output. Cocoa is nearly as important as coffee and accounts for about one-third of the ivory Coast's exports. The acreage under cocoa is about 660,000 and the main growing areas are in the east. Production is again almost entirely in African hands. Cocoa has become a popular, easily grown crop and from 1940 to 1953 the acreage under cocoa doubled.

The small cocoa plots are widely dispersed, three-quarters of the trees being over 10 years old and nearly one-third over 25 years old. Swollen shoot disease is a serious problem. The age of the trees and poor methods of cultivation have reduced the yield to less than a third of the plantation crop. However, new stock has been planted and quality improved, so that by 1961 most of the cocoa sold was first grade. Palm-oil production has declined because of the attraction of cocoa and coffee, the Man Mountains in the north-east being the main producer. Uniform temperatures and humidity and more even distribution of rainfall than in Guinea favour the cultivation of bananas, which are grown almost solely by French planters. Production is only about one-third that of Guinea. There are nearly 10 million Kola trees in the forest belt near Abidjan and Man and there is an extensive trade with the savanna area to the north. The Ivory Coast forests have great potential, and over 10,000 square miles are cut over, the main exports being mahogany and iroko. Exports of wood have increased from about 50,000 tons in 1947 to nearly ½ million tons now.

Mineral exploitation is of increasing importance, the production of diamonds having increased many times since 1948. Manganese is worked at Grand Lahou, and there is a declining production of gold.

Abidjan, the capital of the Ivory Coast and the main port, is one of West Africa's major cities, with a population of 180,000 (Fig. 43). The city is built on the northern side of the Ebrie lagoon, partly on a peninsula and partly on Petit Bassam Island. Abidjan has developed as the Ivory Coast's major port since 1951 when it was connected to the sea by the Vridi Ship Canal, nearly 2 miles long and 49 feet deep. Abidjan has superseded Grand Bassam and Port Bouet and in 1961 handled over 2 million tons of cargo. There is some local traffic by barge along the lagoons which stretch for nearly 200 miles along the coast and a railway extends over 700 miles inland to Wagadugu. Rivers are of little value for navigation so that the railway to the north is a vital factor linking the savanna zones with the coast. However, an increasing proportion of traffic goes by road, and the Ivory Coast has the best developed road network of the ex-French territories in West Africa.

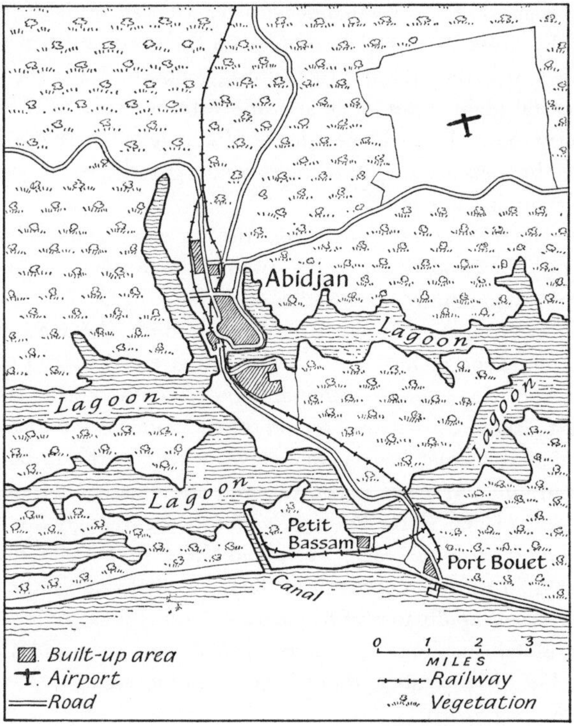

FIG. 43—Abidjan (from *Nouvel Atlas Général*)

Of all the former French West African territories the Ivory Coast has the most promising economic future. The broad forest belt, originally a handicap to growth, now produces large amounts of coffee, cocoa, timber and oil-palm products. The Ivory Coast is also fortunate, taking a long-term demographic view, in that the country's resources can support a far greater population than there is at present, although the lack of labour has proved a problem in the past.

SOUTHERN GHANA

The V of the Middle Volta and its tributary the Afram cuts like a spear into the forest and coastal region of Ghana. To the east of Accra is the line of the Akwapim Togo hills of schists, sandstones and quartzite aligned in a north-easterly direction. Although below 2,000 feet in altitude, they resemble miniature mountains as a result of overfolding and thrusting. Trending north-west from Accra is the triangular *Ashanti plateau* of pre-

Cambrian rocks. Northern Ashanti, however, consists of the Primary sandstones of the Volta Basin which impinge on the pre-Cambrian rocks in an impressive scarp from Kwahu to Gambaga. South of the pleateau is the narrow coastal plain about 15 miles wide with outcrops of pre-Cambrian and Primary rocks. This widens into the Volta Delta where the older rocks are covered by Tertiary and Quaternary sands, gravels, and clays and where there are many inselbergs.

Ghana, which derives its name from the ancient Sudanese Empire, was until recently the Gold Coast, an indication of the former value of the trade in gold which was rapidly surpassed in importance by the slave trade during the seventeenth century. This part of the Guinea Coast attracted a far larger concentration of forts and castles than elsewhere along the coast of West Africa, as, for example, Elmina, founded by the Portuguese in 1481, Cape Coast Castle, Christiansborg and others. A major physical reason for this was that the coast instead of being backed by the usual mangrove swamps and rain-forest has a strip of scrub and grassland stretching west, nearly as far as Cape Three Points. *Rainfall* is surprisingly low and is less than 40 inches on an average, Accra and the Lower Volta Plain indeed having less than 30 inches per annum. To the west of Cape Three Points the rainfall increases rapidly to over 80 inches at Axim. The régime is equatorial with two rainfall maxima, the heavier in May and June and a subsidiary from October to November. Rainfall becomes increasingly seasonal north of the Mampong Scarp.

Apart from the coastal plain east of Takoradi the *vegetation* consists of forest, much of it semi-deciduous. Little more than 5% of Ghana is true forest or forest reserves, the climax vegetation over most of southern Ghana being secondary bush rather than forest.

The southern third of Ghana, approximately 30,000 square miles in area, supports 80% of the *population*. Some 5 out of Ghana's 7 million people live here. The Akans are the largest tribe, comprising the people of Ashanti and the Western and Eastern Regions, with the Guans to the east and north, the Ewes near the Togo border and Ga-Adangmes along the coast and in the Volta delta. High densities of population are found along the coastal plain east of Cape Three Points with densities of over 200 per square mile near Accra.[16] Densities are also high in eastern Ashanti, particularly round Kumasi and the Akwapim-Togo hills.

Cocoa, more than any other economic product, symbolizes Ghana's prosperity. Nearly 5 million acres are under cocoa. It is pre-eminently a

---

[16] According to the 1960 census about 17% of the population of Ghana lives in towns of 10,000 or more.

peasant crop, grown by 1 million farmers on small holdings usually 1 to 3 acres in size. Some of the cocoa farms are larger and owned by absentee landlords. Production first started in Akwapim at the end of the nineteenth century, but had declined greatly by the 1930s because of soil erosion and swollen shoot disease. There are three main areas of production now, two in Central Ashanti and the third north of Accra. Cocoa is a characteristic cash crop of the high forest zone; it needs shade and constant high humidity and a well distributed rainfall of from 50 to 70 inches per annum. Soils should be well drained so that cocoa trees are usually planted on hill slopes, although some cocoa grows in valley bottoms. As the cocoa trees take more than five years to mature, food crops—yams, cassava, peppers, egg-plants, okra, onions, tomatoes, and fruit—are planted in the interim. After the cocoa crop has been harvested for the first time food crops are shifted to areas of poorer soils, perhaps at the edge of the cocoa land.

Cocoa farms are a permanent feature of the landscape and are sometimes over five miles from the village while the food plots are moved periodically using a system of bush fallowing. Little attention need be paid to the cocoa trees once they have been established. There are problems of swollen shoot (Plate X) and other diseases which caused production to fall after 1936. Vigorous attempts have been made by the government to eradicate swollen shoot and by 1961 over 100 million trees had been destroyed, new areas replanted and trees sprayed extensively.

The large yellow-green cocoa pods hang awkwardly from the mottled trunks of the trees. The pods contain the cocoa beans which are harvested mainly from October to February, after which a short dry period from December to March enables the beans to be dried on wide trays. Most of the work on the cocoa farms is done by hired labour which has migrated from the overcrowded and economically backward savanna areas of northern Ghana and Upper Volta.

The large-scale production of cocoa over the last 50 years has been a great stimulus to Ghana's economic development. At times Ghana has been responsible for half the world's production and high prices since the war have helped pay for development projects such as the Volta River Scheme. Exports are usually over half the total value of exports. The pre-war problem of fluctuating prices has been overcome to some extent by the Cocoa Marketing Board which paid the farmer a guaranteed sum independent of world prices.[17] The large decaying family houses of Aburi and

[17] Since November 1961 the name of the Board has been changed to the Ghana Agricultural Produce Marketing Board, which has taken over the responsibility of marketing other cash crops as well.

Akropong are witness to the initial burst of cocoa prosperity in this area. The general effects of the cocoa boom can be seen in the developed communications and settlement pattern of southern Ghana. It is, however, an uneasy prosperity as over half Ghana's revenue is derived from a tropical crop, cocoa, which is subject to the whims of fluctuating prices and the ravages of disease.

Besides cocoa and the basic food crops, kola nuts, coconuts, palm oil and kernels, rubber and limes are produced on a small scale. There has been a considerable attempt to diversify agriculture since independence. In 1963 there were over 100 state farms, most of them being experimental farms taken over from the former Department of Agriculture and the Agriculture Development Corporation. There are in addition 35 Workers' Brigade farms with over 20,000 acres under crops such as rice, vegetables, tobacco, cotton and maize.

Timber exports in 1964 were worth nearly £15 million, about one-third being sawn timber and two-thirds logs. All the trees, including the comparatively soft Wawa, are hardwoods. There are nearly 6,000 square miles of forest reserves, mostly in the region of heavy rainfall west of Takoradi. The bulk of the timber production is still from areas outside the forest reserve that are agricultural. When the ground is cleared for cultivation the tall trees are left, and if commercially valuable are cut down at a later time. Only about one-tenth of the 200 species of trees are exploitable at present, two-thirds of the log exports being Wawa, the rest being mostly cedars and mahoganies.

Southern Ghana is also fortunate in having large exploitable deposits of gold, diamonds, manganese and bauxite, as shown by the table giving the value of exports for 1963.

	£ *million*
Gold	11
Diamonds	3·3
Manganese	4·0
Bauxite	0·5

From 1945 to 1954 the mining industry exported £130 million worth of *minerals*. Exploitation of gold in the Gold Coast dates back a long time, although it is unlikely that this was the source of gold for the Sudanese empires of Mali, Songhai and Ghana. Gold is now obtained mostly from deep mines near Tarkwa, Obuasi and Prestea. In 1960 several mines facing closure because of rising costs were taken over by the Government. Diamonds are all alluvial and mostly industrial stones. There are two main areas of exploitation: first in the Birrim field where rich deposits are worked mechanically, mainly by European companies, and second in the Bonsa field which is exploited by Africans. Here a considerable proportion of the diamonds are lost through inefficient methods; pits are dug haphazardly and diamonds are panned and picked out by hand. Exports worth about £4 million in 1953 had more than doubled in value by 1960.

In contrast the value of manganese production fell from nearly £9 million in 1953 to about £4 million in 1963. The main worked deposits, which are large and with a high content of manganese, are at Nsuta. Mining is opencast and mechanized, and exports are railed to Takoradi. Bauxite is of great potential value and the main centre of production is now at Kanayerebo, nearly 50 miles north-west of Dunkwa. Other deposits are at Yenahin and Ejuanema which will be worked when the aluminium smelter planned for the Volta River Scheme is established.

Alongside southern Ghana's rapid economic development has been a marked growth of *urbanization*; for example, in 1931 only 22% of the population was to be found in settlements with a population of 3,000 and over, but this had risen to 31% by 1948. It is important, however, to distinguish between a town and a village in Ghana where social and not economic function as in Europe is the main criterion for differentiating a town in the traditional sense. A town is the seat of a chief and the political and social centre of the tribe, whereas a village is usually an offshoot from a town. Thus Aburi, with a far smaller population than Nsawam, is a town and Nsawam is only a village belonging to Aburi. This distinction is tending to break down with economic growth and a new type of commercial town has grown up, enjoying nodality of position, such as Nsawam, New Tafo, Asamankese, Nkawkaw and above all Accra. These commercial centres have a shopping area, flourishing market, large lorry park, and separate residential quarters with better housing for higher commercial and administrative classes and poorer housing for immigrant workers in the Zongo.

There are considerable contrasts in building and house styles between north and south in Ghana. In the north huts are round and usually thatched, a number of small huts, grain bins and cattle enclosure forming the homestead or compound. In the south houses are square and usually made of swish or concrete block walls and corrugated iron roofs with a number of rooms grouped round an inner courtyard.

Typical of the older town in the south is *Aburi*, approximately 1,000 feet up in the Akwapim hills and 20 miles from Accra. It forms a kind of hill station for the capital and has a large botanical garden which attracts many visitors. The Basel missionaries settled at Aburi and established it as a girls' educational centre, and it now has two middle schools, a secondary school for girls and a teachers' training college. The missionary quarter with its churches and schools has grown up alongside the non-Christian tribal nucleus containing the chief's house. It was a thriving commercial town when the cocoa boom was at its height, but it has since declined in importance and many large 'prestige' houses are now half deserted.

Seen from the air *Accra* covers a great area and stretches out its tentacles towards the airport and university. In 1876 Accra replaced Cape Coast as the capital of the Gold Coast. It has become the country's second port despite the fact that it is only a surf port. Its main function is the unloading and import of cargo; for example, in 1960 it unloaded over $\frac{3}{4}$ million tons compared to less than 100,000 tons loaded. Takoradi is now Ghana's main port and in 1960 loaded nearly 2 million tons and unloaded 1 million tons of cargo. The new harbour at Tema 17 miles to the east of Accra will increasingly take away traffic from the roadstead at Accra, but the coastal spread of the city may eventually absorb Tema. Accra is also an important road and rail centre with a railway to Kumasi and it has an international airport. Its phenomenal growth from about 27,000 in 1901 to 338,000 in 1960 is connected with its selection as capital of the Gold Coast and then independent Ghana, the prosperity engendered by cocoa and its network of communications.

From the original nucleus near James Fort and Ussher Fort the city has spread outwards in a semicircle over the coastal plain and now covers a large area. The original population, composed largely of Gas, has become very heterogeneous with nearly half the population non-Ga and drawn from many parts of West Africa. The original ties of kinship have been replaced by widespread social links based on common interests, place of residence and occupation. The great expansion of population and its cosmopolitan nature has given rise to many social problems. The growing volume of traffic raises familiar problems of traffic congestion, less serious, though, than in other West African cities such as Lagos whose nucleus is on an island. As elsewhere in Africa there are great contrasts between the modern buildings on Liberty Avenue and small overcrowded houses where people live under slum conditions (Plate XI).

Kumasi, in the heart of the cocoa country, is the capital and largest town in Ashanti. It was founded in the seventeenth century and is the seat of the Asantahene, and so shared the fortunes of the Ashanti state with a greatly reduced population after the Ashanti wars at the end of the nineteenth century. By 1960 the population had grown rapidly to nearly 200,000. Kumasi is thus second in size only to Accra which is situated nearly 200 miles to the south-east. Kumasi is an important regional and communications centre with a large market and wide variety of products, and railway links with Takoradi and Accra and roads to the coast and north to Tamale. The great increase of population and its heterogeneity, as in Accra, have caused many problems. Large numbers of immigrants have come from the north of Ghana and ex-French territories, and are housed in the usual zongo. The considerable foreign population of Lebanese, Indians and Europeans is also housed in separate residential areas.

Ghana was the first West African country to achieve independence, in 1957, and has become a champion of Pan-Africanism. Her prosperity is based largely on cocoa, but also on minerals and timber. Her main needs are to lessen this undue reliance on cocoa and develop the backward northern two-thirds of the country which contributes little to the economy of Ghana, apart from exports of labour. Northern Nigeria's exports of cattle, cotton and groundnuts on the other hand play a vital role in helping to balance the economy of Nigeria. The *Volta River Scheme* may help somewhat to redress the imbalance between north and south in Ghana. A dam is being built in the Volta river gorge at Akosombo (Plate XII), and turbines will generate over $\frac{1}{2}$ milion kw of electricity. This will be used partly to smelt aluminium at the Tema Smelter, the bauxite coming from Yenahin,[18] and partly for industries such as breweries, furniture, soap and cigarette-making and agricultural processing. A 200-mile long lake will be dammed up, providing a cheap waterway to convey the agricultural products of the Middle Belt and north to the markets of the south. Abundant water will also be available for irrigation and the development of the livestock industry. 80,000 people are being resettled in 52 new villages.

SOUTHERN TOGOLAND

To the east of Ghana are the two long narrow strips of southern Togoland and Dahomey. Togoland, stretching 400 miles from north to south, has a total area of about 21,000 square miles and a total *population* of about $1\frac{1}{2}$ million. Formerly German territory, it was administered as a mandate and later as a trust territory by Britain and France. The ex-British part of Togoland has joined Ghana, while the ex-French part has become an inde-

---

[18] Extensive bauxite deposits on the Alewa Range 60 miles by road from Tema are now proved suitable.

pendent country associated with the French Community. There are more than 20 different ethnic groups in the territory; the leading group is the Ewe, living in the coastal area astride the Togo-Ghana border, of whom only 40% now live in Togo. Despite the splitting up of the tribe the Ewes of Togoland reacted strongly against a proposal by Dr Nkrumah in 1959 to integrate Togo in Ghana. The recent assassination of Togo's president, whose policy was anti-Ghanaian, may bring about closer links with Ghana.

The *climate* is dry along the coast as in Ghana, but increasingly wet with a double maximum of rainfall in the Togo mountains. There is a *lagoon-fringed coast* with an economy, as in Ghana, based on coconuts and fishing, backed by a low sub-humid plateau with scrub vegetation, baobabs and savanna crops such as groundnuts, maize and guinea corn, but also oil palms. The *Togo mountains* lie to the north of the plateau and are a highly dissected region with narrow ridges, deep valleys and areas over 3,000 feet in altitude. It is a continuation of the forested Togo mountains of Ghana and grows forest crops such as cocoa and coffee. There is a very high population density with up to 1,500 people per square mile of cultivable area in the rugged Kabrai country. Fertile soils derived from granite and gneiss, terraced fields, manuring and irrigation help to maintain such a high density, but resettlement schemes are being carried out in the south. There are very large deposits of phosphates near Lake Togo and over ½ million tons are exported after being washed, dried and purefied.

The main town and capital is *Lomé* with a population of 70,000. It is also the main port in Togoland and has rail connections with Palime and Sokodé in the north and Dahomey along the coast.

SOUTHERN DAHOMEY

Southern Dahomey is wedged in between Togoland and Nigeria and has only a narrow strip of coastline about 70 miles wide. The total area of Dahomey is 44,000 square miles, about twice the size of Togoland, and it has a *population* of about 2¼ million. There is a lagoon coast backed by a sandy plain rising to the characteristic low plateau. Rivers such as the Ouémé have divided the plain and plateau regions into three or four north-south trending strips. *Rainfall* increases along the coast from under 30 inches on the Togo border to about 50 inches on the Nigerian border. During the seventeenth century Dahomey extended its control to the sea, but became subject to the Yoruba during the eighteenth century. The French became involved politically with Dahomey in the second half of the nineteenth century, and it became a French colony in 1892. Dahomey achieved its independence in 1960.

Coconut plantations owned by non-Africans have been established on the seaward edge of the lagoons. There are nearly 30,000 acres of coconut palms along the coast, most of the coconut products being consumed locally. The sandy coastal plain is *cultivated* continuously in places with extensive stands of oil palms and food crops such as maize and cassava. Oil palms are very important in Dahomey's economy, providing over two-thirds of the exports, mostly in the form of palm kernels. Oil palm products come from wild palm groves which occupy an area of 1,500 square miles. The palm tree is greatly prized by the Dahomeans as it provides cooking oil and a fermented drink, is used for medicinal purposes and also as a substitute for paraffin. However, most of the trees are over fifty years old and are not being replaced in sufficient numbers.

Porto Novo is the capital, with a population of 30,000. It is connected to Lagos by a lagoon waterway, the use of which has declined greatly with the growth of road communications. Most trade now goes via Cotonou, which has a long jetty projecting over the bar and is the main port and rail-centre, with railways along the coast to Lomé and Porto Novo and north-wards to Parakou in the savanna zone. The Ouémé river, flowing into Lake Nokoue, is navigable for 125 miles. There are 360 miles of railways operated by the Benin-Niger Company.

NIGERIA

Nigeria has by far the largest population of any West African country, although not the largest area (Fig. 35). Its total *population* of nearly 36 million (1960 estimate) inhabits an area of 356,000 square miles. Modern Nigeria is little more than 40 years old and it was only in 1914 that the Northern and Southern Provinces were united. Nigeria achieved its independence in 1960, and since the war a federal form of government has gradually replaced the former unitary state. Four contrasting main regions and many minor regions are linked together, their economies being complementary, their social and political systems divergent. In the south the Eastern and Western Regions, with the small Mid-Western Region, occupy less than one-fifth the total area of Nigeria, but have nearly half the total population. The following table gives the 1960 population estimate in millions, with the 1963 census in brackets.

Region	Area (Sq. miles)	Population (millions)	
North	281,800	19·5	(29·8)
East	29,500	8·3	(12·4)
West	30,400	7·1	(10·3)
Mid-West	14,900	1·5*	(2·5)

*1953 estimate

The core of the *Eastern Region*[19] is Iboland, a triangular area with its base along the Niger delta and its apex near Enugu; this is an area of extremely poor sandy, leached *soils*, high humidity and a well distributed *rainfall* of over 60 inches per annum and yet, because of the porous soils, a paradoxical lack of water. The area was originally covered by rain-forest which has nearly disappeared except in the river valleys and has been replaced by oil palms which give Iboland the appearance of a vast oasis. *Population* density is very high, second only to the Nile Valley, with densities of over 1,000 per square mile in parts of Owerri and Onitsha. The Ibos, politically active and with a great thirst for education, form the majority of the population of the Eastern Region. The Ibibios number about $\frac{3}{4}$ million and live in Calabar Province west of the Cross River. Nearly $\frac{1}{4}$ million Ijaws live in the Niger Delta. A major reason for this great density of population in an

[19] This was proclaimed as the breakaway republic of Biafra in June 1967.

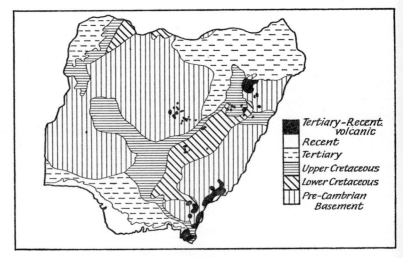

FIG. 44—Nigeria: geology (after Buchanan and Pugh)

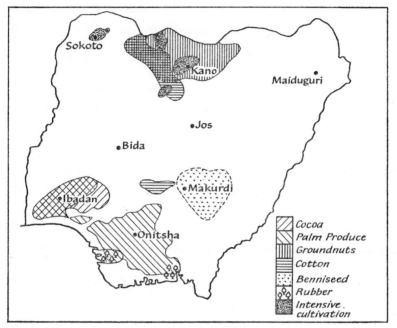

FIG. 45—Nigeria: export crops

unfavourable environment is the protection from attack afforded by the Niger river and the forests. In contrast to the Western and Northern Regions there was no strong tribal organization or urbanization in pre-colonial days, and the social system was based for the most part on clan and family life and dispersed settlement.

Recently, however, there has been a marked growth of towns such as Port Harcourt, Aba, Onitsha and Enugu. The effect of mission enterprises in the form of schools, churches and roads has been to provide a focus for village life and a seemingly continuous line of villages is found along some of the main roads, as for example from Aba to Port Harcourt.

By far the most important *cash crop* of the Eastern Region is palm-oil production (Fig. 45) over 90% of which is from small African farms, the rest from plantations, particularly those of the United Africa Company. The oil palm will tolerate poor soils, but needs an adequate well distributed rainfall of over 60 inches per annum and high humidity. Oil is obtained from the pericarp of the fruit and from the kernels, the latter being exported and crushed in the importing countries to obtain the oil. More than twice as much palm kernels are exported as palm oil. Over the period from 1953 to 1960 an average of over 400,000 tons of palm kernels was exported compared with less than 200,000 tons of palm oil. Exports of palm products were worth £31·6 million in 1964. Large amounts of palm oil are used locally, and its quality is improving with greater use of oil mills and better methods of refining.

There are a number of different types of land use with farm land of two to three acres in scattered strips occupying most of the area. In addition there are the compound lands or kitchen gardens of about half an acre, scattered oil palms close to the village, plantations of oil palms and reserves of woodland and waste. In the congested areas, such as Aba, the plots are becoming permanent and fenced and the fallow greatly reduced with resultant lower yields. Ibo farming techniques are changing, however; manure is used, cultivation in the kitchen garden is continuous and land may be pledged indefinitely for a loan and rented. The men plant and look after crops and trees needing attention such as yams, pineapples, oil palms and coconut palms, but the women still do the bulk of the farm work. The main food crops are roots such as yams and cassava; the latter have increased in acreage recently with the rise in food prices and because they are easier to cultivate than yams.

The discovery of *petroleum* in 1956 by the Shell BP Petroleum Development Company at Oloibiri in the Niger Delta is having a great effect on the economy of the Eastern Region. The size of the oilfields has been extended by other finds adjacent to Oloibiri. Oil pipelines have been built to Port Harcourt and an oil refinery came into production in 1965. An oil terminal has also been completed at Bonny to take 18,000-ton tankers, and the bar of the Bonny river has been dredged to permit the passage of tankers of this size. In 1964 production was nearly 6 million tons.[20] Natural gas has been found in considerable quantities and piped to a power station at Afam for industrial and domestic use.

The Eastern Region's other main mineral resource is *coal* from the Udi-Enugu coalfield, which has reserves of about 30 million tons and a current production figure of less than $\frac{3}{4}$ million tons per annum. Quality is poor as the coal is sub-bituminous, and production is declining as the railway, until recently the main user, is turning to oil. There are possible future uses for Enugu coal if chemical or other industries are developed nearby.

The main towns of the Eastern Region are Enugu, Port Harcourt, Onitsha, Aba and Calabar. *Enugu*, at the foot of the Udi Enugu scarp,

[20] A major problem is that Nigerian oil is very expensive to produce and the proving of oil resources has been a long and costly process. There are also physical problems of a hot and humid climate, swampy delta, complex geology and the depth of the oil-bearing strata.

owed its origin to coal and has grown rapidly recently (1960 population—63,000), as it has become the capital of the Eastern Region. It has wide tree-lined streets with a gridiron pattern, a commercial core, a separate administrative sector and a College of Technology on the outskirts.

Port Harcourt is the main port of the Eastern Region and second to Lagos in importance. It is situated on the estuary of the Bonny River and is the terminus of the railway to Kaduna. It was originally built for coal exports from Enugu, but now exports tin and large quantities of palm produce. The discovery of oil in the delta has caused the growth of industries and a marked increase in the population which is now over 100,000.

The *Niger delta* is a maze of distributaries, large and small, providing a fan of waterways and creeks along which it is possible to move for considerable distances by launch or dug-out canoe. The banks of the creeks are lined with mangroves whose stilt-like roots seem to squat on the mud. Periodically a higher patch of red or yellow sand above tidal and flood level supports a small fishing village. Fishing is done from dug-out canoes, by using fish traps across the mouths of small creeks or by casting a square net. Historically this area was of importance as a centre for slavery before the Pax Britannica was established and later for anti-slavery patrols by the British Navy. Small states such as Bonny throve on the slave trade and now only rusting cannon, cannon balls and sagging Victorian corrugated iron frame houses serve as a reminder of the past.

A number of small ports have developed, such as Forcados, Burutu, Warri and Sapele, the last the site of a large United Africa Company sawmill. Their use is limited by the sand bars across the mouths of the major Niger distributaries which, unless constantly dredged, only permit the passage of small ships. This muddy world of oily tidal waters and fishing villages has been invaded by oil-drilling equipment and rigs, and a great economic transformation has occurred, particularly in Port Harcourt.

The *Western Region*, with a *population* of 7·1 million within an area of 30,400 square miles, is far less congested than the Eastern Region but nevertheless has nearly four times the population of Dahomey within a smaller area. *Soils* are more fertile than in the East and are derived from the Pre-Cambrian rocks of the ancient dissected plateau forming the core of the Western Region. *Rainfall* is less heavy and continuous and the vegetation consists of a dry deciduous forest much cut into for farming land. In the pre-colonial era many strong and well organized states grew up west of the Niger delta. The largest and most important of the forest states in West Africa was Oyo, possibly based on the more ancient state of Ile Ife. The impact of slavery and the use of firearms caused an increase in warfare culminating in the Yoruba wars and Fulani invasion of the nineteenth century.

The *Yoruba*, numbering 5 million, are the largest ethnic group in the Western Region. A remarkable feature is the high degree of *urbanization* of nearly a half of the population, compared to only 14% and 9% in the Eastern and Northern Regions respectively. Although urban concentrations were found before the nineteenth century, it was the necessity for defence in the wars of the nineteenth century that was a prime factor in the growth of a town such as Ibadan. The tradition of urbanization goes back for a thousand years. Its origins are not clear, but are possibly the product of favourable physical factors and ideas of kingship which filtered through the forest belt from Egypt and the Sudan. The early Yoruba towns grew up round the royal palaces with their courtiers' and craftsmen's houses. The pre-nineteenth century town consisted of an agglomeration of large compounds enclosed by mud walls and each divided into a series of courtyards surrounded by small living-rooms. A group of compounds formed a quarter, these in turn being separated by waste spaces and grouped round the palace and market; the whole was surrounded by a defensive mud wall pierced by gateways.

There are two or three main types of *Yoruba towns*. The first group consists of large towns centrally situated, and forming the capital of a state, with farm lands up to ten miles from the central town and a ring of hundreds of hamlets to accommodate seasonally those who usually reside in the capital. The second group consists of smaller towns with farmlands close at hand beyond which are numerous villages, each with well defined boundaries. A third type of settlement is tending to develop now as people stay for longer periods in the hamlets and on farms. The hamlets have developed as centres for tax collection, as markets, and as road centres with the growth of motorized transport.

The largest Yoruba town is *Ibadan*, a population in 1960 of 600,000 making it the biggest city in West Africa, and indeed in intertropical Africa. It is the capital of the Western Region and a major commercial and industrial city. Its ability to protect a population swollen by refugees from the Fulani conquest meant that by the end of the nineteenth century it already had a population of over 100,000. Further growth occurred after the railway reached it in 1901.

The town has developed round a north-south trending ridge, the highest points being the Aremo and Mapo hills which served as useful defensive positions. The town is best seen from the Bower Memorial Tower on Aremo hill. On festival nights swaying and dancing processions of people and zigzagging cars and lorries weave chaplets of many coloured lights round the central ridge, while in the distance the flickering lights of bar, stall, hut and house indicate the wide extent of the city. By day, as elsewhere in a West African town, the market is the great attraction. Ibadan has three large markets selling a wide variety of goods and attracting hordes of gaily dressed men and women in flowing robes, voluminous dresses, turbans and scarves.

The crafts and industries of Ibadan are a mixture of old and new, of traditional and modern, the latter being now more important. The weavers, dyers and metal workers are to be found along with the tailoring and wood-working establishments, the workshops, repair shops and garages scattered about Ibadan. Small shops, stalls and houses are mixed up together along the streets and there is no regular layout in the older part of Ibadan. The oldest part lies within the city walls and near Mapo hill, where the density of population reaches over 50,000 per square mile. Architectural styles range from the simple compounds with mud walls and corrugated iron roofs to debased Brazilian copied from Lagos, and modern concrete with flat roofs, projecting eaves, verandahs, latticed walls, louvred windows and liberal use of colour.

The centre of Ibadan is expanding rapidly; there are incipient skyscrapers and traffic congestion is already a problem. The commercial core, with a planned layout and rectangular pattern dominated by large buildings housing firms such as John Holts and U.A.C. and with many smaller Indian and Lebanese stores, is very different from the sprawling indigenous town with its hundreds of small shops and stalls lining the streets. Book-shops, pools offices and patent medicine stores representing a great interest in education, wealth and health seem to be the most popular, in that order of precedence!

On the outskirts of Ibadan are various residential suburbs, while the University College, medical school and hospital form a small town in their own right with well laid-out grounds and attractive buildings.

In the smaller Yoruba towns there is the same air of vitality and sense of colour, and there are more large double-storey houses with the Brazilian arch and fewer modern buildings than in Ibadan. The lorry park, market and Oba's palace form the economic and religious centres of the typical Yoruba town, which is at its most colourful at festival time. Oshogbo's climax of the year is the Oshun festival during August which is attended by large crowds watching the drumming orchestras and dancing cult groups.

Surrounding these towns are the *farms* whose land is cultivated under the traditional form of bush fallowing, although the rotation has become shorter with growth of population and growing of cash crops. Most farms are small, from 1–2 acres in size, the Yoruba farmer often living in the town, a third to a half of whose population may be agricultural. Cultivation tends to be more vertical than horizontal, with a diminishing tier of crops ranging from the tall oil palm and other perennial tree crops such as cocoa, kola nuts, rubber, plantains and various fruits to the root crops of yams and cassava and the annual grain crops. The main food crops are yams, cassava and maize, and cash crops are cocoa, kola nuts, yams, cassava and fruit.

The main cash crop is *cocoa*, Nigeria ranking third as a world producer after Ghana and Brazil. Production is almost entirely from the Western Region, particularly in the Oyo and Ondo districts, and is limited in the north by lack of humidity, in the south by poor soils and near Ibadan by diseases such as swollen shoot and black-pod fungus. Over 300,000 farmers plant cocoa and production in 1960 was 154,000 tons, worth £35 million, and in 1965 was worth £43 million.

To the south and east of Yorubaland is a *forested region* with a higher rainfall of over 100 inches and acidic soils not unlike that of the Eastern Region, but with a sparse population forming a frontier zone between the Yoruba and Bini peoples. Palm-oil products, rubber and rice are the main crops. *Rubber* is an important cash crop of fairly recent growth, the main areas being Benin and Sapele. Estates have been recently established and nurseries set up to supply high-yielding clonal seedlings to farmers. Exports in 1960 were valued at over £9 million, of which about 40% was crêpe rubber worked up locally. In 1965 rubber exports had increased in value to £11 million. The *timber industry* is also important, the Western Region being the main timber-exporting region. There are over 10,000 square miles of closed forest which provide large exports of logs and sawn timber each year; obeche is the most popular type of tree and supplies two-thirds of the log exports. At Sapele there is a large factory which produces plywood and veneers, and there are many other sawmills. *Benin*, the main town in this area, was at the height of its importance during the fifteenth and sixteenth centuries when it had a well-developed culture of considerable artistic merit, its bronzes and ivory carvings being well known. It was visited by the Portuguese in the late fifteenth century and in 1500 was five miles in width, surrounded by a deep moat, and contained many buildings and wide streets, the Oba's Palace being an imposing building. It became a victim to its own excesses caused by the slave trade, and a decline set in after the seventeenth century. There is now a population of some 50,000, the town of Benin being the centre of a state comprising about 1 million Edo-speaking people and the nucleus of the *Mid-Western* region formed in 1963, consisting of Benin and Delta provinces west of the Niger. The Mid-Western is the smallest and least populous region in Nigeria.

To the north of Yorubaland the forest stops and the prevailing vegetation is derived savanna. Ilorin marks the southern limit of Fulani influence and conquest, yet the Ilorin Emirate has a Yoruba majority. The Western Region therefore claims that Ilorin should be included within its boundaries and excised from the Northern Region.

South of the Yoruba plateau is the *coastal plain* with a lagoon and sand-bar coast similar to that of Dahomey. About sixty miles east of Lagos the coast curves south to meet the great bulge of the Niger delta. Here palms change to mangrove swamp and sand becomes mud. The sandy beach forms a rectilinear line caused by the eastward drift of material by long-shore drift and is only broken occasionally by a channel. Behind the sand bars are lagoons which provide a sheltered waterway to the Niger delta. Coconut plantations and fishing villages are found on the line of beach and sand bar.

The main interest of this region is the town of *Lagos*, Nigeria's main port and the capital of the Federation (Fig. 46). The island of Lagos was a leading slave centre until the mid-nineteenth century, and the Yoruba wars supplied many thousands of slaves. Because of British anti-slavery activities Lagos was annexed in 1861 and became a colony in 1886. As with Accra and Cotonou, the town grew with the extension of the railway to the north and with the growth of port facilities. The nucleus of the city is the unplanned, densely populated African quarter of Lagos Island. Later expansion was to Ikoyi and via Carter Bridge to Iddo Island and the main-land, with residential areas at Ebute Metta, Surulere and Yaba and south-west to Apapa harbour and the industrial area. Lagos now stretches north-wards almost 12 miles to the airport at Ikeja, where the Western Region plans to have an industrial estate.

The main commercial and administrative area is in the southern part of Lagos Island near the Marina and facing the lagoon. There are great con-trasts of wealth and poverty with skyscrapers and modern office blocks next to single-storey houses, shacks and slums. Along Victoria Avenue which leads to Carter Bridge there are a large number of Indian and Lebanese shops. And in the maze of tiny streets between these two sections of the commercial quarter there are a multitude of petty traders, tiny shops, shacks and houses. The density of population is over 100,000 per square mile here and the twisting streets with open drains pulsate with people, while meat and fish markets are in full swing and smell. The Lagos Execu-tive Development Board is moving people out of this highly congested

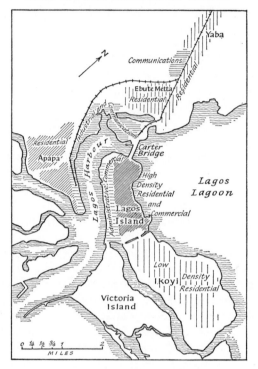

FIG. 46—Lagos

area, and rehousing them in new housing estates on the mainland at Surulere. Redevelopment of the slums is hampered by an agreement whereby slum-dwellers retain their freehold rights to tiny family plots, but cannot pay the cost of the redevelopment.

Ikoyi is a great contrast to the slums of Lagos Island and is a smart residential area housing important government officials, commercial executives and foreign diplomats living in splendour in Italo-American types of houses. Wharves at Apapa harbour first built in 1926 were extended in 1955 and the channel dredged to maintain a depth of 32 feet. Lagos is the most important port in Nigeria and in 1960–1 handled about $1\frac{1}{4}$ million tons. An industrial area which is expanding rapidly is being developed near Apapa. There are a variety of industries—vehicle assembly, light engineering, soap and margarine works, a brewery, agricultural processing factories and factories producing textiles, furniture and other consumer goods.

Lagos, like Accra, Ibadan, Dakar and many other West African cities, has grown rapidly in recent years. In 1921 the population was 100,000 and in 1964 nearly ½ million. The rapid growth of these towns is connected with the quickening pace of commercial and political development and the panoply and prestige of independence has caused a riot of new buildings. Lagos faces planning problems with the great congestion on the island and its lack of access to the mainland (there is only one bridge), with drainage and sanitation due to the high water table, and there is also the political problem of the federal capital extending its boundary into the Western Region.

Despite Nigeria's economic growth—with 1965 exports worth nearly £270 million and over £50 million up on 1964—her political stability was shaken by two military coups in 1966, a change from unitary to federal status and the large-scale exodus of Ibos from the Northern Region because of Hausa-Ibo conflict. In July 1967 civil war broke out in Nigeria when Federal forces invaded the Eastern Region, which had earlier proclaimed itself the Republic of Biafra.

In Ghana Dr Nkrumah was also toppled from power early in 1966 and was replaced by a military regime.

CHAPTER NINE

The Congo Basin

THIS great semicircular region mainly of low altitude consists largely of the drainage basin of the river Congo and its tributaries, but includes a small area comprising the basins of the Sanaga and Ogowé rivers to the east of the Cameroon Highlands and the Lake Chad drainage. The basin-like formation and low altitude are in marked contrast to the three high plateau regions of the Eastern Horn, East and Southern Africa; the Congo Basin is linked more with the lowland zone of West Africa and has a westerly rather than an easterly orientation. It fronts on to the Atlantic, but has a coastline of less than a thousand miles, the Congolese Republic having only a narrow oceanic corridor, while the Central African Republic and Chad are entirely landbound (Fig. 47). It is the most equatorial of Africa's regions, being roughly bisected by the equator, and stretching north to about 9° and south to about 13° of latitude. The region also stretches more than half-way across Africa and is a pivotal area round which the other regions are grouped. Within the region the massive drainage system of the Congo, with its many thousands of miles of navigable waterway, could be a major integrating feature were it not for the interruptions to navigation and the restricted outlet to the sea. The Congo would have been the gateway to a continent but for the major physical barrier of the plateau edge across which it tumbles in a cataract course only a hundred miles from its mouth.

Development has been further hindered by the steamy blanket of rain forest, the continuously hot and humid climate, the miasmic fevers and tropical diseases, and general problem of communications over vast areas away from the rivers. Major developments have occurred on the encircling mineralized rims of the plateau, particularly to the east and south. Here, too, population has tended to concentrate on the healthier uplands rather than in the hot and soggy bowl of the central Congo plain. Political control of the rims has generally passed from the powers controlling the basin; Rwanda and Burundi, formerly linked to the Belgian Congo, are now independent, the Central African and Middle Congo republics are members of the French Community, Angola is now under Portuguese rule and Katanga has striven (if unsuccessfully) to follow a separate course.

Much of the *history* of the region is bound up with the Kingdom of

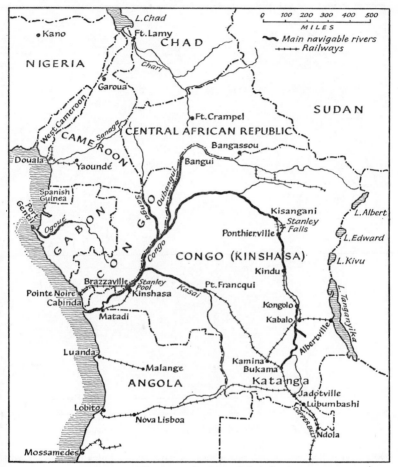

FIG. 47—The Congo Basin: political divisions

Kongo which flourished in pre-European days. There has been long Portuguese contact with the region, and much was learnt about Congo drainage. The contact was, however, generally a bitter one as the region supplied millions of slaves for Brazil and the Americas. European interest in the region was rekindled with Stanley's and de Brazza's explorations towards the end of the nineteenth century. It was Stanley who, on his second journey to Africa in 1874, finally solved the problem of Congo drainage by sailing down the river to Stanley Pool, which bears his name. The Congo Independent State was set up by King Leopold of Belgium in

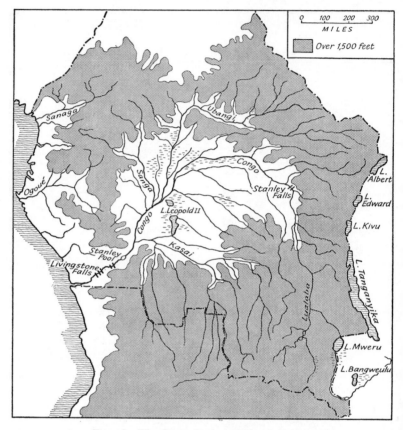

Fig. 48—The Congo Basin: relief and drainage

1885 and it became in essence a vast private estate. Palm oil, rubber and other wild plant resources were collected by enforced labour until 1907 when the Congo Independent State was taken over by the Belgian government and it became the Belgian Congo. The northern rim of the Congo basin became incorporated into the French sphere of influence as French Equatorial Africa, the names of the rival protagonists being retained in the capitals of the two Congo Republics, Leopoldville (now renamed Kinshasa) and Brazzaville, which face each other across Stanley Pool. German intervention in the wedge of country between the Niger and Congo drainage caused most of this area to be proclaimed as the German territory of Kamerun, which subsequently became a mandated territory administered by France and Britain and is now independent.

276

RELIEF AND STRUCTURE

The Congo basin consists of a *great depression* 1,000–1,500 feet in altitude, largely uniform in surface, let down into the African plateau and surrounded by the ancient crystalline rocks that rise to 3,000 to 5,000 feet generally, except in the Western Rift Highlands where individual peaks, as in the Ruwenzori range, may rise above the snowline (Fig. 48). Sedimentary rocks, mainly the soft Lubilash sandstones, similar to the Karoo deposits of South Africa, and alluvium occupy the floor of the basin. The Kundelungu beds, older sedimentaries consisting of conglomerates, limestones and schists of Permian and Triassic age, are found on the fringe of the basin. Surrounding the sedimentary basin are the rims of ancient plateau rocks composed of Archaean and early Palaeozoic rocks. Important economically are the Kambove series of dolomitic rocks which contain the valuable Katanga copper deposits (Fig. 49).

DRAINAGE

The *Congo and its tributaries* form a great fan of waters that converge on Stanley Pool and plunge over the Crystal Mountains in a series of 32 cataracts over a distance of about 280 miles. The Congo is nearly 3,000 miles long and drains a basin of nearly 1½ million square miles. It rises in

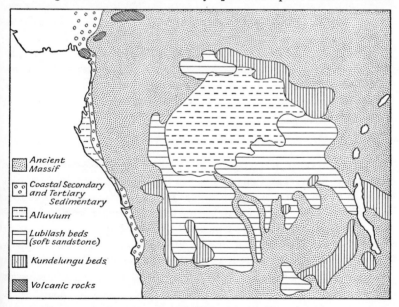

FIG. 49—The Congo Basin: geology

the highlands of Katanga at 11° 45′ S. and flows northwards as the Lualaba to Kisangani (formerly Stanleyville), where there are 7 cataracts and the river falls 200 feet in 60 miles, then curves west and south-west to receive its two great tributaries the Ubangi and Kasai near its Channel course before debouching on Stanley Pool and the plateau rim. Below Matadi the river flows for 100 miles in a wide estuary up to 10 miles wide to enter the Atlantic in a great rush of waters and without the usual delta. Contrary to most African rivers, it has two periods of flood in May and December as it crosses the equator, and also two rainfall maxima. The mean discharge of the Congo is about 2½ million cubic feet a second.

CLIMATE

The climate is largely dominated by its equatorial location and the enclosed nature of the basin. There is heavy rainfall: most of the region has over 60 inches and the West Rift Highlands and part of the central Congo plain has over 100 inches. Temperatures are continuously high with a very slight range and average 77–80° F. (25–7° C.) throughout the year in the basin itself. This is probably Africa's most unhealthy and enervating climatic region as no relief is brought by sea breezes. To north and south the climate becomes more tropical with an increased temperature range and lower, more seasonal rainfall. Temperatures are lower on the high plateau at 10° S., with average annual temperatures below 70° F. (21° C.) in Katanga, and a seasonal range of 15° F. (8° C.). The rainfall is also about 50 inches a year and far more concentrated, with a dry season from May to October. The West Rift Valley highlands have the coolest temperatures and Tshibanda at an altitude of nearly 7,000 feet has mean daily minima of 51° F. (11° C.) compared with mean daily minima of 69° F. (20° C.) Kinshasa (Leopoldville).

VEGETATION

Vegetation varies from equatorial forest to alpine vegetation on the Ruwenzori range. The Congo basin has the largest stretch of equatorial forest in Africa. There is a great variety of trees (which include silk cotton, red cedar and African mahogany), some of which reach a height of nearly 200 feet, (many of these are supported by great buttresses). A mass of lianas clings to the trees, but there is little undergrowth. In the western half of the Congo basin there is a great deal of flooding, and swampy forest occurs with fingers of gallery forest lining the major rivers and forming an impassable green wall. Much of the area is occupied by secondary forest which is denser than primary forest and has numerous small trees and considerable undergrowth. The savanna belt almost encloses the rainforest and occupies the southern half of the Congo basin and the encircling plateau slopes. It is found in areas with a marked dry season and where annual firing of the grasses occurs. Grasses vary in height from 3 to 4 feet, with species such as *Sporobolus* and *Eragrostis*, to elephant grass which reaches a height of 16 feet. Lalang grass 4–7 feet in height is a feature of areas of abandoned cultivation and forms a continuous growth which is difficult to eradicate.

POPULATION

The population numbers only about 24 million and is sparsely settled over a large area of over 2 million square miles, with an average density for the whole region of only 12 per square mile. This ranges from large areas of forest desert, which are almost uninhabited, to parts of the Cameroons and West Rift valley highlands where the density may be over 100 per square mile. The shortage of manpower is serious in Gabon and the Congo (Brazzaville). The majority of the people of the region live in the Congo (Kinshasa) and numbered over 15 million in 1964.

A remarkable feature of the former Belgian Congo has been the rapid spread of urbanization, the rural percentage of the population having dropped from 91% in 1941 to 77% in 1953, nearly 3 million Africans then living in towns and industrial areas. Leopoldville grew from 22,000 people in 1933 to 420,000 in 1961. The population is largely composed of Bantu who form about three-quarters of the total, most of the remainder being made up of Sudanese negroes. About 100,000 nomadic Pygmy hunters have survived in the forests.

ECONOMY

The Congo basin is predominantly *agricultural* in contrast to the pastoral-agricultural economy of East and Southern Africa (Fig. 50). Over large areas shifting subsistence cultivation is the rule with crops such as cassava, rice and maize. A cash economy has spread sporadically in plantation, mining and urban areas. There are extensive oil palm plantations in the western Congo lowlands, cotton and rice are widely cultivated in the northern and southern sectors of the basin, and coffee and tea in the eastern highlands. In the early years of this century wild rubber and palm oil were collected and with ivory formed the main exports. After 1908, however, state and private enterprise in the form of large commercial companies, such as the Union Minière du Haut Katanga and Lever Brothers, exploited mineral resources and established plantations of oil palm and other crops.

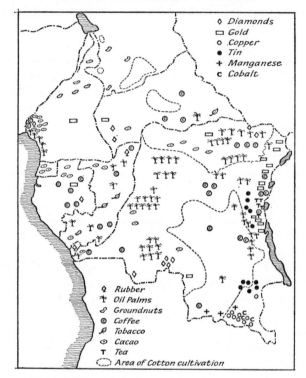

FIG. 50—The Congo Basin: chief minerals and crops

Cooperation between state and private enterprise has been a feature of the economic development of the Belgian Congo. The mineral resources have been systematically exploited for more than fifty years, the state holding the mineral rights, and in an average year exports of copper, diamonds, tin, zinc, gold and other minerals account for two-thirds of the total exports. The tables below show the production of some of the main agricultural products by weight in 1963, and of minerals by weight in 1955 and 1963.

Production 1963		Minerals	1955	1963
(thousand metric tons)			(million tons)	
Palm kernels and		Copper	0·23	0·27
Palm oil	101·6	Zinc		
Rubber	37·6	concentrate	0·11	0·15
Coffee	66·0	Diamonds	12·4	14·5
Cotton seed	26·0		million carats	million carats

In 1961 because of the effects of internal unrest the production of tea, cotton and pyrethrum, and of minerals such as cassiterite, wolfram, diamonds and tin fell markedly. Agricultural exports fell from about 700,000 tons in 1959 to 400,000 tons in 1961, while African export crops declined by two-thirds from 1959 to 1962.

A quarter of the world's potential *hydro-electric power* lies in the Congo basin and a number of schemes have been developed in Katanga, Kivu and near Kinshasa (Leopoldville). A vast project for the future is the Inga scheme on the lower Congo planned to supply power for the processing of aluminium and uranium. Political uncertainty and the development of other major African schemes, such as the Volta River Project, have caused the Inga Scheme to be shelved for the time being. Over most of the Congo Basin there is, however, a noticeable lack of economic development. Exports from former French Equatorial Africa and Cameroon were only 3% of total African exports in 1961.

REGION 1—KATANGA

The Congo basin is divided into a number of minor regions of Katanga, Kivu, the Ubangi-Middle Congo lowlands, the Ogowé-Sanaga lowlands, the northern Congo and the south-western Congo.

The *Katanga* is a distinctive region on the southern uplands which are composed of Archaean and Palaeozoic rocks and act as a major gathering ground for Congo waters. The altitude is over 4,000 feet; the rainfall is seasonal with a dry period from May to October and temperatures are lower and with a greater range than in the central Congo. The vegetation consists of savanna woodland similar to that of Zambia.

The main economic significance of the Katanga lies in its *valuable mineral deposits*, consisting of copper (the most important), zinc, manganese, cobalt, uranium, silver and coal. With the Copper Belt of Zambia the Katanga accounts for nearly one-sixth of the exports from tropical Africa, making this area the most significant economic node north of the Republic of South Africa. Copper comes from the Kambove series and consists of oxide ores of a very high copper content and sulphides which are found at depth in the Kipushi mine. Deposits are mainly superficial and easily mined by opencast means, except for the deep mine at Kipushi. The copper concentrate from Kipushi is smelted and then treated in converters to produce blister copper before export to Belgium. As in Zambia there are ancient copper workings scattered over the area and these encouraged the early exploitation of copper by the Belgians. Mining began in 1912, twenty years before the opening up of the Zambia Copper Belt.

Initially the great problems were isolation, lack of labour supply and lack of communications, as the region is in the heart of Africa nearly 1,500 miles from the sea. Transport problems were solved with the development of the Congo's road and rail network (Fig. 51) and by the mid-1930s there were rail outlets via Port Francqui on the river Kasai, Lobito Bay in Angola and via the Rhodesian railway to Beira. Mining is controlled by the great Union Minière du Haut Katanga (U.M.H.K.). Some 30,000 Africans are employed on the Katanga mines and, in contrast to the migrant labour system with hostels for men in South Africa, a permanent labour force with comfortable family houses has been developed in the Katanga, thus avoiding some of the social evils of the migrant labour system. In addition, the U.M.H.K. controls metallurgical plants and workshops, produces explosives and generates hydro-electric power.

Mining started near Elisabethville (now renamed Lubumbashi), but has spread outwards now, the copper mining area stretching approximately north-west to south-east for a distance of nearly 200 miles. The mines are divided into Northern, Central and Southern: the Ruwe, Musonoi, Kamoto and Kolwezi mines in the north, the Kambove mine in the centre and the Kipushi and Ruashi mines near Lubumbashi in the south. The Katanga copper-mining area is, therefore, not as concentrated as that of the copper belt of Zambia. Other minerals mined within this area are cobalt from the Star of the Congo mine near Lubumbashi; cobalt, manganese and zinc from Kambove; uranium from Chinkolobwe,[1] and coal from Luena near Bukama. Tin is also mined near Manono about 150 miles north-east of Bukama. Coal resources of low grade are very small compared with those of Rhodesia and South Africa, and production in 1963 was only about 70,000 tons. The Katanga is fortunate in possessing adequate hydro-electricity from power stations on the Lualaba and Lufira rivers. Power is even supplied from Le Marinel power station on the Lualaba to Kitwe in Zambia. Copper is smelted at Jadotville and Lubumbashi.

The Katanga has also developed its *agriculture*. Ranching and dairying are carried on, stimulated by the urban and mining markets, while vegetables and fruit are grown near Lubumbashi and the main towns. The tsetse-free uplands make cattle raising possible and the Katanga contains the bulk of the Republic's 800,000 head of cattle.

Lubumbashi (formerly Elisabethville) with a population of over 180,000, including some 14,000 Europeans, is Katanga's chief town and was for a

[1] Until 1961, when mining of uranium ceased.

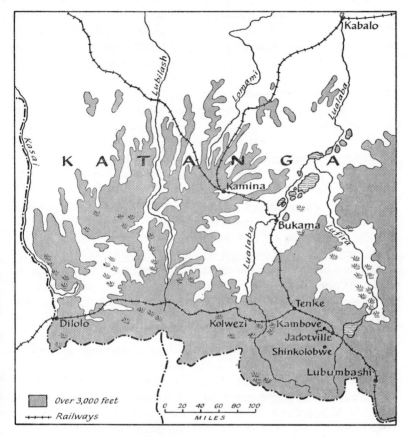

FIG. 51—Katanga: relief and communications

while the capital of President Tshombe's separatist state of Katanga. It was founded in 1910 and named after the Belgian queen. It is considerably larger in size than the Zambian towns, but like them is spaciously planned with many wide tree-lined streets and imposing buildings. By rail it is little more than 100 miles from Ndola in Zambia, but over 1,300 miles from Lobito Bay and over 1,600 miles from Beira.

REGION 2—KIVU

The second region of *Kivu* is north of the Katanga and comprises the lower eastern slopes of the plateau to an altitude of 5,000–6,000 feet. It stretches 800 miles northwards to the Uele river and includes the provinces of Kivu and the higher eastern part of the Eastern Province of the Congolese Republic. Although the faulted blocks, volcanoes and rift valley lakes are linked physically, and to some extent economically, more with East Africa than with the Congo basin, they will be referred to here where necessary.

Most of the rocks are Archaean, but in the south there are exposures of sandstones, shales and limestones of the Lubilash and Kundelungu beds. As the region lies athwart the equator it experiences the climatic variations common to the whole of the Congo. Vegetation too varies greatly from the gallery forest of the Congo tributaries to savanna and a kind of steppe vegetation in the higher areas with grasses, aloes and euphorbias, and above 10,000–12,000 feet a sub-Alpine vegetation of tree groundsels and giant lobelias.

Population is concentrated more on the high plateau than in the lower reaches of the Congo basin and in parts of Kivu is over 100 per square mile. The high grasslands round Lake Kivu are comparable with the Kenya Highlands in their suitability for European settlement. Many Europeans settled here in the post-war decade and grew crops such as coffee and tea. *Bukavu* (Costermansville) is the capital of Kivu Province and has grown up on the western shores of Lake Kivu at an altitude of nearly 5,000 feet. With a European population of 4,327 in 1959, it has become an important health resort.[2]

Agriculture varies greatly according to altitude. Arabica coffee is grown at altitudes above 3,000 feet and comes mainly from Kivu Province. Tea cultivation started in 1946 on a small scale to supply the local market. A certain amount of pyrethrum is grown in the higher parts of Kivu as in the Kenya highlands. Cattle are also concentrated on the high tsetse-free grasslands of Kivu, the second main cattle-rearing area in the Congolese Republic.

Mineral production is of some importance along the plateau rim. Some 40,000 Africans are employed in the mining industry in Kivu. Tin is mined over a large area from Katanga to Kivu; production in 1963 was 16,500 tons of tin ore and 1,400 tons of refined tin, and was much lower than in 1959 and 1960 because of local troubles. Both reef and alluvial gold are mined in the scattered workings of the Kilo-Moto area which supplies two-thirds of Congolese output, but which represents only about 1% of world production.

Transport links are to both east and west, a considerable proportion of the traffic before 1960 coming via the Kasese terminus of the Kenya-Uganda railway, and by road to Kivu and Ruanda-Urundi. Lake Tanganyika offers north-south links, but despite its length of over 400 miles it is of

[2] The European population has declined greatly and Bukavu's role as a health resort has dwindled since Congolese independence.

little significance in the communications system of either East Africa or the Congo basin. However, with the completion of the Kamina-Kabalo railway it is now possible to go by rail form the Cape to Albertville and after a short journey on Lake Tanganyika to resume the journey across Africa by rail to Dar es Salaam.

REGION 3—UBANGI-MIDDLE CONGO LOWLANDS

The third region consists of the Ubangi-Middle Congo lowlands and stretches north of the equator towards the Sudan belt and the Nile water-shed. The northern part of the region consists of ancient plateau rocks, much of it being above 2,000 feet in altitude. The area between the Middle Congo and Ubangi is below 1,500 feet in altitude and consists largely of alluvium and Kundelungu sendimentaries. Away from the equatorial régime of the Central Congo with its double maximum of rainfall, the rainfall becomes increasingly seasonal and limited to the period from April/May to October, with a 5–6 month long dry season on the northern fringe of the region. Vegetation in the southern third of the region consists largely of tropical rain-forest grading off to Guinea and Sudan savanna in the north.

The boundary of French Equatorial Africa and the Sudan was only fixed in 1924 and the area divided administratively into the territories of Gabun, Middle Congo, Ubangi-Shari and Chad only in 1946. They became independent in 1960 as members of the French Community and are now renamed the Gabon Republic, Republic of the Congo, the Central African Republic and the Chad Republic respectively.

THE CENTRAL AFRICAN REPUBLIC

With a population of little more than 1 million, including about 6,000 Europeans, in an area of about 240,000 square miles, the Central African Republic has an average density of only 5 per square mile. Reasons for this low density, which is characteristic of much of the northern half of the Congo basin, are the extremely unhealthy nature of the hot lowlands, the high mortality rate with a life expectancy of only 35 years, the isolation and the former ravages of the slave trade.

As the region possesses little in the way of mineral resources *agriculture* is of paramount importance, with cultivation of crops in the southern areas of high rainfall and pastoralism to a minor extent in the few tsetse-free areas of the north. Cotton is the chief crop with nearly ½ million acres under cultivation, an acreage nearly as much as that of the Congolese Republic, but with a production of only about 30,000 tons of cotton seed and lint in 1963 representing about half the value of exports. However, the high costs of production and transport make it uneconomic to sell on the world market, and it can only be sold at artificially high prices in France. The area of cotton cultivation stretches south of the Ubangi river to about the line of the middle Congo. The Eastern Province is the leading producer in the Congolese Republic with a production of about 50,000 tons of cotton, and higher yields than in the Central African Republic. Cotton is also produced in the Southern Congo, and by 1960 nearly ¾ million farmers were growing cotton. Total production was, however, only half that of the Sudan. Some 80% of the cotton is exported, the rest being used by local spinning factories. Exports fell markedly in 1961 and 1962. Robusta coffee is produced in areas of less than 2,000 feet in height, total production from the Central African Republic being over 6,000 tons in 1962 from an area of about 72,000 acres. In the savanna areas of the north groundnuts are the main crop, production, however, being very small compared with the great groundnut areas of West Africa. Sisal is grown near Kovango in the east, but its price is not competitive on world markets. Cattle still number less than ½ million, but the savanna regions could be developed more intensively for cattle rearing.

The production of alluvial diamonds has declined greatly since 1955 and in 1962 about 175,000 carats were produced, representing less than 2% of the Congolese Republic's production.

The middle *Congo and its main tributaries*, the Ubangi and the Tshuapa, are of great importance and they provide some 2,000 miles of *navigable waterway*. Barges of nearly a thousand tons in size ply between Kinshasa (Leopoldville) and Kisangani (Stanleyville) on a stretch of over 1,000 miles of the middle Congo. Above Kisangani rapids interrupt navigation at two points: between Kisangani and Ponthierville and Kindu and Kongolo; the gaps are bridged by railway. A short stretch of navigable waterway on the Itimbiri river leads to Aketi from where a railway goes nearly 400 miles eastward into the coffee, cotton and gold-producing areas of the Eastern Province.

The Central African Republic is handicapped by its interior location and lack of seaboard. Outlets are either by water from Bangui down the navigable Ubangi to Brazzaville and thence by rail to Pointe Noire or by road through the Cameroon Republic to Douala. A railway linking Yaoundé and Bangui is projected, providing a shorter link without break of bulk.

The main towns in the region are *Kisangani* and *Bangui*. The latter has a population of about 100,000, is the head of navigation on the Ubangi river and the capital of the Central African Republic. Kisangani is an important river port and provincial capital.

REGION 4—OGOWE-SANAGA LOWLANDS

West of the Ubangi and south of the Bamenda highlands is a region almost circular in shape fronting on to the Atlantic. Excluding the northern savanna belt of the Cameroon Republic, which is considered separately with Chad, the area of the fourth region of the Ogowé-Sanaga lowlands is about 350,000 square miles and the population was estimated at just over 3 million in the late 1950s. The region is divided politically between the republics of Cameroon, Gabon and the Congo, and the small enclave of Rio Muni. Control of the region was divided largely between France and Germany until the First World War. French interest in the region goes back to the first half of the nineteenth century and Germany acquired control over part of the Cameroons at the end of the nineteenth century in the face of British interest in the area. Spanish Guinea, which includes the islands of Fernando Po, Annobon, the Corisco Islands and Rio Muni, remains as a survival of early European slaving interest in the region. With the loss of German control during the First World War most of the Cameroons became a French Mandated, later Trust, Territory which achieved independence in 1960.[3] Two small strips of territory along the Nigerian border, the Northern and Southern Cameroons, were made British Mandates, the northern strip now being incorporated into northern Nigeria and the southern strip into the Cameroon Republic as a result of plebiscites. Gabon and the Congo Republic also achieved their independence in 1960.

The region consists of the drainage basins of the Sanaga, Ogowé and Sanaga rivers, the last a tributary of the river Congo. The Guinea ridge, part of the African plateau, is the major watershed. The coastal lowlands are narrow, varying in width from twenty to over a hundred miles wide, and are composed of *sedimentary rocks* of Cretaceous to Recent age. These lowlands have been built up by rivers such as the Sanaga and Ogowé, the latter having constructed a delta 50 miles wide. There is a steep rise to the plateau from the lowlands, the dissected edge reaching a height of over

---

[3] 100,000 square miles ceded by France to Germany in 1911 were taken by France and did not form part of the Mandate.

3,000 feet in the Massif du Chaillu. North of the Sanaga river the Archaean platform is widely exposed with many large intrusions of igneous rocks. South of the Sanaga the geology is more complex and the relief rugged because of the high rainfall. Away from its upturned and dissected edge the plateau forms a horseshoe of highland, the Guinea ridge, most of which is above 2,000 feet in altitude and carved by deep river valleys. The highest part, the Bamenda Highlands from 3,000 to 4,500 feet with small areas above 7,000 feet, forms part of the boundary between the Congo basin and West Africa.

Most of the region has an equatorial type of *climate* without a well defined dry season and consistently high temperature and humidity. Daily mean maximum temperatures vary from 80° to 90° F. (27°–32° C.) and mean minima from 65° to 75° F. (18°–24° C.), the monthly range being about 6° F. (3° C.). Rainfall varies from about 60 to over 100 inches per annum and there are two maxima from March to June and October to December. Along the coast south of Cape Lopez the effect of the cool Benguela current is felt, causing a diminution of both temperature and rainfall. Altitude also causes a decrease of temperature in the Bamenda highlands.

Vegetation consists mostly of tropical rain-forest which is particularly well developed in Gabon. There are patches of savanna in the south and along the coast. The forest resources of the region are of great economic importance.

Population is scanty with a density of less than 5 per square mile over large areas. The forest areas are almost uninhabited, the most densely populated areas being in the coastal towns and on the high plateaux. Factors to account for the low density of population are similar to the Congo basin, namely the unhealthy climate, high incidence of disease, dense vegetation and ravages of slave raiding. Ethnic diversity is considerable as here Bantu meets Negro Africa. The bulk of the population is Bantu and there are many small tribes, Gabon alone having 40 different tribes. There are small numbers of Pygmies in the forest and to the north are Sudanese negroes.

The region has been further subdivided into three minor regions which correspond approximately to the political units of Gabon, Congo and Cameroon Republics.

GABON

Gabon, consisting largely of the drainage basin of the Ogowe river, is the smallest with an area of 101,000 square miles and estimated population of only 460,000. It is a prosperous country with an expanding economy based

IX. Yam mounds in the Middle Belt of Ghana.

X. The effects of swollen shoot on a cacao tree, Ghana.

XI. Accra, Ghana.

XII. Akosombo Dam, Volta River Scheme, Ghana.

on forestry and mining. *Forestry* products are responsible for over half the exports. A softwood, okoumé, used for veneer and plywood is the main tree exploited and production has quadrupled to over 600,000 tons in ten years. Attempts are being made to develop plantations of okoumé which would greatly increase the yield. There is a large plywood plant at Port Gentil with an output valued at nearly £3 million in 1961.

In future *mineral* exports should increase considerably. Petroleum from the Port Gentil peninsula south of Cape Lopez has been produced since 1957, production in 1963 being nearly 1 million tons. There are great reserves of manganese and iron ore. Manganese deposits at Moanda are estimated at 200 million tons, one of the largest deposits in the world, while reserves of iron ore near Mékambo may amount to 1 billion tons, much of it with over 60% iron content. There are, however, considerable transport difficulties to overcome before these deposits can be exploited and over 600 miles of railway will have to be built over difficult terrain. Gabon also has a valuable potential source of energy in her numerous rivers and rapids from which hydro-electric power can be generated. As yet, however, there is little in the way of industrial development.

Over half the population of Gabon are engaged in agriculture which is still mainly backward and subsistent with crops such as bananas, manioc, maize, taro and yarns. Attempts to increase the amount of cash crop farming have met with some success and exports of cocoa and coffee have increased greatly since 1940. Gabon has considerable natural resources, but her economic growth is handicapped by the tiny size of the population, lack of manpower, high infantile mortality, insidious effects of disease, wide ethnic differences, the rugged nature of the country and dense vegetation.

THE CONGO (BRAZZAVILLE)

The Congo (Brazzaville) lies to the south of Gabon and north of the river Congo which forms the main drainage. It has an area of 180,000 square miles, almost exactly the same as the Cameroon Republic, but an estimated population of less than 1 million, about a quarter that of the Cameroon Republic. The Kongo are the largest ethnic group forming nearly half the population. They were of historical importance as they built up the Kingdom of Kongo.

The economy is primarily *agricultural*, although only 20% of the exports are agricultural products. The main cash crops are palm kernels, with exports of 9,700 tons in 1963, and palm oil, with 1962 exports of 6,500 tons. The main producing area is in the south. Groundnuts are also produced in the south, other cash crops on a minor scale being rubber, tobacco, cacao and coffee. The main subsistence crops are cassava, plantains, sweet potatoes, yams and rice. A major project is the Niari Valley Industrial and Agricultural Corporation which operates four plantations employing 10,000 Africans. The Niari Valley has considerable physical advantages with two rainy seasons, a dry spell from May to October, and deep soils. It is also situated along the line of the Brazzaville-Pointe Noire railway. Local crops such as sugar-cane and oleaginous products are processed, the Corporation operating a sugar refinery and an oil mill. There are a variety of holdings, ranging from European estates of 2,000 acres to the paysannat schemes in which African farmers are encouraged to grow oil palms and coffee. The main crops grown are groundnuts, sweet potatoes, urena, manioc, maize, coffee, and green crops.

Forestry, as in Gabon, provides the country's major source of income, and timber exports, consisting mostly of logs, account for over half the total exports. Only a small proportion of the timber is processed locally and made into veneer and plywood. The main commercial varieties of trees are limba, mahogany and okoumé. *Mining* is of less importance than in Gabon and lead was the only mineral exploited on a considerable scale, but the M'Passa mines in the Niari Valley produced only 330 tons in 1963. Petroleum has been exploited at Pointe-Indienne near Pointe Noire since 1960. As in Gabon there are great potential sources of hydro-electric power particularly along the Kouilou river.

The Congo (Brazzaville) stretching north-east for 700 miles fulfils an essential transit role between the Atlantic coast and the landlocked republics of Chad and Central Africa, and has the best developed system of *communications* in former French Equatorial Africa. There is a 320-mile long railway from Pointe Noire to Brazzaville which was completed in 1934. Freight traffic has doubled since 1948. The artificial harbour at Pointe Noire, the main port for former French Equatorial Africa, has been completed and is being equipped to handle exports of manganese from Gabon. Its traffic has grown rapidly since 1950 and in 1960 the port handled nearly 800,000 tons of cargo. The Ubangi river is navigable for nearly 400 miles and, combined with the Congo, provides an all-water route of over 700 miles from Brazzaville to Bangui.

Brazzaville is the capital of the Congo (Brazzaville) and faces Kinshasa, the capital of the Congolese Republic, across Stanley Pool. With a total population of over 100,000, including over 5,000 Europeans, Brazzaville has a French atmosphere with its sidewalk cafés, restaurants and hotels. It has an international airport of growing importance used by over 2,000 planes a year and situated almost midway between the major regions of

East, West and Southern Africa. *Pointe Noire*, with a population of over 50,000, is the only port and the terminus for the railway to Brazzaville. It handles a growing amount of trade, much of it transit to the interior. It also has a small industrial sector with woodworking, light metal, ship-building and repair industries.

SOUTHERN CAMEROON

The third sub-region of the southern half of the Cameroon Republic has an area of about 80,000 square miles and population of about 2 million (Fig. 52). The total population of the republic is now nearly 5 million, which is by far the largest of the four republics north of the river Congo. Population in the north has been much influenced by Fulani incursions and Islam. Among a number of tribes in the south the Fanti are well known ethnologically and occupy a large area astride the Cameroon Gabon border. Hunters and shifting cultivators, they move their villages every five to eight years. The villages are of interest as they are built in the form of a street with a terrace of huts on each side and a blockhouse at each end.

The Cameroon-Bamenda highland, consisting of volcanoes such as Mount Cameroon rising from crystalline and volcanic plateaux and much of it over 4,000 feet in height, with fertile volcanic soils, proved attractive for German settlement before 1914 and nearly 300,000 acres were alienated. There is a comparatively dense African population, the Bamiléké being the major tribe with a density of over 700 per square mile in places; consequently, there is much overgrazing and soil erosion. The Germans initiated the *plantation* production of cocoa, coffee and bananas and studied scientifically the problems of tropical agriculture. This area produces 80% of Cameroon's coffee. West Cameroon, the former British Cameroons, has a population of nearly 1 million and an area of about 16,800 square miles. The Cameroons Development Corporation is the main commercial organization and basis of the economy, managing nearly 20 plantations and producing 33,000 tons of bananas, nearly 900 tons of palm oil and palm kernels, and 4,000 tons of rubber in 1962. Palm-oil products, timber and rubber are produced in the coastal belt. Cocoa, with a production of 85,000 tons in 1963, is the main export, accounting for over one-third of the total and followed by coffee with a 1963 production of over 50,000 tons. Aluminium smelted at Edea on the Sanaga river accounts for about one-sixth of the exports.

The granitic Adamawa massif, 2,500–4,500 feet in height, extends in an east-west line across Cameroon, forming a marked barrier between north

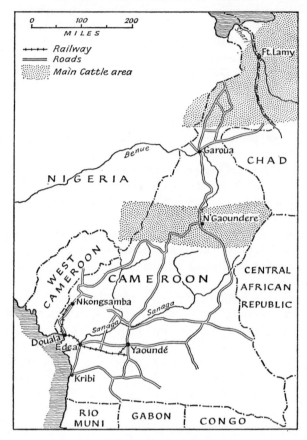

FIG. 52—Cameroon

and south. There is only a small population, with densities of less than 10 per square mile. This is a savanna region and cattle rearing is the main economic activity. There are over ½ million cattle and some are sold in the south. The area has considerable economic potential as there is no over-stocking, or rinderpest and foot-and-mouth diseases.

Douala, with a population of 120,000, is the largest town and main port. It is situated on the left bank of the navigable Vouri river and, with quays for ocean-going ships, handles most of the country's trade. It is the terminus of the main railway to Yaoundé and Mbalmayo on the Nyong river. A secondary railway goes from the small port of Bonaberi opposite Douala to Nkongsamba and the coffee plantations of the Cameroon Highlands.

Yaoundé with a population of 60,000 is the capital and an important communications centre with a railway to Douala and road links to Bangui and Fort Lami in Chad Republic.

REGION 5—NORTHERN CONGO

The fifth region comprises the territory of Chad and the northern part of the Cameroon Republic.

CHAD

This republic is landlocked and inaccessible, lying north of the Central African Republic and a thousand miles from the sea. It was part of former French Equatorial Africa, and achieved its independence in August 1960; it now forms part of the French Community. It has an area of nearly $\frac{1}{2}$ million square miles and a population estimated in 1962 at only $2\frac{3}{4}$ million, the bulk of which lives in the better watered area south of Lake Chad. About half the territory is desert and has a population of about 50,000 nomadic pastoralists.

The republic extends from about 8°–23° N. and thus varies considerably in climate and vegetation. North of the Ubangi-Shari plateau there is a very large depression drained by the Shari and Logone rivers flowing into Lake Chad. Lake Chad is situated at an altitude of 800 feet and is only a slight remnant of its size in Quaternary times. The area of the lake varies greatly, in times of flood covering 10,000 square miles. The western shore of the lake is very flat and a rise of a foot in water level will increase the area of the lake by several hundred square miles. The lake can be divided into two main areas: the northern area, which is fed by only one major river, the Komadougou from Northern Nigeria, is far shallower and dries out more than the southern half fed by three major rivers, the Yadseram, Logone and Shari. In July all but the eastern half of the lake dries out. The south-eastern half has far more permanent water, with a depth of 5–12 feet near the delta of the Shari river. Here there are numerous islands, many of which are inhabited. The Shari brings down floodwaters in December and January, causing the lake to spread out over a wide area; it retreats again in May or July.

Over half the country has less than 10 inches of rainfall a year and is, therefore, only suited to extensive pastoralism. South of about 14° N.

rainfall increases to about 25 inches and near the southern border to about 40 inches.

Because of its geographical position Chad has been a meeting place for the nomadic, Islamic peoples of the north and the sedentary Bantu population of the south. It was the focus of states such as Bagirmi and Wadai and came under French control at the end of the nineteenth century.

In the Shari-Logone basin in the south-west, cotton is the main export crop, exports in 1959 being worth over £4 million and accounting for 80% of the total exports.[4] It has been grown as a cash crop by dry farming methods since 1929 and yields have risen considerably since 1946, the area under cultivation now being 700,000 acres. Groundnuts are grown on a considerable scale in rotation with cotton. Rice is grown on about 50,000 acres in river valleys such as the Logone. The main food crops are millet and sorghum.

Pastoralism is important in the arid north and Chad is estimated to have about 4 million cattle and 4 million sheep and goats. Sheep are reared largely for the local market, about 1 million being slaughtered each year. Nearly 150,000 cattle, about 130,000 sheep, and hides and skins are exported each year, mostly to Nigeria. Attempts to improve the livestock industry include the export of meat by air, the provision of refrigerated abattoirs at Fort Lamy and Fort Archambault, and the digging of wells. Fishing is also important, and over 70,000 tons of fish are obtained each year from the Logone-Shari-Lake Chad basin. Large quantities of smoked fish are exported to Nigeria and dried fish are sent to Southern Chad from the Lower Shari.

Communications are vital to such a large and isolated region and have been developed greatly since the war. Air links are important and other links are by road, rail and river to ports such as Pointe Noire, Douala, Port Harcourt and Lagos. There are no railways, and many of the roads are impassable for much of the year owing to the clay surface of the Chad basin and the floods from January to July. Rivers are navigable for part of the year, but the Shari from Fort Lamy to Chad is navigable the whole year and is used for the export of dried fish and natron to the south. Another important water route is from Garoua via the Benue and Niger to Burutu, and although only possible for three months of the year it is used for the transport of cotton, salt and other goods.

Aridity and isolation are major problems that confront the new republic of Chad.

The remote region of *North Cameroon*, which has a high density of

[4] Production in 1963 was 93,000 tons of seed cotton.

population, consists of a long narrow strip of territory stretching from the Adamawa Massif to Lake Chad. The physical background is similar to neighbouring parts of Chad. Cotton and groundnuts are the main cash crops and rice and millet the main food crops. There are also nomadic pastoralists with considerable numbers of livestock. Garoua on the upper Benue is used as a port from July to September for the import of petroleum and other products through Nigeria and the export of cotton, groundnuts, hides and skins.

REGION 6–SOUTH–WESTERN CONGO

The last region in the Congo basin is a square of territory bounded to the west by the Cuango river and the Crystal Mountains, to the east by the main stream of the Congo and south by the Congo-Zambezi divide, and includes the north-eastern plateau slopes of Angola. It stretches from the equator to approximately 10° S. The region therefore descends in broad terraces from the rim of the encircling plateau to the great curve of the forest-clad basin of the Congo, much of it below 1,500 feet in height, and crossed by the wide meandering arcs of major tributaries such as the Kasai, Sankuru and Tshuapa. The central plain is covered by dense rain-forest clinging thickly to the banks of the Congo, its tributaries and the thousands of islands. Great areas are water-covered or marshy and there are few villages in this morass-like area.

Relief and drainage

Ascending from the central plain on to the first terrace and from the alluvium on to the Lubilash sandstone, the forest thins out and retreats to a thin line of gallery forest along the main streams, like fingers pushing into the savanna glove. The basin was formerly the bed of an ancient inland sea whose remnants are to be seen now in Lakes Leopold II and Jumba and the marshy land at the junction of the Congo and its tributaries. Relief is uniform and rivers flow in wide meanders entrenched in the alluvium. The Congo is over 5 miles wide in places, but narrows where a major tributary, such as the Ubangi, joins the main river. The Congo and its great tributaries finally bunch together near Stanley Pool. The river plunges through the western barrier of the plateau, known here as the Crystal Mountains, in a deeply entrenched cataract course. Between Stanley Pool and Matadi there are 32 cataracts in a drop of nearly 1,000 feet. From Matadi to the sea the wide estuary of the river is again navigable for nearly 100 miles. Behind Stanley Pool there are nearly 9,000 miles of navigable waterway.

Climate

The climate of the Congo basin is mostly equatorial with constant high temperatures and humidity and abundant well distributed rainfall. Much of the area has over 80 inches of rainfall with the characteristic double maximum. The atmosphere is steamy with rising mists, stagnant water, sodden ground, dripping leaves and great evaporation, and seems almost to generate its own cycle of rainfall. Kinshasa (Leopoldville) has an average relative humidity of over 90% at 7 a.m. The climate is even more uncomfortable than that of West Africa. Where the plateau slopes rise out of the plain ameliorating conditions set in, temperatures become cooler, rainfall decreases and the climate becomes more seasonal.

Population and agriculture

The great majority of the population are Bantu, but there are small nomadic groups of Pygmy hunters in the rain-forest. Most of the forest Bantu live in clearings along the banks of the rivers which act as highways for trade and travel and provide fish. Shifting cultivation is ubiquitous, and the ground is abandoned after three to four years. Cassava is the main crop; sweet potatoes, yams and plantains are also grown. Forest products are collected particularly palm oil, and fishing and hunting are important subsidiary occupations. Tribes such as the Bateka on the Congo and the Balunda on the Kasai are great fishermen and trade smoked fish for cassava and forest products. Long villages line the banks forming a kind of riverine ribbon development. Huts in the west and centre are rectangular (the influence of the west coast), those in the north-east are round (the influence of the east coast).

Oil palms are the most important cash crop: the main areas are in the western half of the basin between the lower Kasai and the Kwango. Extensive plantations were established by Huilerie Belge, a subsidiary of Unilever, which in 1961 had 150,000 acres of oil-palm plantations. Exports increased from 90,000 tons of palm oil in 1938 to about 220,000 tons in 1963. Palm oil is produced mainly on plantations and palm kernels on African plots, which may be up to 100 acres in size. The proportion of palm oil has increased greatly. Quality is high compared to other parts of Africa, partly because production is from cultivated and not wild trees. Most palm kernels are pressed in the Congo for the extraction of the white palm-kernel oil, in contrast to West Africa which sends its palm kernels to Europe for the extraction process. Palm-oil exports are handled in bulk shipments. Although a decline set in after independence, the Congolese Republic normally supplies about one-fifth of world palm oil and one-sixth of palm-kernel production. Cotton is another cash crop of increasing importance in Kasai Province and is widely grown by Africans.

The former Belgian administration carried out an energetic programme of *agrictultural improvement*. INEAC (*Institut National pour l'Etude Agronomique du Congo Belge*) has developed higher yielding tree and other crops and demonstrated to Africans the value of crop rotations and compost. A particular feature in the former Belgian Congo has been the growth of the corridor system of agriculture whereby fields are systematically cultivated, then allowed to revert to bush on a twenty-year cycle. This attempt to intensify the prevailing methods of shifting agriculture has been partly successful. The Belgian paysannats scheme has been carried out on a large scale and farmers have been carefully instructed in the type of crop to plant and the kind of rotation to adopt. By 1958 nearly $\frac{1}{4}$ million people had been settled in various areas, which total over 2 million acres. Crops such as groundnuts, rice, urena and coffee are produced and the cash income very greatly increased. However, since 1960 the schemes have run down and their continuance is doubtful.

Mining

The Congolese Republic is the world's largest producer of *industrial diamonds;* the diamond fields in the Kasai Province are located in two areas, round Tshikapa on the river Kasai and Bakwanga on the Bushimae. The diamonds, from a source in the pre-Cambrian basement complex, are found partly in Triassic conglomerates, partly in creek and terrace gravels. They were first discovered in 1909 and in the early 1950s production was about 12 million carats, which by 1961 rose to 18 million carats. Operations at Tshikapa virtually ceased in 1962. 95% of production is industrial diamonds, nearly all from Bakwanga, only 5% being gem stones.

Communications

The communications pattern of the Congolese Republic has been conditioned by the extensive system of the river Congo. Roads and railways have often been built to supplement the waterways. The Congolese Republic has over 9,000 miles of *navigable waterway*, the longest stretch being on the middle Congo for over 1,000 miles from Kinshasa (Leopoldville) to Kisangani (Stanleyville). During the period 1940–60 waterways carried nearly half the total traffic. The Kasai is navigable to Port Francqui, from where a railway leads to Katanga. Tugs pull long strings of barges and river steamers stop periodically at small ports where clearings have been made in the dense forest. Steamers drawing a maximum of about six feet can ply along the main river, which is hampered by shifting sand bars and numerous islands. The spread of water hyacinth has been a problem on some of the tributaries, but its growth has been controlled on the main river by chemical spraying.

Urban settlement

Kinshasa (Leopoldville) is the capital and is a booming African city, the largest in the republic, with a rapidly expanding population which has grown from 41,500 in 1939 to over 400,000 (Plate XIII). There are a number of clearly differentiated sections. One part contains the main river port, from which a railway leads to *Matadi*, by-passing the cataract section of the Congo (prior to July 1966 this section bore the name Kinshasa, now extended to cover the whole city). The main commercial section overlooks Stanley Pool and faces Brazzaville only a mile away. The European residential area stretches for miles along the banks of the river. On the southern outskirts of the town are the African suburbs with neat white houses and great numbers of wattle and daub huts fanning out across the savanna. The city has important administrative functions, catering for a growing civil service. It is also an industrial centre with a number of food-processing and consumer industries, including shipbuilding, and the manufacture of textiles and chemicals.

Economic future

The future of much of the Congo basin, particularly that of the Congolese Republic, is uncertain. The precipitate granting of independence and with-drawal of the Belgian administration left such a state of confusion that United Nations aid had to be sought on a massive scale.[5] Tribalism and regional separation movements threatened to break the unity of the state. However, Congolese leaders such as Tshombe realized the urgent need for foreign advisers and financial aid on a large scale. The region has great potential with the mineralized rims of the south and east added to the tree crops of the central basin, and economic development was carefully organized by the Belgians with a combination of state and great commercial concerns, although political growth was hampered. The threads of this great web of Belgian economic activity and administrative control were

[5] The U.N. force had nearly 19,000 men at the beginning of 1963.

broken in 1960 and have not been reknit as yet. Outlying regions such as the Katanga and Kivu are 1,000 miles from Kinshasa (Leopoldville) and even loose links are difficult to organize. The Katanga has tried to pursue an independent existence, but has now been brought back into the Congolese Republic.[6] Added to the problems of distance and ethnic diversity are the environmental problems of an unhealthy climate and large expanses of marsh and forest in the hot steamy depression of the central plain.

[6] The former President of the Katanga became the Prime Minister of the Congolese Republic, a striking change in roles. Later Tshombe was forced to flee the country and in July 1967 was arrested in Algeria.

CHAPTER TEN

The Eastern Horn

THE Eastern Horn of Africa, with an area of about $\frac{3}{4}$ million square miles and estimated population of 24 million, consists of the greatest expanse of highland over 7,000 feet in altitude in Africa (Fig 53). This great northern block of highland, the culmination of the eastern plateau of Africa, rises sharply from an encircling border of sea, desert scrub, and bush. The region is now divided politically between Ethiopia (Abyssinia), French Somaliland, Somalia and the northern frontier zone of Kenya.

It derives its *regional distinctiveness* from the long-continued existence of the Ethiopian-Coptic culture of the highlands amidst a sea of Moslem culture and religion. The precipitous highlands, rising 7–8,000 feet from the arid plains, form extremely strong natural defences whose only access until recently was by means of watercourses, dry for only part of the year, which led up the walls of the escarpment to the high plateau above. Although two of the world's oldest routeways, the Nile Valley and the Red Sea, lead past the Ethiopian plateau it has remained aloof and inviolate by virtue of its great natural barriers. Its freedom from outside control, except for a very brief period of Italian rule, has earned it the admiration of ex-colonial Africa and Addis Ababa has become the centre for many Pan-African political conferences.

The region is one of Africa's major *watersheds*, the source of the Blue Nile and the major Nile tributaries of the Atbara and Sobat, which together supply the bulk of Nile water. Control of this watershed is thus of vital concern to the Sudan and Egypt. The region is also an ethnological divide between the northern Semitic and Hamitic lands and Bantu and Negro Africa.

RELIEF, STRUCTURE AND DRAINAGE

The landscape is marked by a number of youthful geological forms, with fresh scarps, steep valleys, disturbed drainage patterns and few alluvial depositional features. The highlands of Ethiopia consist of a number of *plateaux* deeply dissected by a series of gorges and down-faulted valleys. The base of the plateau consists of Archaean rocks overlain by basalt, and other volcanic rocks, such as dolerites and trachytes, at altitudes of over

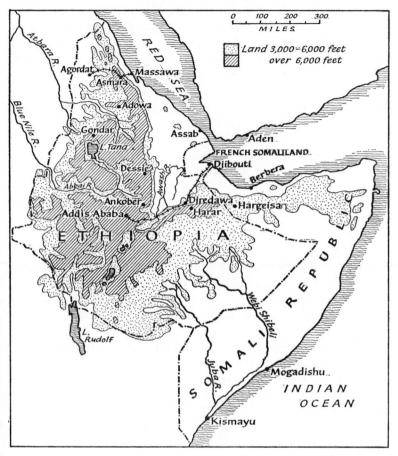

FIG. 53—The Eastern Horn: relief and political divisions

8,000 feet, which tend to form terraced ridges or fantastically shaped volcanic peaks. In Somalia the basement rocks are covered for the most part by terrestrial Cretaceous sandstone and shales. There are two main plateau systems, divided by the eastern and western arms of the Great Rift Valley which form a Y shape meeting in the Awash valley. The northern section is the more significant and higher, while the southern section trending east-west has a limited area of highland which soon descends to lower and drier plateaux and plains.

The northern plateau stretches for approximately 600 miles from the Awash valley to Asmara and is, on the whole, over 7,000 feet in altitude;

from it rise numerous volcanic peaks and mountain ranges from 12,000 to over 15,000 feet high, Ras Dashan in the Simien Mountains, for example, being 15,158 feet. In some places the plateau surface has been preserved, while in others remants of former erosion cycles are preserved in the many mesas and buttes known as ambas. These have played a prominent role in Ethiopian history and, as in Basutoland, have acted as natural fortresses. The eastern edge of the plateau is sharply defined by the western arm of the Rift Valley and persists as a dramatic scarp for 400 miles in a north-south direction, rising almost sheer for 7,000–8,000 feet above the arid lowlands to the east. The western edge of the plateau is also precipitous, but far more dissected by the headwaters of the Nile which have cut gorges thousands of feet deep into the plateau.

The plateau is tilted mainly to the west, causing most of the *drainage* to flow westwards to the Nile, the Takkaze (Atbara), Blue Nile and Sobat taking 80% of Ethiopia's drainage. The southern and eastern arm of the plateau, stretching from the Awash Valley to Lake Abaya north-east and then east via the Chercher hills into Somalia, forms a minor watershed with the Awash draining northwards into an inland drainage basin and the Juba and Webbe Shibeli south-east to the Indian Ocean. The northern edge of the eastern plateau is also marked, although not so dramatically, by an arm of the Rift Valley.

The arms of the Rift meet south of Ankober and slice across the plateau at a height of about 6,000 feet in a well defined trench, as in Kenya, with scarp walls rising to 13,000 feet and a chain of lakes such as Lake Zwai, Abyata and Abbaya (Margherita). On all sides there are spectacular changes of relief and climate, great horizontal changes being swiftly confined within a narrow vertical compass. From the cold, misty plateau edge at 9,000 feet near Ankober the traveller looks down to the sun-scorched scrub 7,000 feet below, or from the torrid heat of the Blue Nile gorge 5,000 feet up to the serrated edge of the cool plateau above.

CLIMATE AND VEGETATION

The *climate* varies greatly from the heat and aridity of the Danakil desert to the frost of the dega above 8,000 feet and occasional snow on the mountains above 13,000 feet (Fig 54). The Ethiopians themselves distinguish three zones based on altitudes the first is the *quolla*, below 5,000–6,000 feet with average annual temperatures over 80° F. (27° C.) in places and vegetation varying from savanna to desert. The second is the *voina dega*, from 6,000–8,000 feet, comprising the bulk of the plateau with higher rainfall and lower temperatures below 68° F. (20° C.) and with varied

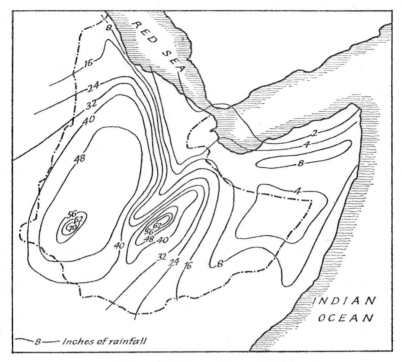

FIG 54—The Eastern Horn: rainfall

vegetation and crops. The third zone is the *dega* consisting of small scattered areas above 8,000 feet and with temperatures from 50°–60° F. (10°–
16° C.). The highlands intercept the Indian monsoon and thus have
abundant precipitation with the little rains from February to March and the
great rains from June to October. Most rain occurs in the south-west,
amounts decreasing to the north and east. *Vegetation* tends to be more
luxuriant in the higher parts of the southern plateau, but thins out to
grassland with scattered trees, such as junipers, and bushes on the northern
plateau and tree euphorbias and giant senecios in the highest zone. The
Somaliland plateau consists of acacia savanna which degenerates to a
meagre scrub in the drier areas.

HISTORY

The diverse physical environment has had a great effect on the historical
and ethnic background of the region. The main cultural influences have
come from the north and spread successively southwards down the great

303

western plateau. The kingdom of Axum flourished in Tigré from the first to the seventh century A.D., and was allied with the Byzantine Empire and the Greek world and maintained trading relations with many countries through the port of Adulis. The Ethiopian Coptic culture has been kept alive by its continuous, although at times tenuous, contact with the Coptic Church in Egypt. The spread of Islam down the Nile Valley and along the Red Sea isolated the Christian state, but the great height of the plateau and its precipitous walls protected it from overthrow, as happened with the kingdoms of Kush and Meroe in the Sudan.

The successful survival of Christianity in Ethiopia gave rise to the legend of Prester John. The Portuguese in the course of their expansion in the Indian Ocean maintained contact with Ethiopia for a considerable period in the sixteenth and seventeenth centuries and were able to help crush the Somali conquest of the plateau by Muhammad Gran. The Portuguse link had considerable architectural influence on the castles and churches of the north.

Internal unity has been difficult to achieve except under strong leaders or in modern times with the aid of communications. The broken nature of the country with its small plateaux isolated by deep gorges encouraged the survival of the historic kingdoms of Tigré, Amhara, Shoa and Gojjam, the last being cut off for months at a time by the flood waters of the Blue Nile gorge which almost encircles it. Centuries of isolation were only broken by the construction of the Djibouti-Addis Ababa railway which, commenced in 1894, only reached Addis Ababa in 1917. With the colonial scramble for Africa Ethiopia was a rich prize, but managed to foil Italian attacks at the battle of Adowa in 1896. The Italians were successful in their conquest of Abyssinia for a brief period from 1936 to 1941, since when the country has again been independent. Eritrea, which was Italian from 1889 to 1941, was federated with Ethiopia in 1952. There is a complex feudal system and the country is now divided into provinces ruled feudally still by rases who, as the local potentates, have considerable authority. The Emperor Hailie Selassie, who put down the 1960 revolt, has been a major force in the unification and modernization of the feudal states comprising Ethiopia, many of these states being independent or semi-independent until recently. The Coptic Church has for long been a significant source of national culture and a rallying point for many of the people. But it is a very con-servative organization and owns nearly a quarter of Ethiopia's land; most of the rest is owned by the Emperor and the large landowners.

[*Embassy of the Republic of Congo* (*Kinshasa*)

XIII. Kinshasa (Leopoldville) and the river Congo.

XIV. Zebu cattle, Ethiopia.

[*Imperial Ethiopian Embassy*

XV. Tigre Province, Ethiopia.

XVI. Dar es Salaam and the harbour.

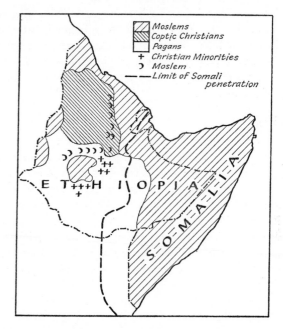

Fig. 55—The Eastern Horn: religions and Somali penetration

POPULATION

The peoples of Ethiopia vary greatly in language, ethnic origin and religion (Fig 55). There are some 100 tribes, 70 languages and 2 important religions. The word 'Abyssinia' means 'mixed'. The Amharas, numbering some 2 millions, are Coptic Christians and mainly of Hamitic origin. They form the dominant racial stock in the central highlands and the main political force in the country. The Gallas form over half the population and are also of Hamitic origin. They first appeared in Ethiopia in the sixteenth century and became important as troops in the dynastic wars of the eighteenth century. They have little political consciousness compared to the Amharas and are partly pagan, partly Christian and Moslem. In the desert areas the Gallas have become nomadic pastoralists. Many of the Gallas have acquired a considerable negroid strain. Other smaller groups are the Tigraians of the north, also a mixture of Hamitic and Semitic stock, the Falashas or Black Jews (Jewish influence dating perhaps from the pre-Christian era), Somali and Danakil nomads of the desert area, and Nilotic tribes such as the Nuer.

The total population of the area is about 24 million,[1] the bulk of the population being situated on the Ethiopian plateau between 5,000 and 8,000 feet in altitude. The desert areas below 3,000 feet have a sparse population consisting largely of nomadic pastoralists. Similarly the areas above 8,000 feet are underpopulated largely owing to the coldness and inclemency of the weather.

AGRICULTURE

Over 90% of the population is *agricultural*. With such a wide range of rainfall, altitudinal zone and soil there is a wide range of *crops* from wheat, millet, teff, rubber, bananas and coffee to vines, apricots and figs and many other fruits and vegetables. Agricultural methods are backward and the land is only scratched by a primitive plough, while grain is thrashed by oxen or with flails. The areas above 8,000 feet contain some of Africa's finest grasslands and there is a great potential for commercial stock-raising. Only extensive pastoralism is possible in the desert and semi-desert areas and concern is far more with quantity than quality of stock. (There are few exploitable minerals as yet.)

The region has some of Africa's most valuable potential farming land and a wide range of agricultural products could be exported on a comparatively large scale. Until recently, however, the great bulk of the population has remained outside the money economy and there were few exports until after the war.

REGIONS

The Eastern Horn may be divided initially into two minor regions, largely on grounds of altitude and rainfall. The areas above about 5,000 feet are usually well watered and have a considerable temperature range, so that crop cultivation is widespread. Areas below 5,000 feet are hotter and drier and the major activity is generally nomadic pastoralism. The northern plateau areas are also the seat of the Coptic Christian culture, whereas the southern and eastern lowlands are inhabited by Moslems. There is thus a strong contrast between the areas above and below 5,000–6,000 feet in altitude.

The plateau region may be further subdivided into the northern and south-eastern plateau blocks. Indeed because of the extremely dissected nature of the plateau and varied physical conditions a large number of further minor regions may be enumerated. However, in an African setting

---

[1] Estimates of the population of Ethiopia vary widely from 17 million to nearly 25 million.

with an immature population pattern and uneven economic development the geographer is more interested in the broad sweep of the region than in the minutiae of detail which have often still to emerge.

The plateau

The *northern plateau*, stretching north-south for about 600 miles from Asmara to Addis Ababa, is the most important region. This is the *voina dega* of the Ethiopians with patches of plateau above 6,000 feet in altitude, often isolated by great gorges containing Nile drainage and possessing the old kingdoms, the heart of the Amharic domain and the Coptic Christian church. Here are found the old capitals of Aksum and Gondar, the rock churches of Lalibela, Lake Tana and many mountain ranges with peaks over 13,000 feet high. Above 8,000 feet it is a cold, almost uninhabited area, but below this height the population clusters thickly in places.

The canvas, as elsewhere in Africa, is so wide that only a few of the general patterns and basic colours can be picked out. The great rains come in June or July when the land is ploughed and grain crops are sown, with barley first followed by wheat and other crops. By mid-October the fields are green with young crops, barley being half grown. With the cessation of the great rains the country dries up and grass turns from green to brown. Crops tend to be harvested during this period, the grain being trodden out by the oxen and winnowed with wooden shovels. Cattle graze on the stubble whose short spikes turn gold and brown in the early morning and late evening sunlight. Fresh land is prepared for planting and the ox-drawn plough turns the new grass over while worn-out land lies fallow. With the lesser and more variable rains of February to March the grass may revive and afford much-needed grazing for the thin zebu cattle. Shifting cultivation is the rule, a primitive plough being used rather than the hoe. It is estimated that land under cultivation is less than 10% of the total area. About half the cultivated area is planted to grains.

The main *food crops* are various forms of sorghum and millet which are made into thin, round sheets of bread called injera and eaten with a highly peppered sauce known as wat. Cereals are also used for fodder crops and beer. Wheat, barley and oats are grown at higher altitudes and are less common, although wheat has been an export crop since 1945. Various types of beans, peas and lentils are grown, and there are many varieties of fruit —oranges, dates, figs, bananas, peaches, apricots—and vegetables. Oil seeds such as castor oil, peanuts, cottonseed and linseed are important. Cotton of poor quality has long been grown by peasants and is now being produced on a small scale as a plantation crop. Sugar estates were established after 1950 at Wonji on the Awash river and in 1958 at Shoa. Output is now sufficient to supply the Ethiopian market, with a small surplus for export.

The *major cash crop* is coffee, which is grown in the lower and warmer areas of the south, and provided about half the exports in 1961. Its name is derived from the Kaffa province in the south which was the original home of coffee. Much of the crop is still from wild trees, but there are an increasing number of plantations in the south-west. Production, which was only 14,000 tons in 1945, had increased to 45,000 tons in 1959 and 65,000 tons in 1962, worth £5 million, and could be increased greatly in future but for the prevailing low prices.

Estimates of the numbers of *livestock* in Ethiopia vary widely, but there may be up to 20 million cattle (Plate XIV), 20 million sheep and 15 million goats, which would make Ethiopia of major importance as a pastoral country with regard to numbers. There are wide areas of good natural grazing and greatly increased quantities of fodder crops could be produced. At present, the standard of livestock rearing is low and cattle, in the characteristic African fashion, are regarded as more valuable for social and ritual than for economic reasons. Pastoral products account for little more than 5% of total exports. If stock diseases such as rinderpest could be controlled and the quality of the small zebu type cattle improved, there are considerable prospects for the export of meat and pastoral products other than the present hides and skins.

Little is known about Ethiopia's mineral wealth. Gold occurs in placer deposits to a limited extent, production in 1963 being only 27,000 ounces. Iron, mica and other minerals have been found and oil prospecting is being undertaken in Ogaden. Potash has been discovered in the Danakil desert and may soon be exported.

The *settlement* pattern varies considerably, too; the Amharas tend to live in round huts forming isolated homesteads perched on ridges and hill tops and surrounded by groves of eucalyptus trees. There are large numbers of churches, many thatched and like oversized huts and some constructed of stone. They are also found on hilltops and form the pivot of rural life, although seldom the focus of a village. In contrast, the Tigrean villages are stone built and contain substantial stone houses, sometimes with an upper storey. The majority of these houses are square in shape and flat-topped with tiny windows and an outside staircase leading on to the roof which may shelter sheep and goats at night. Some of the double-storey circular stone houses look like Border keeps and are used for defence against Shifta attacks. Stone built village churches are conspicuous on rising ground. At Lalibela and elsewhere there are remarkable rock churches, carved in red volcanic rock, some dating back to the thirteenth century, and in their conception not unlike the Nabatean rock-hewn buildings in Petra.

Of the old capitals Gondar, Aksum and Makallé are of considerable interest. *Gondar* was the capital during the seventeenth and eighteenth centuries and was at this time of large size, with over 40 churches. It contains the castle palaces of Fasilidas and other kings, the building of which was aided by Portuguese craftsmen, although the style is more Arabian in motif. The city has many metal workers and craftsmen and has for long been a centre for handicrafts. It now has a population of 22,000. *Makallé* is the capital of Tigre and has one of the largest markets in the northern part of the plateau and is the centre of a great salt trade. *Asmara*, the former capital of Eritrea, is situated at an altitude of nearly 8,000 feet and has a population of over 100,000. It is connected by railway with Massawa and by road with Addis Ababa. It has the air of an Italian city, and there are still about 20,000 Italians living there.

By far the most important town in Ethiopia now is *Addis Ababa* (meaning New Flower) which was deliberately chosen by Menelik as his capital at the end of the nineteenth century. The town straggles along the foot and slopes of the Entotto hills at an altitude of 8,000 to 8,500 feet. The huts, houses and buildings of the widespread town are hidden in extensive groves of eucalyptus trees which were first planted 50 years ago and have now largely replaced the slow-growing native juniper. This Australian gum tree is of great benefit to the Amharas as it is quick-growing and coppices freely when cut down. It supplies fuel and building poles to the capital and has been instrumental to a considerable extent in enabling the location of the capital to be fixed here instead of having to be moved elsewhere with the exhaustion of local fuel supplies. The capital is situated in a nodal position at the southern edge of the northern plateau and commanding the Rift Valley and eastern plateau of the Danakil and Ogaden country and routes to the east. The population is now nearly $\frac{1}{2}$ million and is expanding rapidly, with a mixed indigenous population and very varied foreign community, Armenians and Greeks handling most of the retail trade. There is an extensive market and on Saturdays donkeys come in from the surrounding countryside with great loads of hay and earthenware vessels. Another feature of the capital and its immediate hinterland is the large number of churches.

One of the major barriers to economic growth on the plateau is the lack of roads. There is a rudimentary pattern of nearly 3,000 miles of main roads centring on Addis Ababa but most of Ethiopia still has to rely on animal transport. There are about 5 million horses, mules, donkeys and camels.

The *second plateau region* is smaller and lies to the south and east of the main northern plateau, separated from it by the north-east to south-west trending Rift Valley containing a number of lakes. The highlands to the east of the Rift Valley rise to over 13,000 feet in places, but are generally lower and have a more luxuriant vegetation than the northern plateau while landforms are rounder and more undulating than the angular ridges and mountain peaks of the north and centre. The plateau gradually narrows to the Chercher hills where there are red soils and much Mocha coffee is grown. *Harar* is the main centre of the eastern plateau and is well situated at the border of Galla and Somali territory. It is the second city of Ethiopia but has a population of only about 25,000 and until 1887, when it was captured by Menelik, was the capital of an independent Moslem emirate. It is a Moslem town in contrast to Christian Addis Ababa and is picturesque with narrow streets and defensive walls reminiscent of the Middle East. Situated some 6,000 feet up, its climate is temperate and cooler than Dire Dawa some 30 miles away situated at a lower altitude on the Djibouti-Addis Ababa railway.

The eastern lowlands

The second main region comprises the great eastern swathes of low plateau, plain and coastal strip, forming over half the area of the Eastern Horn, but with only a small population, most of the area having a density of less than 4 per square mile. The characteristic Archaean Metamorphics are covered by Cretaceous sandstones in Somalia. The Chercher hills and the continuation of the higher parts of the plateau into Somalia divide the small triangle of Danakil and French Somaliland from the far larger southern triangle of Ogaden and Somalia. From this central divide the land falls fairly steeply to the Danakil desert in the north, but more gradually in the south. The most spectacular change of altitude is to be found in the abrupt rise from below sea level in the Danakil depression to the wall of the eastern escarpment at over 10,000 feet near Adigrat.

Rainfall is slight and very variable, most of the area having less than 15 inches, although locally in a few areas over 40 inches may occur. Temperatures are also high, particularly along the Red Sea coast at Massawa and at Berbera, which has a mean temperature of over 80° F. (27° C.) for the three months June, July and August, but decrease considerably in the mountains overlooking the Gulf of Aden.

The only permanent rivers are the Awash in the north which flows into an inland drainage basin, and in the south the Juba whose major tributary the Webi Shibeli flows parallel to and near the coast for over 200 miles.

before joining the Juba near its mouth. Most of the drainage consists of ephemeral rivers.

SOMALIA

The former territories of British and Italian Somaliland have been united to form the independent country of Somalia which has a population of about 2 million in an area of 270,000 square miles. The total Somali population in the Eastern Horn is about 3 million, although some estimates place it at 5 million (Fig 55). The population of the second region consists mostly of nomadic or semi-nomadic Somali and Danakil pastoralists, possessing large herds and flocks, estimated to number over 2 million camels, $1\frac{1}{2}$ million cattle, $4\frac{1}{2}$ million sheep and nearly 4 million goats. The number of cattle is considerable in view of the semi-arid nature of the environment. Over most of Somalia life is dominated by the necessity to find grazing and water for *livestock*, the only means of existence. When water and grazing are scarce competition for these resources leads to inter-tribal warfare. For much of the year cattle tend to be pastured near the reliable water supply of the Awash and the Juba. Camels only needing water once a week can range far more freely over the dry plateau. Camels are the mainstay of the local economy and provide transport, meat, milk and hides. The quality of livestock is poor and stock losses are frequent. Hides and goat and sheep skins, after bananas, are the main items of export. Somalia is the world's main exporter of frankincense, and gum arabic is collected from the acacia trees. Cultivation is only of minor importance, and only 0.3% of the land is cultivated. Millet and maize are the main subsistence crops, while in the valleys of the Juba and Webi Shibeli, under Italian influence, irrigated crops of sesame, cotton, bananas, sugar and dates are produced. The main crop on the Webi Shibeli is sugar. Bananas supply about two-thirds of the exports and are grown on Italian-owned plantations along the Juba river.

The nomadic population live in hemispherical huts, consisting of grass mats over a framework of wooden hoops, that can be taken apart quickly and moved by camel. As water and pasturage are of vital importance tribal rights are carefully regulated and in many cases the stock is moved to extra-tribal areas. Political frontiers mean little to pastoralists and recently tension has been created on the borders of Ogaden and Somalia when movement across the frontier was disputed.

The only permanent population is in the coastal ports and the inland town of Hargeisa. *Mogadishu*, founded over a thousand years ago, is the capital and is situated near the Webi Shibeli valley. It has an estimated population of over 70,000 and, despite the handicaps of a poor harbour which is only suitable for lighters and dhows, it is the country's chief port. The other main ports are *Berbera*, 400 miles west of Cape Guardafui, and an outlet by road for Harar and Hargeisa, and *Kismayu* at the mouth of the Juba river and to the south of Mogadishu.

In 1966 *French Somaliland* still remained aloof from Somalia and was the only colonial enclave left in the region. With an area of only 8,500 square miles and population of less than 70,000 its only economic significance lies in the port of Djibouti, the terminal of the Addis Ababa railway. Besides acting as an important centre for bunker fuel, it handles the bulk of Ethiopian trade, although the port of Assab[2] in Eritrea has a good natural harbour and with Massawa should handle an increasing proportion of this trade. During the eighteenth century *Massawa* was the terminus of the caravan track from Ethiopia, and was revived by the Italians to serve as Eritrea's main port. It is connected by railway with Asmara. The coastal strip of Eritrea is very hot, Massawa having one of the highest mean annual temperatures in the world. The population is Moslem and consists of nomadic pastoralists with herds of camels, sheep and goats, and a way of life similar to that of Somalia. The production of salt is important near Massawa and gum arabic is collected.

Somalia is a poor, arid region whose budget still has to be subsidized to a major extent by British, Italian and World Bank aid. There are no minerals, little agriculture and only extensive pastoralism to provide an economic basis for the new state. Illiteracy is widespread and political unity is difficult to achieve among nomadic clans constantly at strife and in a region where clan warfare is a universal practice. The Somalis are also averse to manual labour and other occupations such as trade, leaving this to Arabs, Pakistanis, Goans and Italians. With a high rate of illiteracy the Somalis are further handicapped by an arid environment and thus have a very low income *per capita*, estimated at only about £15 per annum.

---

[2] Assab has been developed to accommodate three large vessels, but is not being used to capacity, partly because of the high cost of transport up a steep road to Dessie.

CHAPTER ELEVEN

East Africa

EAST AFRICA is a fairly well defined region astride the equator and stretching from nearly 5° N. to about 12° S. (Fig 56). It has an area of nearly ¾ million square miles and a population of about 30 million and includes territories that have been under British, Belgian and German influence in the last sixty years, namely Kenya, Uganda, Tanzania[1] Rwanda, Burundi, and the offshore islands of Pemba and Mafia. The colonial association with Britain has produced a certain similarity of economic and political development, particularly with the grouping of common service under the East African High Commission, now known as the Common Services Organization.

RELIEF AND STRUCTURE

The *major boundaries* of East Africa are largely physical, lines of drainage being prominent. The northern boundary is composed partly of high ground south of the Sudd, but mainly of a wedge of desert and semi-desert from Lake Rudolf eastward to the Indian Ocean. The southern boundary, which is more a political than a physical barrier, consists of the Rovuma river, south of which lies Portuguese East Africa. A more suitable physical boundary might, perhaps be the Zambezi river. The eastern boundary is the Indian Ocean, while to the west the Rift Valley highlands and lakes form a well marked crescent-shaped boundary which is a crenulate line at a height of about 3,000 feet threading the eastern limits of the Congo Basin.

East Africa consists largely of a *great plateau*, mostly above 4,000 feet in altitude, whose main relief features are aligned north-east to south-west compared to the east-west trending features of West Africa.

CLIMATE

Because of the high altitude and variety of relief there is a considerable diversity of climate. Rainfall varies greatly in seasonal amount and distribution and much of the region suffers from aridity, only 2% having over

[1] Tanganyika and the island of Zanzibar have united to form the Republic of Tanzania. As the union is purely political the two areas may still be treated as distinctive units from other points of view, with Zanzibar included in the coastal region.

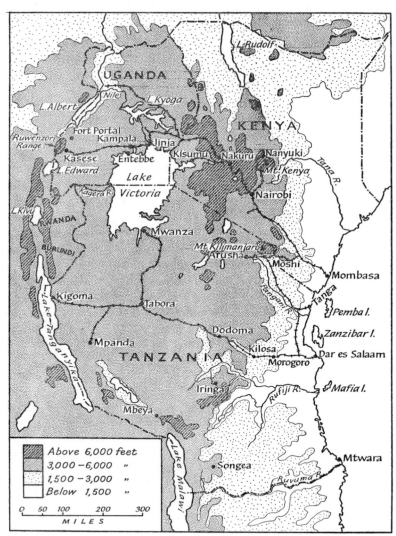

FIG. 56—East Africa: political divisions

2 inches of rain in every month, while nearly two-thirds of the area has a drought of six months or more in the year. Compared to West Africa, temperatures are usually low because of the considerable height of the plateau. Soils are generally poor and porous except where enriched by volcanic action.

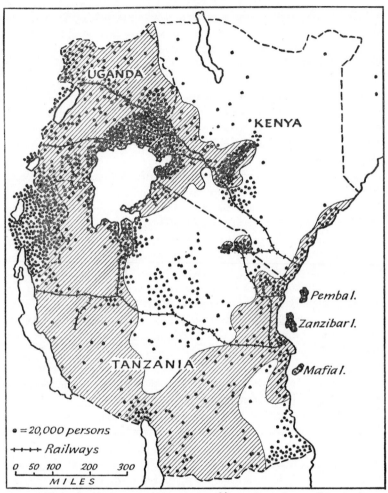

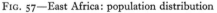 The shaded area has over 70% probability of 30 inches of rain per year

FIG. 57—East Africa: population distribution

POPULATION

The physical background over much of the region is harsh and this is reflected to a considerable extent in the distribution of population. There are over 200 different tribes, varying in size from the tiny groups of Pygmies and Wanderobbo to the 2 million Baganda. There are also great ethnic variations; these range from Hamitic pastoralists such as the Galla, Boran and Masai, the Nilotes—the Luo of Kenya and Acholi of Uganda—and the Nilo-Hamites such as the Kipsigis and Nandi of Kenya, to the Bantu who form the bulk of the population. The Bantu are represented by major tribes such as the Kikuyu and Kamba of Kenya, the Chagga and Sukuma of Tanzania and the Ganda of Uganda.

There are islands of dense population in favoured areas of high rainfall, but the general pattern is one of sparse population in a wide expanse of low and fluctuating rainfall. In those regions where historical factors have caused the spread of bush and tsetse fly as well there is usually no population. Two-thirds of Tanzania is uninhabited.

Population density is high in Kigezi, the northern shores of Lake Victoria, parts of the Kenya Highlands, and in scattered areas of Tanzania, such as the Uluguru mountains and the northern end of Lake Malawi (Fig 57). The density may be over 500 persons per square mile and pressure on resources considerable, causing soil deterioration and decreased crop yields. Indeed, in the early 1950s, many observers thought that parts of East Africa were overpopulated. The 1953–5 East Africa Royal Commission Report takes the opposite view, however, and considers that there are too few rather than too many people in East Africa for the development of her economic resources. The considerable non-African section of nearly $\frac{1}{2}$ million people, two-thirds of them Asians, have played a major part in this economic (European administrators are being replaced rapidly by Africans) development. Their future is uncertain and many of the white settlers have left Kenya. The Asians who control much of East Africa's commerce and provide many of her artisans fear African competition.

Asians and Arabs have long been attracted by the prospects of trade along the coast of East Africa and a string of trading settlements has been established from the Eastern Horn of Africa as far south as the Zambezi delta since pre-Christian days. The presence of large areas of high plateau has among other factors attracted European settlers since the early years of this century.

Nevertheless, the lack of exploitable minerals such as gold and silver and of other valuable commodities, the difficulty of penetrating East Africa from the north and west with barriers to southward movement such as the Sudd in the upper reaches of the White Nile and the rain forests of the Congo basin, and the greater economic attraction of the Far East, has meant the isolation of much of the interior from world contact until the late nineteenth century. Attention was focused more on the Sudan belt, because of comparative ease of movement, and on the mineral resources of south-east Africa than on East Africa.

AGRICULTURE

Agriculture is of paramount importance in the economy of East Africa; mining and industries play only a minor role. The islands of economic activity and high population density are, however, separated from each other by great distances. Communications are poorly developed and costs of transport to world markets are high. Partly because of the patchy nature of economic development and population distribution and the general lack of correlation between areas of dense population and of economic activity, migrant labour has developed on a large scale. The great bulk of East Africa's population is still rural, but the urban percentage is increasing rapidly, while labour mobility in urban and plantation areas is high and output low. Farming methods are usually backward and must be viewed against a background of poor soils, variable rainfall, conservatism and lack of education.

POLITICAL STRUCTURE

Politically East Africa is divided into Kenya, Uganda, Tanzania, Rwanda and Burundi and the offshore islands of Pemba and Mafia. Kenya has an area of about 225,000 square miles and a population in 1962 of about $8\frac{1}{2}$ million, including about 50,000 whites, 180,000 Asians and 40,000 Arabs. Uganda has an area of about 92,500 square miles, a population of about 7 million, but with a much smaller non-African element than Kenya, consisting of 80,000 Asians and 10,000 Europeans. With an area of about 362,000 square miles and a population of 9 million including 90,000 Asians, 18,000 Arabs and 21,000 Europeans, Tanzania is the largest of the East African territories. The most densely populated areas are, however, Rwanda and Burundi with a combined population of nearly 5 million and an area of only 27,000 square miles. The offshore islands have a population of about 300,000, the majority being in Zanzibar.

SUBREGIONS

The whole region of East Africa has an area of nearly $\frac{3}{4}$ million square miles and a population of about 30 million. The accompanying map shows its subdivision into five minor regions consisting of the plateau, plateau foreland, the coastal plain and the islands, highlands and Lake Victoria borderland (Fig 58).

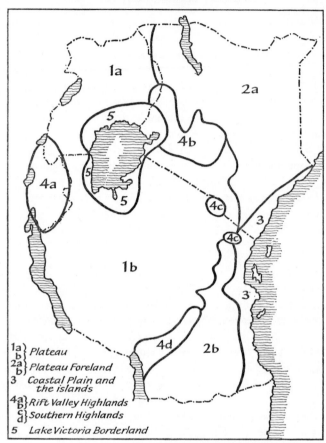

1a} }
b} Plateau
2a} }
b} Plateau Foreland
3   Coastal Plain and
    the islands
4a} }
b} Rift Valley Highlands
c} }
d} Southern Highlands
5   Lake Victoria Borderland

FIG. 58—East Africa: geographical regions

REGION 1—THE PLATEAU

RELIEF, STRUCTURE AND VEGETATION

The plateau, a great almost circular region, is the core of East Africa and contains within its bounds the East and West Rift Valley highlands and the Lake Victoria lacustrine strip. It is about a thousand miles long from north to south and is flanked to the west by the Congo basin and to the east by the terraced foreland of the Nyika. The plateau is a continuation of the South African plateau, and is composed of *ancient crystalline rocks*, mostly gneisses, of the basement complex, mantled by great lava flows and other

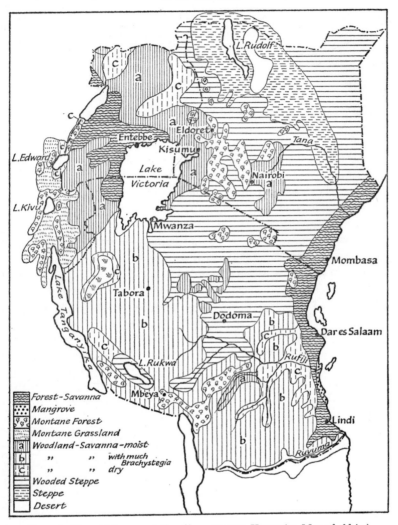

FIG. 59—East Africa: vegetation (from UNESCO *Vegetation Map of Africa*)

deposits. The average altitude is about 4,000 feet and its surface is level or undulating forming an immense peneplane diversified by granitic outcrops and abrupt hill masses. Its widest extent is south of Lake Victoria, where south of Tabora the level plains, clothed in the main with dry deciduous woodland of *Brachystegia-Isoberlinia* and known as miombo woodland, seem to stretch endlessly to the horizon. Burton describes the dense vegetation of the Manyoni District: 'The general aspect is a dull uniform bush, emerald-coloured during the rains, and in the heats a network of dry and broomlike twigs'. *Miombo woodland* covers over half of Tanzania and wooded savanna about one-quarter (Fig 59).

CLIMATE AND SOILS

The plateau experiences a considerable variety of climate, ranging from the equatorial type on the northern shores of Lake Victoria to an almost temperate climate in parts of southern Tanzania. The great bulk of the plateau is south of the equator and therefore affected by the south-east trades in summer and the north-east trades in winter. It has a seasonal rainfall with a maximum from October–November to April–May and a dry winter. North of Lake Victoria the extent of plateau is much smaller and rainfall away from the lake is still seasonal in régime, but higher in amount and more reliable. The rain-bearing south-easterly winds cause heavy precipitation on the south-eastern fringe of the plateau and highland masses and peaks such as the Rungwe and Uluguru Mountains and Mount Kilimanjaro, the Rungwe mountains, for example, receiving over 100 inches per annum. There is a marked rain shadow on the eastern half of the plateau which has less than 30 inches of rainfall. Partly because of its location on the eastern and leeward side of the continent the plateau rainfall, which is largely convectional, is extremely variable and in areas with below 30 inches of rainfall cultivation may be hazardous. Dodoma, with an average annual rainfall of 22·24 inches, had annual totals varying from less than 13 inches to 26 inches in the period 1953–6. Temperatures are lower than on the coast and Tabora at about 4,000 feet in altitude has a mean annual temperature of less than 73° F. (23° C.), compared to Dar es Salaam with nearly 78° F. (25° C.).

Soils are very varied and along a line from Arusha to Dodoma they vary from heavy red clays and rich volcanic ash to the infertile plain soils characteristic of a great part of Tanzania.

HISTORY

Widespread stone ruins and terracing on the plateau testify to ancient settlement and cultivation. Later the plateau was a major migratory route for Bantu moving southwards in search of fresh grass and water, and in the early nineteenth century for northward movement of Nguni and other peoples fleeing from the disastrous Zulu wars of the 1820s. During the nineteenth century Arab slave caravans controlled by the Sultan of Zanzibar operated along a major route from Bagamoyo, the mainland base, via

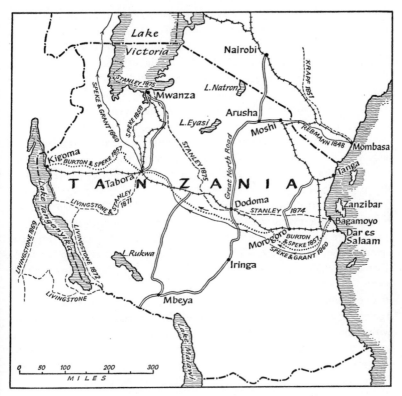

FIG. 60—East Africa : routes taken by some of the European explorers; and modern communications

Dodoma and Tabora to Ujiji on Lake Tanganyika. Slaves and ivory proved a lucrative combination. This was the route followed by the explorers Speke, Burton and Livingstone and later by the Central Railway (Fig. 60). In the late nineteenth century German occupation was resisted strongly by the Hehe, Angoni and other people. The combination of slave raiding and warfare has caused depopulation, bush invasion and tsetse infestation over wide areas of southern Tanzania.

POPULATION

The population pattern of the plateau is very patchy. There are local concentrations in the better watered areas such as Sukumaland and the northern tip of Lake Tanganyika, and great uninhabited areas in the wastes of the central and eastern plateau, reflecting the variability of rainfall, lack of water, infertility of soil and widespread distribution of tsetse fly. The importance of physical factors in the distribution of population in East Africa is brought out clearly in the admirable maps of the East African Royal Commission Report 1953–5 and articles by Gillman published in the 1930s. The majority of the people inhabiting the plateau and East Africa as a whole are Bantu, for example the Hehe of southern Tanzania.

The *northern section of the plateau* is about one-third the area of Uganda, and of small extent compared to the southern section. It consists of wooded savanna with over 40 inches of seasonal rainfall in some areas. It is inhabited largely by Nilotic pastoralists practising seasonal cultivation and growing millet, maize and other crops, following a very different way of life from the Bantu people inhabiting the lake shore.

The Karamoja area in north-eastern Uganda is backward and the Karamojong are mainly pastoralists. Soil erosion is serious in parts of the area. There are nearly 4 million cattle in Uganda, providing an offtake of about 700,000 head in 1960. There are also nearly 4 million sheep and goats. Tsetse infestation is far less serious than in Tanzania and livestock are widely distributed in northern Uganda.

The *southern section*, excluding Sukumaland, has an area of nearly 150,000 square miles. Much of it, especially the eastern half, is semi-arid with some areas receiving less than 20 inches of rainfall. It is therefore not suited to the cultivation of subsistence or cash crops. Also, owing to the ravages of trypanosomiasis, large areas are unoccupied by livestock or man. Economic exploitation of the region, is, however, somewhat of a problem. Around Tabora there is a cluster of population with groundnuts and castor beans the main cash crops. At Urambo land is leased to the tenants by the Tanganyika Agricultural Corporation, some holdings being over 1,000 acres in size.

AGRICULTURE

An attempt in 1947 by the Overseas Food Corporation to grow *groundnuts* on a commercial scale hitherto not tried in tropical Africa proved a dismal failure. The main area was at Kongwa with subsidiary areas in the Southern and Western Provinces and by 1949 less than 15% of the $\frac{1}{2}$ million acre target had been cleared at Kongwa. The planners had not taken into account the uncertainties of the physical environment such as the highly variable rainfall, limited areas of fertile soil and dense vegetation, and the technical problems of maintaining machinery in a tropical environment. After the loss of about £25 million the scheme was greatly restricted in 1950 and in 1954 placed under the Tanganyika Agricultural Corporation.

The Corporation has, however, profited to some extent from the lessons learnt and is experimenting with ranching and settlement schemes. Three ranches have been developed at Kongwa, Ruvu and near Handeni, with a total of about 25,000 head of cattle. African tenant farmers have been allotted 10–15 acre plots and grow groundnuts, maize, sorghum, sesame and other crops.

Experiments to produce higher grade beef cattle have been carried out at Mpwapwa, the headquarters of the Veterinary Department, which is near Kongwa. The problem is to breed a hardy animal that will stand up to the difficult environment and yet mature rapidly and be of comparatively large size. A Boran-Aberdeen Angus cross has been tried out with some success at Kongwa. Unless the carrying capacity of the savanna, which is very low, is improved it may be uneconomic to clear large areas of tsetse fly. However, the only economic future for large parts of the plateau would appear to be extensive pastoralism. At Urambo in the Western Province the cultivation of groundnuts has been suspended and flue-cured tobacco produced instead. The African Tenant Farming Scheme, somewhat similar to that at Nachingwea, is being expanded.

TOURISM

East Africa has a wide variety of wild life that is proving a great tourist attraction. In Kenya tourism is second only to coffee as a source of foreign exchange, and national parks cover an area of nearly 9,000 square miles. The attitude of the African farmer, who formerly only regarded wild game as a source of meat, has changed, and he has come to realize increasingly that wild animals properly conserved are an inexhaustible source of wealth. Economic development over much of the plateau may thus lie in game conservation rather than pastoralism.

MINERALS

Minerals may prove to be of considerable importance when the plateau has been more thoroughly surveyed geologically. The United Kingdom Atomic Energy Authority has opened an office in Dodoma to aid in the prospecting for uranium deposits in East Africa as a whole. Lead, copper, silver and gold were obtained from the Mpanda mine which is connected by a branch line 131 miles long to Kaliva on the Central Railway. Lead was the most important mineral produced, but production has now ceased because of the exhaustion of lead ore.

REGION 2—THE PLATEAU FORELAND

The second minor region of the plateau foreland consists of two wedge-shaped lobes meeting near Tanga. Situated to the east below the plateau, altitude is from 1,000 to 3,000 feet. The *northern section* comprises over half the area of Kenya, with about 600,000 people, and is composed largely of volcanic rocks of Tertiary to Recent age. This area, known as the *Nyika*, consists mainly of semi-desert and desert, much of it with less than 10 inches of rainfall. Rainfall is also very variable and temperatures are high. Vegetation consists of thorn scrub. It is a region of difficulty and is very sparsely populated by *nomadic pastoralists*, such as the Galla, Somali, Turkana and Boran, owning herds of cattle. As this is a frontier zone there are frequent border raids by the Rendille tribe who inhabit southern Ethiopia, and the Somali are pressing for incorporation in the Somali Republic. Rainfall increases in south-eastern Kenya but population is slight except along the banks of the Tana river and in the Teita Hills.

The *southern area* is better watered, parts of it such as the Matengo Highlands having over 40 inches per annum. At altitudes of 3,000 feet temperatures at night are cool for much of the year. Soils vary widely from grey, sandy soils of low fertility to rich loams and clays. The population is surprisingly sparse considering the favourable nature of the environment, and is probably due to recent depopulation caused by events such as the Maji-Maji rebellion, when the Angoni rose against the Germans. A wide variety of subsistence *crops* is grown, such as maize, sorghum, rice, cow-peas, beans, cassava and groundnuts, while cash crops are sesame, tobacco, coffee, beeswax, cotton, cashew nuts and castor; tobacco is produced by Africans in the Songea district, while sugar is planned for the Kilombero Valley. An interesting and efficient form of hillside farming has been developed by the Matengo tribe; earth is piled at the side of holes in the hillside and crops such as maize are grown on the spoil. Weeds and residue are thrown into the holes which are then filled with the previous season's soil and new holes dug, so maintaining soil fertility and preventing erosion.

There are large *coal* deposits at Ngaka in the Songea District, unexploited as yet owing to their remoteness and the high cost of transport.

REGION 3—THE COASTAL PLAIN AND THE ISLANDS

RELIEF

The third subregion is the coastal plain, generally below 1,000 feet in altitude, a long, narrow, fairly well watered strip stretching from the mouth of the Tana river to the Ruvuma river in the south. It consists of sand and coral limestone with Jurassic limestones and shales to the west in Kenya, and in southern Tanzania a sandy or coral shore, mangrove-fringed in places, and backed by a dissected scarp which rises to about 1,200 feet.

CLIMATE

Much of the region has over 40 inches average annual rainfall, and a modified equatorial type of climate and well distributed rainfall with maxima in April–May and November–December. Temperatures tend to be high, although modified by sea breezes in areas such as the new district station at Kilwa Masoko. The average annual temperature range is also small, for example Mombasa and Dar es Salaam with mean annual temperatures of 78–80° F.(25°–27° C.) and a mean annual range of 7°–8° F. (4° C.).

HISTORY

The coastal area of East Africa has had contact with the outside world since pre-Christian days and a large number of trading settlements such as Lamu, Malindi, Mombasa and Kilwa Kisiwani were set up. Kilwa probably dates from the tenth century A.D. and became the most important port on the East African coast until the sixteenth century. Gold, ivory, tortoise-shell, ambergris and slaves were exported to India, Arabia and even as far as China. Much of this trade network was destroyed after the Portuguese occupation in the sixteenth century. The Swahili people, an intermixture of Bantu, Arab, Persian and other immigrants, inhabit the coastal strip and speak a distinctive language, Swahili, which has become the lingua franca of East Africa. Much of the coastal region has a fairly high density of population.

AGRICULTURE

The most important crop is sisal, particularly in the Tanga Province of Tanzania, produced on European and Asian owned plantations. Tanzania is the world's largest producer, with exports worth £22 million per annum (1963 and 1964). Sisal was first introduced into Tanganyika in 1893 and later spread to Kenya. It is a very hardy plant and will grow at altitudes of up to 6,500 feet and with a rainfall of below 15 inches. It will grow on black cotton soils if well drained, but prefers lime soils. The sisal plants, not unlike the crowns of giant pineapples, form serried, spiky ranks on the hillsides, myriads of them to each estate.

There are few estates in the coastal area of Kenya, but in Tanzania estates are found in three main areas, in Tanga Province, along the coast or the Tanga-Moshi railway, for over 200 miles inland from Dar es Salaam and at Lindi in the south. Tanga Province produces nearly two-thirds of Tanzania's sisal. It is best grown as a plantation crop, as there is a period of 2–5 years before the first fibre is produced, and complicated processing and expensive machinery are needed. Overheads are very high, as a large labour force is needed, and a light railway is required to transport the leaves to the factory. Capital requirements for a small estate are estimated to be £300,000. Transport costs are high, so that a coastal location is advisable. The main factor governing the size of an estate is the decorticator, which with a capacity of 1,500 tons of fibre per annum, presupposes an estate with an average size of 3–5,000 acres. One large estate of 80,000 acres employs 20,000 Africans. Another important factor governing location of a sisal estate is an adequate water supply for decortication. Recent research carried out at the Sisal Growers' Research Station near Tanga has perfected a new hybrid plant which produces nearly 600 leaves compared to the 220 leaves of the average plant. This will be of great significance to the sisal industry and Tanzania as much of the land at present under sisal could be released for other purposes. Sisal prices have fluctuated greatly since 1945, ranging from about £50 to £150 per ton.

Coconut palms are numerous and copra and coconut oil are produced by Africans while matting and ropes are made from the coir. Cashew nuts are produced in Tanzania on a fairly large scale, exports in 1962 being worth £2 million. Oranges are also grown and bananas and other tropical fruit command a ready market in the Tanga Province.

TOWNS

The coastal towns have a cosmopolitan air and interesting architecture, a blend of Arabic, German and colonial styles which might be termed 'coastal crenellated'. Mombasa and Dar es Salaam are the most important towns in the coastal area, and the main ports in East Africa. *Mombasa* had an estimated population in 1960 of 189,000 with 32,000 Asians, 27,000 Arabs, nearly 5,000 Europeans and the rest Africans. The main town is on Mombasa island with the old dhow harbour on the northern side and the modern harbour of Kilindini (meaning deep water) to the south. There is an extensive anchorage and eleven general cargo berths. In 1964 the port handled over 4 million tons of cargo. Mombasa was an important trading centre before the sixteenth century and, until the eighteenth century, formed one of the major Portuguese ports on the East African coast, as shown by the massive Fort Jesus. The building of the railway to Kisumu and later to Kampala gave it added importance and it is now the main outlet for the trade of both Uganda and Kenya. A causeway connects the island to the mainland and the town has now spilled over onto the mainland in the direction of the airport.

Ocean-going dhows of 200–300 tons size come to Mombasa from December to April on the north-east monsoon, bringing dried fish, dates, Arab furniture and Persian carpets, and return to India and Arabia in June on the south-west monsoon. Mombasa is developing as a minor industrial centre with food-processing and light engineering industries, an oil refinery and a cement works at Bamburi.

Dar es Salaam, about 200 miles south of Mombasa, means the Haven of Peace (Plate XVI). It was only a fishing village before being developed by the Germans after 1884. It has become the capital of Tanzania, the largest town, main port and terminus of the Central Railway which runs to Kigoma on Lake Tanganyika and has a branch line from Tabora to Mwanza on Lake Victoria. The town is situated on the northern shores of an almost landlocked bay which provides a wide harbour. The port, which had only lighterage facilities until 1956, now has three deep-water berths, but is handicapped by the narrow harbour entrance. The dhow traffic of Dar es Salaam is far more important than that of Mombasa, as shown by the following figures:

	Steamships		Sailing Vessels	
	Number	Reg. Tonnage	Number	Reg. Tonnage
Mombasa	1,330	4·4 m.	323	11,357
Dar es Salaam	1,163	3·6 m.	1,116	45,915

Dar es Salaam is also handicapped by the lack of good road outlets and, as the site of the city is low-lying, widespread flooding may occur during the wet season. In 1957 the population of the city was about 131,000, including nearly 100,000 Africans, 27,000 Asians and 9,500 Europeans. The Asian population controls the commerce of Dar es Salaam and owns much of the property. The African population is polygot and made up of many tribes. They live in Swahili-type houses build of mud and coral blocks with rooms opening off a central passageway.

Minor ports along the coast of Tanzania are Tanga, which handled $\frac{1}{4}$ million tons in 1961, and Lindi and Mtwara in the south. Mtwara was developed for the abortive Groundnuts Scheme and has a large surplus capacity. The railway line to Nachingwea has now been removed because of the lack of economic development, but Mtwara may be revived as a possible port for Zambia.

The *islands of Zanzibar and Pemba* are fairly close to the coast and the town of Zanzibar is only 26 miles north-east of Bagamoyo. It has long been a centre of Arab influence apart from a short period of Portuguese rule. In 1840 Sultan Seyyid Said transferred his capital from Muscat to Zanzibar. He expanded the cultivation of cloves and made the city and port the most important commercial centre in nineteenth-century East Africa. It became the centre of an extensive slave trade which was only suppressed in 1873 at the insistence of the British. The climate is equatorial in type, as on the neighbouring mainland, with well distributed rainfall of 60–80 inches and high temperatures throughout the year.

Zanzibar, almost entirely made of coral, has an area of about 640 square miles and a population of about 165,000, the majority African but with large Arab and Asian minorities. The majority of the population lives in the west where the soils are more fertile. *Pemba*, about 50 miles north of Zanzibar, has an area of 380 square miles and a population of 133,000. Zanzibar and Pemba together produce 80% of the world's supply of cloves, the area under cultivation being about 50,000 acres. The plantations, formerly owned by Arabs, Indians and Africans, have now been taken over by the government which seized power in December 1963. Coconuts are also important, and coconut oil, oil-cake and coir fibres account for one-fifth of the total exports. The islands face serious economic problems as in 1963 the price of cloves was only one-seventh of the 1958 price. The economy is in a state of stagnation and the inhabitants are reluctant to attempt growing suitable new crops such as nutmegs, coffee and limes.

REGION 4—THE HIGHLANDS

RELIEF AND STRUCTURE

The fourth region consists of a number of separate blocks of highland that rise above the plateau level to an average altitude of over 6,000 feet, but with individual peaks such as Mounts Kilimanjaro and Kenya (Plate XVII) and those of the Ruwenzori range rearing above the snowline. The most important highland areas are the East and West Rift Valley Highlands, the Kilimanjaro-Meru and Usambara mountains in north-eastern Tanzania and the Southern Highlands from Mbeya to Iringa in southern Tanzania. The Kenya Highlands and the East Rift Valley Highlands form a compact block compared to the widely separated highland masses of Tanzania.

The highlands are usually composed of *plateau basalts* and phonolites and other volcanic rocks which were poured out, probably in Cretaceous times, over the ancient plateau rocks. Peaks of volcanic ash and lava rise from the lava plains, the greatest in Kenya being Mount Kenya. The great cones of Mounts Kilimanjaro and Meru are an impressive sight on the eastern side of the rift valley. Viewed from the air Mount Kilimanjaro looks like a great inverted pudding basin crowned with white icing and resting on a sloping green base. Faulting and vulcanism have occurred at various periods from pre-Cambrian times on, but the present appearance of the eastern Rift dates only from the middle Pleistocene. Block faulting, possibly of Tertiary age, has caused the complex fault system of the eastern zone, as in the Pare and Usambara mountains of Tanzania. The two great cracks representing the eastern and western arms of the *Rift Valley* diverge from the Ethiopian plateau and with their striking assemblages of highlands, fault blocks, volcanoes and lakes cut across the plateau to reunite near Mbeya.

The *eastern arm of the Rift* passes almost north-south through Kenya and in its course through the Kenya Highlands is very clearly defined (Plate XVIII). The highest part of the rift floor is at Naivasha (6,000 feet) where the bordering mountainous ramparts reach a height of nearly 14,000 feet in the Aberdare ranges to the east. There are a large number of cones and necks of volcanoes such as Mount Menengai and Quaternary lake deposits and lakes much reduced in size since Quaternary times on the floor of the rift. The Quaternary lacustrine deposits correspond to some extent with former lake shores and may be connected with pluvial periods during the Ice Age. Lakes Naivasha, Nakuru, Magadi, Natron and others were far more extensive in Quaternary times. Ancient lake-beds contain much evidence of early animals and man at the Olduvai Gorge. In Tanzania the shrunken Lakes Natron, Eyasi and Manyara from an area of inland drainage.

The *western arm of the rift* begins north of Lake Albert, where there is a great scarp nearly 5,000 feet high, and curves in a southerly and south-easterly direction to join the eastern arm just north of Lake Malawi. Rainfall tends to be higher along the Western Rift, the lakes larger and the highlands more dissected. In the north are Lakes Albert, Edward, George and Kivu and to the south the 400-mile long Lake Tanganyika with depths of over 3,000 feet. The Ruwenzori range (Mountains of the Moon) are a great upthrust mass of non-volcanic, crystalline rocks with a number of snow-clad, glacier-fringed peaks more than 16,000 feet high. Surrounding Lake Kivu is high, dissected and faulted country with the high plateau of

Rwanda, the volcanic peaks of the Mfumbiro group, some over 14,000 feet high, and a host of small scoria cones on the valley floors.

CLIMATE

The climate has been modified considerably by the high altitude of this subregion, which induces heavier rainfall on the southward-facing slopes and lower temperatures. The Kenya Highlands are crossed by the equator, so that two rainy seasons occur with short rains from October to December and long rains from March to May, with a fairly dry period from November to March. In the highlands of southern Tanzania rainfall is seasonal with the maximum from about October to April. Rainfall may be up to 100 inches per annum, but often is less than 40 inches as the following figures illustrate:

Station	Altitude (feet)	Mean Ann. Temp. °F. (°C.)	Lowest Mean Temp. °F. (°C.)	Av. Ann. Rainfall (Inches)
Eldoret	6,863	62·5 (17)	37·0 (2·8)	40·86
Equator	9,062	55·8 (13)	38·6 (3·5)	46·93
Nairobi	5,495	67·3 (19·6)	44·0 (6·8)	34·33
Kabale	6,138	61·9 (16·7)	37·0 (2·8)	38·67
Iringa	5,380	66·3 (19·2)	41·8 (5·4)	29·57

Compared to stations on the coast with mean annual temperatures of 78–80° F. (25–27° C.), areas at over 5,000 feet have temperatures 15–20° F. (8–11° C.) lower. The climate at these altitudes tends to be cool and bracing, although the range of temperature is small for a station like Nairobi, which is nearly on the equator.

VEGETATION

Vegetation varies greatly with altitude and aspect. On mountain slopes from about 6,000 to 12,000 feet various types of forest occur. Rain-forest with camphor, yellow-wood and other trees is found on southward and eastward-facing slopes, and a drier type of forest with cedar as a characteristic tree are found on westward and northward-facing slopes. In Kenya cedar forests are found at altitudes from 6,000 to 9,000 feet, camphor forests from 7,000 to 9,000 feet and bamboo forests from 8,000 to 10,000 feet. The forests alternate with much grassland, and above 12,000 feet is the moorland zone of tree heaths such as the lobelias and giant senecio, shrubs, plants and mountain pastures, stretching up to the snowline at 15,000 to 16,000 feet.

There are *four main subregions* consisting of: the western highland block comprising Ruwenzori, Kigezi, Rwanda and Burundi; the eastern block of the Kenya Highlands; Mounts Kilimanjaro and Meru and the Usambaras; and the southern block of the southern Tanzania highlands.

A. WEST RIFT VALLEY HIGHLANDS

The western highlands are politically divided between Uganda, which contains about half the Ruwenzori range and Kigezi, and the now independent states of Rwanda and Burundi, formerly under the tutelage of Belgium.

Uganda

Kigezi is a most attractive region, much of it over 6,000 feet in height, highly dissected, with deep river valleys draining towards Lake Edward and steep terraced hill slopes. The cones of Muhavura, Singida and other volcanoes of the Mfumbiro group rise steeply from a tumbled foreground of hills and mountains, while Lake Bunyonyi, a lovely crater lake, winds round the convolutions of the hills. Mists cling to the valleys giving them an autumnal look early and late each day.

The *Bakiga* are the main tribal group and are cultivators. Beans, peas and millet are the main crops and they are grown on broad bench terraces which help to reduce the danger of soil erosion on the steep hillsides. Population density is high and there is great pressure on the land. The *Kigezi Resettlement Scheme* is an ambitious attempt to solve this problem. By the end of 1953 nearly 24,000 people had been resettled at a comparatively small cost. Since then the rate of resettlement has slowed down and the population has increased so that the problem has become as serious as ever again and there is less prospect of easing the situation by moving people away from the heavily congested areas. More intensive agriculture, farm planning and land consolidation are necessary as in Kenya.

Rwanda and Burundi

Further south are the ex-trust territories of Rwanda and Burundi which, with an area of only 21,000 square miles, have a population of over 5 million. Originally forming part of German East Africa they were administered by Belgium until 1962 when they gained their independence. The bulk of the people are Hutu, but the pastoral Tutsi, who formed the ruling class, are of Hamitic origin. In the political ferment that has arisen since the independence of the Congo, power has tended to pass into the hands of the formerly subservient Bahutu. Rwanda is controlled by the Hutu, the Tutsi having been driven out, whereas Burundi is still under the control of the Tutsi minority. Rwanda has been invaded unsuccessfully three times by the Tutsi refugees, thus imperilling the lives of some 200,000 Tutsi who still live in Rwanda. The economic effects of insecurity and failure to maintain the high standard of cultivation enforced by the Belgians caused the major export, coffee, to drop in 1963 to only about one-third of the value of the 1962 crop. As in Kigezi, pressure of population on the land is severe and this is met in part by the migration of Banyaruanda to Buganda, parts of Tanzania and the former Belgian Congo. Rwanda has the highest population density and lowest *per capita* income in Africa south of the Sahara. Slopes are steep and soils are mostly light and highly eroded. The administration has strongly advocated the growing of food crops to counter-act the periodic famines which used to affect the country. Over 40% of the area is arable, with leguminous crops and potatoes in the valleys, maize and cassava on the slopes and bananas around the homestead. Most of the area under cultivation is terraced or contour-ridged. Coffee is the main cash crop, Robusta being grown up to 4,500 feet and Arabica above this; cotton, sisal, tobacco, tea and pyrethrum are minor cash crops.

In contrast to Kigezi, here there are over 1 million cattle and more than 2 million goats and sheep; livestock and hides are exported. The region is badly overstocked, but reduction of numbers is difficult as cattle are valued more for their long horns than as a source of milk and meat.

B. EAST RIFT VALLEY HIGHLANDS

The second region of the Kenya Highlands is the largest continuous area of high plateau in East Africa (Fig 61). The Kenya Highlands form *two great plateau lobes* like the wings of a butterfly, separated by the eastern Rift Valley. The western wing is generally higher, reaching 10,000 feet in the Mau plateau.

The Mau plateau

It is the home of the Nandi people, formerly warlike pastoralists, who are now turning vigorously to mixed farming. *European farmlands* are found on the Mau plateau and the Trans-Nzoia and Uasin-Gishu plateaux

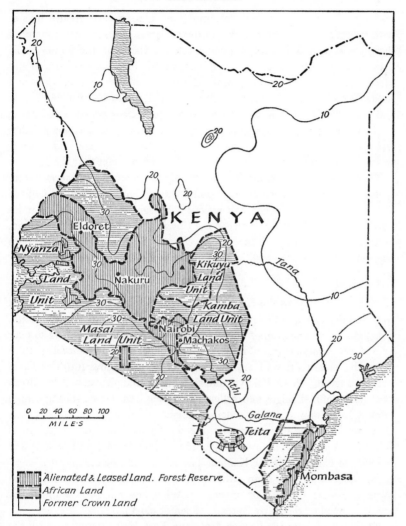

FIG. 61—Kenya: land and rainfall

where rolling grasslands from 6,000–7,000 feet high are used for mixed farming and the cultivation of wheat, maize and other crops. High, healthy uplands encourage the rearing of livestock. The Grasslands Research Station at Kitale has been conducting research into the use of different types of grasses such as Molasses, Rhodes and Guinea grass and leguminous plants and herbs for mixed farming. Livestock products and crops such as

333

wheat, maize and pyrethrum are important. Farms vary greatly in size with ranches up to 20,000 acres and mixed farms of 1,500 acres; many of these could be subdivided. A 2,000-acre farm visited in the Trans-Nzoia district and situated at an altitude of about 6,000 feet was farmed well and carried 500 livestock—cattle, poultry and pigs—and grew maize, wheat and fodder grasses in a seven-year rotation. Rainfall is nearly 50 inches a year and the volcanic soil is fertile. A wide variety of fruit is grown for domestic purposes and peach, pear, apple, pawpaw, banana, mango, guava and other trees surrounded the farmhouse. Almost all the temperate and tropical fruits can be grown together at altitudes of over 6,000 feet near the equator.

There is a large Afrikaans farming population centred on *Eldoret* which, with its rambling corrugated iron roofed houses, Dutch Reformed Church and wide streets, is rather like a South African dorp.[2] *Kitale*, 40 miles north-west of Eldoret and between the Cheringani Hills and Mount Elgon, is the other main town on the western lobe of the Kenya Highlands. Both Eldoret and Kitale are small in size and act as local market centres.

Kikuyuland

The eastern lobe of the Kenya Highlands stretches for nearly 200 miles from the dry Laikipia Plateau in the north to the Ngong hills in the south. The Aberdare Range rises to nearly 14,000 feet and Mount Kenya to over 17,000 feet. Drainage from Mount Kenya is radial and a remarkable pattern of parallel streams flows from the slopes of the Aberdares into the Thika and Tana rivers. This is Kikuyuland, a deeply scarred plateau with hundreds of sharply defined ridges and valleys. The deep, red, fertile soils and high rainfall have long attracted a dense indigenous population, and recently white settlers. The dominant tribe are the Kikuyu, forming one-sixth of Kenya's population and numbering over 1 million people, while 30 miles to the west of Nairobi live the Kamba. The Kikuyu have a complicated social organization with two main patterns, one being closely linked with land ownership—the sub-clan or mbari—the other being territorial and based on the control of ridges by councils of senior elders.

The Kikuyu expanded south from the Fort Hall district to near the present site of Nairobi towards the end of the nineteenth century. This was the frontier with the Masai who roamed and raided over much of what is now the European farming area. But in the 1890s four major disasters, smallpox, rinderpest, locust invasion and resulting famine, crippled the Kikuyu in the southern area and the population was greatly reduced. The

[2] Many of the Afrikaners have gone back to South Africa since the independence of Kenya.

result was that when the first white settlers arrived in 1902 and 1903 this seemingly empty land in Kiambu, Kabete and Limuru was taken up for white farming and alienated. However, nearly all the areas of great potential with over 45 inches rainfall per annum, deep fertile soil and high temperatures that permit two crops to be raised a year, are owned by Africans. The criticism that Europeans have taken over all the best land in Kenya cannot be accepted.

Pressure on land was further increased by other factors common to much of Africa, namely the rapid increase of population in a limited area and the practising of uneconomic methods of farming. The density of population in parts of Kiambu and Fort Hall is over 1,000 per square mile. Before *land consolidation* fragmentation of land was widespread[3] and soil erosion a serious problem. Valiant attempts were made by the government to reclaim land and provide alternative land, but without tackling the social barriers to agricultural improvement. In 1945 the African Land Development Organization was set up and in 1954 a more intensive plan, the Swynnerton Plan, was adopted for the purpose of improving the standard of African agriculture. The administration soon realized that a possible solution lay not in the provision of more land, but in the change of Kikuyu farming methods. Partly because of the Emergency and the removal of leaders who before had opposed the government reclamation programmes, the Kikuyu, an intelligent and energetic people, saw the necessity for re-grouping land and farming scientifically. The people were taken from their scattered farmsteads, or shambas, and collected into villages during the Emergency. Traditional patterns of settlement and land use were broken, there was a change of emotional climate and after the Emergency the Kikuyu pursued the game of land consolidation with as much vigour as they had used formerly in opposing the administration.

The scene has been transformed in Kiambu, Nyeri and Fort Hall; large villages are perched on every hilltop, forming a great contrast to the former sprawling landscape of scattered shambas. Land use is more intensive and there are small farms in a regular layout with fields grouped round the homestead instead of, as before, being scattered in many strips at considerable distances from the homestead. Orderly rows of crops and trees march along the contour, broad-bench terraces climb leisurely up the steep hillsides, the red, gashed landscape is healing rapidly and vegetation and crops are creeping over the red soil. Consolidation has been virtually completed over most of the Kikuyu territory. The minimum economic size of holdings is 4 to 5 acres composed of scattered strips which have

[3] The average number of strips was 8 to a 4-acre holding.

been grouped together.[4] As most of Kikuyuland consists of long, narrow, parallel ridges, farms have been allocated stretching from valley bottom to ridge top. A typical planned holding of 10 to 20 acres may have fenced fields with cash crops such as coffee, pyrethrum and pineapples, part reserved for food crops and part under grass and fodder crops for livestock. With consolidation and the consequent revolution in agriculture the value of African subsistence crops has grown considerably and by 1959 the recorded sale of African produce amounted to over £9 million. By the end of 1959 over 1,300,000 acres had been enclosed and consolidated, over half of this being in the Central Province. By mid-1962 nearly $2\frac{1}{2}$ million acres had been enclosed and almost 300,000 farms created, and by the end of 1965 over $1\frac{1}{2}$ million acres had been registered.

Another resettlement scheme which has met with considerable success is the *Makueni Settlement* amongst the Kamba people. This is in a drier area than Kikuyuland with a rainfall of only 20 to 30 inches and is situated at a height of 3,000–4,000 feet. As in Kirkuyuland the land was being wrecked by donga erosion, but following on work started by the Land Development Board with contour-ridging, livestock improvement and control, and resettlement of people, agricultural methods have improved and a surplus of food and cash crops is being produced. By the end of 1960 the Makueni Scheme was full and had nearly 2,200 registered settlers.

The Mwea-Tebere irrigation scheme on a tributary of the Tana river accommodates over 1,000 families, each cultivating a 4-acre plot. Rice is grown with much success and yields are more than 2 tons an acre.

Economy

It is hoped that, by broadening the agricultural base and diversifying the economy, service and secondary industries may develop in the villages absorbing the landless, decreasing the incidence of migrant labour and making rural life more stable. There are many problems facing African agriculture such as falling world prices for primary products, the economic suitability of particular crops, indebtedness and the flight of capital and increase of rural unemployment following on independence, failure to register change of title, increase of fragmentation and continuance of skilled supervision.

European settlers have done a great deal to build up the agriculture and economy of Kenya and still provide 70% of Kenya's main exports of coffee, tea, wheat, pyrethrum and sisal, despite the growing importance of the African farmers' contribution. European numbers are, however, small, and

[4] Agricultural experts are of the opinion that a larger holding is necessary.

XVII. Mount Kenya and giant groundsel.

XVIII. The Rift Valley, Kenya.

XIX. Nairobi and the Kenya Highlands.

XX. A pyrethrum plantation on the slopes of Mount Meru, Tanzania.

the total white farming population was probably less than 10,000, but the acreage of European-owned land was large, nearly 12,000 square miles. European land holdings are greatly disproportionate (with an average acreage of over 2,000 in 1960)[5] to African land holdings, the great bulk of Kenya's African rural population living in an area of about 52,000 square miles. Nevertheless, in 1960 about 280,000 Africans were employed on 3,600 European farms.

Coffee is the major cash crop. It is excellent quality Arabica, grown mainly near Nairobi at elevations of 5,000–6,000 feet. Although still largely a European crop, the African percentage of coffee has increased to about a quarter, and there are over 100,000 African growers. Tea is second in importance and Kenya is now the largest tea producer in Africa, with exports worth over £6 million in 1964. The main area of production is around Kericho. Over 100,000 Africans now grow tea, especially near the European plantations which occupy over 80% of the area under tea. Kenya produces more than 60% of the world's pyrethrum which is grown mainly above 7,000 feet in height. Most of the pyrethrum is exported.

The following tables illustrate (a) the relative proportion of African and European acreages of coffee, tea and pyrethrum in 1959 and (b) the percentage of non-African to total production in 1960.

(a)	Coffee	Tea	Pyrethrum
African	26,000	1,541	12,000
European	67,000	36,100	29,000

(b)	Coffee	Tea	Sisal	Pyrethrum
	81·0	99·3	95·2	78·8

Livestock accounted for about one-quarter, crops for three-quarters of European gross farm revenue in 1961, although crops occupied less than one-fifth of the area.

The future of white settlement in Kenya is now in doubt with African political domination and the end of British control. Envious eyes are cast by crowded African farmers and landless people on the broad fertile acres of European estates often, as near Nairobi, almost surrounded by African-owned land. The first considerable body of white settlers arrived in 1903 and they were granted leasehold farms of 640 acres in size. Great difficulties had to be overcome; strange pests and diseases destroyed crops and livestock, African labour was unused to European farming techniques, transport

[5] A third of the European holdings were less than 500 acres in size and about one-fifth were above 2,000 acres in 1960.

was costly and roads non-existent at first, while the long depression in the 1930s ruined many white farmers. Lord Delamere did a great deal to prove the suitability of the Highlands for white farming. After the First World War 1,000 farms were taken under the soldier-settler scheme and by 1931 the land under cultivation was about 660,000 acres, a threefold increase in 10 years.[6] Similar progress was made in production during the Second World War; the production of pyrethrum, for example rose from less than 2,000 tons in 1939 to nearly 7,500 in 1945.

With the increase in African cash-crop production[7] the economic importance of European farming will decrease and much European land will be taken over for *African resettlement*. The government proposes to resettle African landless on nearly 2 million acres in the former White Highlands. So far about 350,000 acres have been bought for smallholder schemes near African reserves. Over 100 projects have been approved for yeoman farmers involving about 100,000 acres with farms of up to 300 acres in size that should provide an income of at least £250 a year. The only *minerals* of significance in Kenya are soda ash from Lake Magadi to the value of

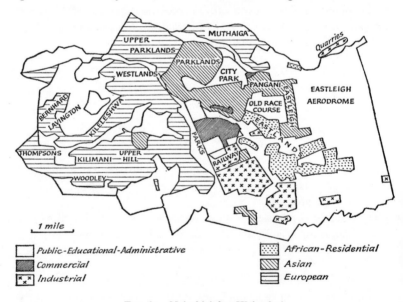

FIG. 62—Nairobi (after Walmsley)

[6] According to the Kenya Land Commission Report of 1933, nearly 12% of the area alienated to Europeans was under cultivation and 41% used for pastoralism.
[7] Since independence the African farmer has increased his share of marketed farm output to nearly one-third.

338

£¾ million in 1964 and copper worth £½ million from the Macalder-Nyanza mine.

The main urban settlement in the Kenya Highlands is *Nairobi*, the capital and largest town in Kenya, the commercial capital of East Africa and, as the headquarters of the Common Services Organization, perhaps also a future capital of a federated East Africa (Plate XIX). It is situated at an altitude of about 5,400 feet at the head of the Athi plains and on the southern fringe of Kikuyuland. It has a population of about 300,000, including about 23,000 Europeans, 100,000 Asians and over 150,000 Africans many of whom are seriously overcrowded.

Nairobi is an important communications centre connected by railway with Mombasa, Kisumu and Kampala, with a branch line north to Nanyuki. The Kenya-Uganda railway, extended to Kampala in 1931, has been, more than any other factor, responsible for the economic development of Kenya and Uganda. It has been able to tap the fertile Kenya highlands and the productive northern shores of Lake Victoria, these regions providing value-able agricultural freight of coffee, cotton, sisal, pyrethrum, tea and other crops for export, in contrast to the infertile, semi-arid, middle stretch of the Central Line in Tanzania. The Great North Road also runs through Nairobi which, with its airport at Eastleigh, is one of the most important air terminals in Africa. Nairobi is the main commercial centre in East Africa and headquarters for the major commercial houses and banks. It is also an important market centre as it lies on the edge of the productive highland region. And further it is the main industrial centre of the region with a fairly wide range of secondary industries, ranging from consumer goods to food processing.

Nakuru, known as the 'Farmers' Capital', is far smaller with a population of about 30,000 and is situated in the Rift Valley on the shores of Lake Nakuru. It is the centre of the white Kenya Farmers' Association and a market centre for much of the Rift Valley. It has a few food-processing industries.

C. THE NORTHERN HIGHLANDS

The Northern Highlands consist of mountains in the Mbulu District such as Hanang and Oldeani, the volcanoes of Meru and Kilimanjaro, and the Usambara Mountains. These highlands are more scattered, less compact and far smaller in area than in Kenya. A considerable amount of land was alienated for German settlers in the north-eastern highlands, but this was confiscated and sold to Europeans and Asians after the First World War. In 1951 the total area alienated was just over 2·3 million acres, representing about 1% of Tanganyika's area. Some of this has been purchased for African settlement and now less than 1% is leased for non-African farming. The main blocks of alienated land are between Mounts Meru and Kilimanjaro, and in Tanga District where sisal is the main crop. Coffee, tea, pyrethrum (Plate XX) and wheat are grown at higher altitudes on the mountain slopes, and sisal and some sugar at lower altitudes and in drier areas. There is a small Afrikaans community on the northern slopes of Mount Meru.

Meru and Kilimanjaro

The great peaks of Meru and Kilimanjaro are a dramatic sight when viewed from Arusha and Moshi. They tower into the tropical air from a tawny arid foreground of thorn scrub, the dark green forest belt partly hiding the cultivated plots on Kilimanjaro, while above the zone of moorland pasture, often obscured by cloud, floats its dome of ice. Over $\frac{1}{4}$ million *Chagga* inhabit the slopes of Kilimanjaro, with a density of 600 people per square mile. Most of the people live in the coffee and banana zone at an altitude of 4,000 to 6,000 feet. The *staple crops* are bananas and coffee, subsidiary crops being sweet potatoes, yams and beans. Below 4,000 feet in the drier area is the banana and maize belt. The vihamba or plots in the upper zone are 2–3 acres in size, those in the lower zone being larger (Fig 63).

Coffee is grown on a large scale by the Chagga and is a lucrative crop,

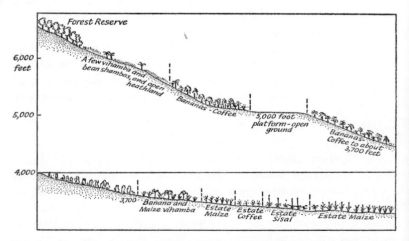

Fig. 63—Tanzania: land-use profiles, Machame road, Mount Kilimanjaro (after James)

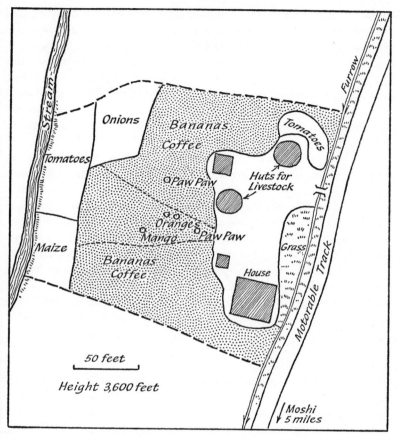

Fig. 64—Tanzania: Romani Thomas's kihamba, Mwasi, Kilimanjaro (after James)

although the return from bananas is actually greater than that of coffee. The *Kilimanjaro Native Cooperative Union* (K.N.C.U.) has played a major part in the successful development of the coffee industry. It is the largest African cooperative society with 35,000 members. The coffee trees grow under the shade of bananas which provide beer, food and mulch, while tall forest trees tower over both coffee and bananas (Fig. 64 and Plate XXI). Older huts may house cattle which are stall-fed on banana leaves and grass, fetched by the women from the plains below. Many new stone houses with corrugated iron roofs bear witness to the prosperity of the Chagga coffee growers.

There are over 2,000 irrigation furrows on the mountain; most of them were constructed before the Europeans came. The water is, however, used uneconomically, over half of it being lost through evaporation on the lower slopes. Congestion in the banana-coffee zone is forcing population on to the lower slopes where settlement schemes will have to be organized using irrigation water. Sugar is grown near Moshi and in the Kilombero Valley, production in 1964 amounting to 64,470 tons. Water supply will present a grave problem in the future, with African and European agricultural demands, the urban needs of Moshi and the hydro-electric power station at Kikuletwa.

From Mount Meru the irrigation channels thread out for 20 miles, creating ribbons of green in the tawny wastes that stretch out onto the plains of Masailand. Cotton, beans and maize are grown on pockets of fertile volcanic soil. Here, too, conflict over the use of water between European and African has arisen, although the Arusha and Meru tribes are far less advanced agriculturally than the Chagga people. The two main towns are Moshi and Arusha; the former, is the regional centre for the European farms and the site for a meat-processing factory; the latter, a flourishing town with many large new buildings, such as the K.N.C.U. headquarters, is the regional centre for the Chagga. Moshi, at a height of 2,750 feet, is lower and hotter than Arusha. Both towns are connected by rail with Mombasa and Tanga.

The Usambara Highlands

There are also European estates on the Usambara Highlands where coffee is the main crop but tea is important, and sisal is grown at lower altitudes. Commercial production of *tea* began at Ambangulu in 1931 and still continues. The tea bushes grow on very steep slopes under shade trees, and the soil is protected by bunds of grass. The bushes take 4 years to mature and last for 60–100 years. Leaves are plucked every 15 days during the dry season, then brought straight to the factory where they are dried, rolled and fermented, redried, graded and sorted and despatched mostly to the London market. About 5 million pounds weight of tea was produced in 1965. There is a high density of population in the Usambaras, and the main cash crop is rice, grown in the swampy piedmont zone. Population density is also considerable in the Uluguru mountains south of Morogoro and there is much soil erosion because cultivation is practised on very steep slopes.

D. THE SOUTHERN HIGHLANDS

The Southern Highlands, a belt of rolling grassland 5,000 to 6,500 feet high, extend from the Rungwe Mountains, which attain a height of nearly

10,000 feet and the Kipengere Range, for about 150 miles north-east to Iringa. Coffee, maize, pyrethrum, tobacco and tea are the main cash crops in the Iringa district. Over half of Tanzania's tea and a third of the pyrethrum comes from the Southern Highlands. The European farming area suffers from remoteness, being 150 to 200 miles from the Central Railway.

The main tribe in the Iringa District is the *Hehe* who form a tribal group consisting of a number of small tribes who have recently been united politically, particularly by the actions of Muyogumba and Mkwawa in the latter half of the nineteenth century. They number over 200,000, are a patrilineal tribe and are both arable and pastoral farmers. The majority live on the plateau at altitudes of 4,500 to 7,000 feet. *Maize* is the staple crop with eleusine, beans, potatoes, ground-nuts and marrows as subsidiary crops. *Livestock* are important as this is a tsetse-free area, and there are large numbers of cattle, sheep and goats.

REGION 5–THE LAKE VICTORIA BORDERLAND

The last region is the Lake Victoria Borderland. Despite the fact that the lake is divided politically by the three territories of Kenya, Uganda and Tanzania it possesses a considerable measure of geographical unity based on a fairly dense population of progressive farmers who grow cash crops such as cotton and coffee on a large scale, who have an advanced form of social organization, as amongst the Baganda, and who are linked to some extent by Lake Victoria transport.

Lake Victoria (Plate XXII) occupies a recent crustal sag between the East and West Rift Valley highlands. It is not caused by faulting and is, therefore, shallow compared to the rift valley lakes of Tanzania and Malawi. It is over 250 miles long and 150 miles broad with an area of nearly 27,000 square miles and is more like an inland sea than a lake, especially when sudden storms cause lake steamers to pitch and toss. High rainfall and the fairly level shoreline penetrated by numerous arms of the lake have attracted a *dense population*, in contrast to the slight population in scattered pockets on the mountainous shores of Lakes Tanganyika and Malawi. The effect of the sleeping sickness epidemic of the early 1900s in Uganda has, however, cleared population away from the northern lake shores. The northern and western shores of the lake receive a *high, well distributed rainfall* of up to 80 inches, whereas the eastern and southern shores are much drier with less than 40 inches. The temperatures near the lake are equable with a diurnal range of less than 15° F. (8° C.) compared to a range of 25°–30° F. (14°–17° C.) away from the lake. Lake breezes tend to modify the high humidity of the lake shore. The climate is, however, hotter and more humid than in the neighbouring highlands. The lack of climatic attraction and the presence of a dense indigenous population are among the main factors which largely preclude white settlement.

UGANDA

Most remarkable of the lake regions is *Buganda*, a rolling landscape of flat-topped Miocene residuals, swampy valleys with sluggish drainage, abundant well distributed rainfall, long grass and patches of rain-forest and rich red soils. There is a characteristic pattern of flat hill tops, averaging 4,300–4,400 feet in altitude, steep valley sides and swampy valley bottoms (Fig. 65). Two crops a year are grown and for a long time a dense cultivating population has inhabited the region, creating a humanized rural landscape. Buganda has a population of about 2 million on less than a quarter of Uganda's area. Ganda comprise about half this number. The Ganda belong to one of the most advanced social and political units in Africa. The original negro cultivators were conquered by Hamitic invaders who now form a ruling aristocracy. During the nineteenth century Buganda grew at the expense of the neighbouring kingdoms of Bunyoro and Ankole.

The head of the state is the Kabaka who is assisted by ministers and a council known as the Lukiko. Under the Kabaka is a strong and well organized system of chiefs with gombololas and murukas. The present Kabaka was, however, forced to flee from Buganda in 1966.

Economy

Under the terms of the *Uganda Agreement* of 1900 the chiefly lands comprising over half of Buganda were broken up into mailo (square mile) blocks and held under freehold tenure. This was a great incentive to the economic development of the region and there are now over 50,000 land-owners. *Cotton* proved an ideal crop and, first planted in 1906, soon formed the basis of the economy. Uganda is the chief cotton-growing country in the Commonwealth and in 1964 the export value of cotton was £16 million. Nearly 2 million acres are under cotton in eastern Buganda and the south and central parts of the Eastern Province, which now produces over half of Uganda's cotton. Uganda cotton has a long staple and fine fibre.

Coffee is the other major cash crop of Uganda and is more popular than cotton as it gives higher yields and only takes about 80 days of work com-pared to 140 days for cotton; from 1960 to 1962 yields averaged over 100,000 tons per annum, worth about £30 million. 90% of the crop is

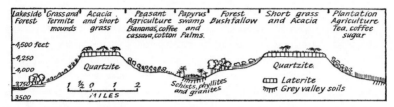

FIG. 65—Uganda: transect across south Kyagwe, showing zones of land use (after McMaster)

Robusta, grown in Buganda, and about 10% is Arabica, grown mainly on the slopes of Mount Elgon.

Both cotton and coffee are grown on small plots or shambas, the cotton being planted after food crops have been harvested. Valuable experiments have been carried out at the Empire Cotton Research Station at Namulonge on rainfall probability and crop moisture requirements. Coffee trees are interplanted with plantains, as in Bukoba and on the slopes of Mount Kilimanjaro, and surround the homestead. Here, too, post-war high prices for cotton and coffee have engendered an atmosphere of great prosperity and new, square, mud-walled, tin-roofed houses replace the older, traditional grass huts. Many farmers carry out simple forms of coffee bush pruning and mulching of bananas, although the majority of farmers in Buganda are still indifferent to the potentialities of their environment. Shambas are usually small, but one farm investigated belonging to a gombolola chief was 25 acres in size, of which 6 acres was under coffee, 6 under bananas, 3 under cotton, sweet potatoes and beans and the remaining 10 acres fallow.

Settlements tend to be concentrated on the deep, red, fertile loams of lower slopes of the residual hills between the steep upper slopes and the swampy grey earths of the valley bottom. Near Kampala hills are capped with a thin layer of laterite and the short grass growing here affords grazing for livestock, while the zone of cultivation is lower down the slope on the tropical red earths. The area of dense population in Buganda corresponds to the zone of elephant grass where *bananas and maize* are the main food crops and coffee and cotton the main cash crops. Further north conditions are drier and millet replaces maize, while cattle become important.

There are two Indian-owned sugar estates at Lugazi and Kakira near Jinja. The Kakira estate consists of 20,000 acres of which 80% is under cane, and there are 8,000 workers. In 1960 a new plantation was started near Jinja as well. Production of sugar in 1964 was over 120,000 tons. Factories on the estate produce sugar, soap, cotton-cake, vegetable oil and sweets. The great fields of cane, roads, large factory and many buildings are a great contrast to the haphazard landscape created by Busoga farming.

Most of the labour on the cotton and coffee shambas is performed by *immigrants*, mainly from the overpopulated highlands of Kigezi, Rwanda and Burundi. In the period 1931–48 the Ganda increased by about 14%, while the immigrant African population increased by nearly threefold to form over one-third of the total population of Buganda. Much of this migration is permanent, thus changing considerably the composition of the African population of Buganda. Employment on Buganda farms forms a contrast to the prevailing pattern of labour migration to areas of European plantations and enterprise in the rest of East Africa.

Towns

The main urban centre in Uganda is *Kampala* (the place of the impala) which is also the commercial capital and increasingly replaces Entebbe, 21 miles away, as the administrative capital. It is a city of many hills, with the Kabaka's palace, cathedrals and other major buildings situated on a number of the hill tops. It is really two towns, the Asian-European municipality and the Kibuga, which contains Mengo, the capital of Buganda, and its environs and encompasses the municipal area on three sides. The 1959 census gives the population as only about 44,000, including nearly 20,000 Asians and just over 3,000 Europeans, but if Mengo and the peri-urban fringe is included the population is probably over 100,000. Kampala is an important route centre and was until recently the terminus of the railway from Mombasa, which has now been extended 210 miles farther west to Kasese to serve the Kilembe copper mines. The town is well laid out with five residential zones and areas for commercial and light industrial development, while former swamps, which were malarial, have been drained and formed into open spaces. The future expansion of the town is problematical, as the Kibuga prevents extension of the municipal boundary on three sides. Another problem is divided control over the built-up area, but with Uganda's independence some form of mixed Baganda-European-Asian municipal control should be worked out.

Entebbe has grown up on the shores of Lake Victoria about 21 miles from Kampala and is the administrative capital of Uganda. The population consists almost entirely of government servants. Increasingly its functions are being taken over by Kampala. It is an attractive town with the international airport located nearby. *Jinja*, situated where the Victoria Nile leaves the lake, is the second largest town in Uganda and has a population of about 30,000. It is the commercial and industrial centre for the Eastern

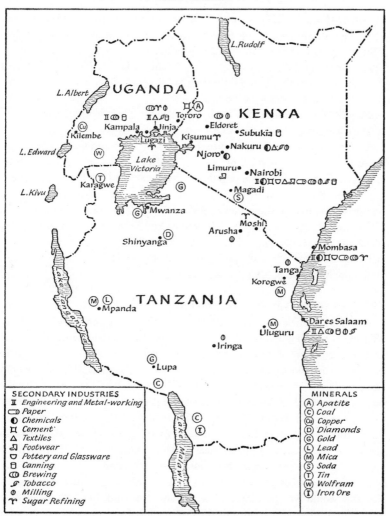

FIG. 66—East Africa: minerals and secondary industries

Province and has a copper-smelting works, a grain-conditioning plant and tobacco, textiles, plywood and cotton-ginning factories, and a brewery. The completion of the *Owen Falls Dam* and hydro-electric scheme has greatly assisted Jinja's growth as Uganda's main heavy industrial centre (Fig 66). Power is now being supplied to Kenya and there is talk of building a second dam lower down the Nile to increase the supplies of hydro-electricity.

347

Copper (Plate XXIII) is mined at Kilembe in the foothills of the Ruwen-zori range production in 1960 being valued at £3½ million and in 1964 at £6·2 million) and is smelted at Jinja before being exported. The ore has 1·8% copper and 0·14% cobalt. The Sukulu Hills near Tororo have large deposits of phosphates, and there is a cement works and an asbestos-cement products factory at Tororo.

Bukoba[8] forms an island of high rainfall south of the Kagera river, on the western shores of Lake Victoria, almost surrounded by the Kagera river, Lake Ikimba and swamps. Despite the infertility of the Bukoban sandstones there is a dense population of Haya who may have come originally from Bunyoro. Some 80% of the population live near the lake. The landscape is not unlike Buganda, with plantains and coffee trees, flat-topped hills and papyrus swamps in the valleys. *Robusta coffee* has long been grown in Uhaya and in 1966 20,000 tons of Arabica and Robusta were exported from Tanzania as a whole worth nearly £7 million. The great bulk of the 70,000 African coffee-growers have joined the Bukoba Native Cooperative Union. As in Buganda the plantain is the staple food.

THE NYANZA PROVINCE OF KENYA

Kenya's Nyanza Province, centred on the Kavirondo Gulf of Lake Victoria, has a population of some 2 million with a density of over 1,000 people per square mile in parts of North Kavirondo. Rainfall is high in North Nyanza, 60–75 inches per annum, and very reliable. The majority of the population are Kavirondo, but there are also Jaluo or Luo of Nilotic origin. Compared to Buganda development is far more backward. A variety of *crops* are produced with a large output of maize and cotton, the main cash crop, while other cash crops such as sugar, coffee and rice are also grown. A large sugar refinery near Kisumu produces much of Kenya's sugar. There are two sugar plantations east of Kisumu.

During the 1930s gold was discovered at Kakamega and it was feared that this would lead to a large-scale alienation of land, but little came of it and the last mine closed down in 1953. *Kisumu* is the main centre in Nyanza Province and is situated at the head of the Kavirondo Gulf. It has a popula-tion of about 21,000, half of whom are Asian. Until 1926 it was the railhead for the Kenya-Uganda railway line. It is an important lake port with a twice-weekly service round the lake and workshops for building lighters and assembling lake steamers.

[8] Bukoba is part of Tanzania, but has been included in the same section as Buganda as it forms part of the same high rainfall region on the northern and western shores of Lake Victoria.

SUKUMALAND

Sukumaland, with an area of about 20,000 square miles, is the last of the Lake Victoria regions. It is situated in a drier region along the southern shores of the lake and is inhabited mostly by the Sukuma who number about 1 million. Much of the natural vegetation has been cleared to create what is known as a cultivation steppe. Wide treeless plains at an altitude of 3,500–4,000 feet, forming part of the plateau and diversified by granitic outcrops, stretch south into central Tanzania. Rainfall is approximately 30 inches per annum: for example, Shinyanga has 31 inches and Mwanza 39 inches. The Sukuma are not predominantly cattle people like the Masai, but are both cultivators and stock-owners. Nevertheless, cattle are an important source of income. Cotton is the major cash crop, the area producing four-fifths of Tanzania's cotton, worth £3–4 million a year. Rice, millet and maize are also grown.

Originally the Sukuma lived in circular fortified villages on granitic outcrops, but recently they have moved away from the congested zone and occupied large new areas under the *Sukumaland Development Scheme*. The object of the Scheme was to reclaim some 10,000 square miles of central Sukumaland and open up 8,000 square miles of empty land in the Geita and Maswa districts. Fly-infested bush was cleared by communal labour, a large number of dams dug and the area settled closely. A great deal has been achieved by the Sukuma themselves, better standards of agriculture practised and greater output of cotton and food crops achieved.

There are many islands on Lake Victoria, some of which are farmed intensively. On Ukara island, because of intense pressure on the limited land available, an efficient form of agriculture developed in pre-European days. Food and fodder crops are grown in rotation and stock are stalled in huts, their manure being used for the fields. Stock numbers are controlled carefully.

Mwanza is the largest town and the main centre for the Lake Province, with a population of nearly 20,000. It is a major lake port and has rail links to the south via Tabora to Dar es Salaam. Mwadui, near Shinyanga, has the former Williamson diamond mine, now owned partly by the Tanzania government. Mining is a fairly simple affair and consists of removing the gravels and sands overlying the diamondiferous pipe. *Diamond* exports in 1964 were worth £6·8 million, the most important mineral export from Tanzania. Geita mine is the largest gold mine in East Africa, with a production of £$\frac{1}{2}$ million in 1956, but it is now operated at a loss. Gold exports for Tanzania were worth £1·2 million in 1964. Tin and wolfram are also produced in the Lake Province.

CHAPTER TWELVE

Southern Africa

SOUTHERN AFRICA occupies the southern half of the great plateau of High Africa (Fig. 67). It projects as a large spade-shaped mass into the southern oceans to 34° S. Although not projecting far enough south to experience the temperate climates of the southern parts of South America and New South Wales, of Tasmania and South Island, New Zealand, Southern Africa, except for the Maghreb, is the most temperate of African regions. The total area is some $2\frac{1}{3}$ million square miles forming nearly a quarter of all Africa and the population is about 42 million (1963).

STRUCTURE AND RELIEF

The dominant physical element is the great plateau which has an average altitude of over 3,000 feet and high upturned rims with altitudes of over 10,000 feet in Lesotho (Basutoland). Almost saucerlike in shape the plateau slopes down towards the Kalahari basin and, where cut into by major rivers such as the Zambezi and Limpopo and their tributaries, the altitude may be less than 2,000 feet. The oldest rocks forming the base of the plateau are Archaean and pre-Cambrian. The Archaean rocks consist of highly metamorphosed sedimentary and igneous rocks and are now exposed only in the eastern Transvaal and north-western Cape and South-West Africa. The pre-Cambrian rocks consist of the Dominion Reef, Witwatersrand, Ventersdorp and Transvaal systems and are of vital economic significance as they contain the gold-bearing quartzites of the Witwatersrand. Where these ancient rocks are exposed a more diversified landscape results than on the Karoo rocks of the plateau, highly dissected in some areas and with bold relief as in parts of the northern Transvaal.

The pre-Karoo rocks of the Cape system, laid down in Palaeozoic times and subsequently folded in Permo-Triassic times, form the distinctive region of the Cape folded mountains. The folded and faulted mountains trend north-south and then east-west more or less parallel with the coast, with longitudinal troughs and mountain-girt basins breaking through to the coast by short transverse gaps known as poorts.

The beginning of the Karoo era was marked by the Dwyka (Carboniferous) glaciation, followed by the swampy forests of Ecca times when the vast

FIG. 67—Southern Africa: political divisions

coal measures of Southern Africa were laid down. In the arid conditions of the Beaufort and Stormberg periods there were great sandy wastes subsequently covered by wide sheets of lava which built up the Lesotho plateau. Southern Africa may have formed part of the super-continent of Gondwana until the late Jurassic when it broke up accompanied by widespread faulting and the formation of the Rift Valley. Several periods of uplift and pediplanation have followed since Cretaceous times, the most important being the warping of Tertiary times when the plateau was updomed around the coast and the Kalahari basin formed, which was subsequently covered by sand in Pleistocene times.

Because South Africa has been dry land since Palaeozoic times and affected predominantly by epeirogenic movements, erosion has been long continued and the Gondwana, post-Gondwana and African cycles have left widespread traces of their course in the form of vast peneplanes on the plateau and terraces on the slopes. Pediplanation, horizontal bedding of Karoo rocks and vulcanicity have caused a monotonous landscape of wide flat plains interspersed with flat-topped and conical hills known as koppies, capped with basalt, dolerite and Cava Sandstone, while doleritic dykes wind sinuously across the country. In Rhodesia the erosion of granite intrusions has caused rounded granitic tors rather than the characteristic koppie. Near the eastern margin of the plateau and the seat of the Stormberg lava flow in Lesotho the terrain is more dissected and angular, but it is on the eastern plateau slopes that dissection is most advanced and there is a wild sweep of deep valley and plunging mountain ridge.

Apart from a narrow coastal plain in northern Zululand, widening out in Mozambique, Southern Africa lacks a true coastal plain. The plateau is close to the coast except in the Southern Cape and southern Mozambique and descends precipitously from its towering fretted edge, known as the Great Escarpment, in a series of broad terraces to the coast. The *Great Escarpment* is a basic element in the physical geography of Southern Africa and varies in altitude from 4,000 to over 11,000 feet in height in theDrakensberg where Stormberg lavas overlie Karoo sediments. The escarpment is least impressive where capped by softer Karoo rocks.

DRAINAGE

The escarpment is an erosional and not a faulted feature, which is being driven back rapidly on the east coast by a series of strong coastal streams, such as the Great Kei, Tugela and Pongola which descend rapidly to base level in the Indian Ocean. The plateau edge is much closer to the coast in the drier west where river erosion is less pronounced. The major rivers, such as the Zambezi and Limpopo, rising on the plateau and draining eastwards into the Indian ocean, have cut great embayments into the plateau. A major peculiarity of Southern Africa's drainage is the fact that the Orange river rising near the east coast in Lesotho then curves south and west to flow into the Atlantic, while the Zambezi rising on the high plateau of Angola flows south and east to the Indian Ocean. The Zambezi is by far the largest river of Southern Africa and with its major tributaries, such as the Luangwa and Shire, has a mean annual flow of 80 million acre feet compared to the 9 million acre feet of both the Orange and Vaal.

Over much of the great interior basin of the Kalahari drainage is interior and does not reach the sea. The major system is that of the Okovango,

XXI. Plantains and coffee bushes, Mount Kilimanjaro.

XXII. Fishermen near Kisumu, Lake Victoria.

[*Uganda Information Dep*]

XXIII. Kilembe copper mine, Uganda.

XXIV. Cape Town and the Cape peninsula.

[*South African Information Service, Pretoria*]

which flows into an inland delta of Lake Ngami, and the Makarikari Salt Pan. Many of the streams that used to flow into this inland system have been abstracted by more vigorous tributaries draining to the Zambezi.

As rainfall over most of Southern Africa has a decided summer maximum, river régimes fluctuate greatly between the floods of summer and the low water of winter. Rivers are of little or no use for navigation as they descend steeply from the plateau edge to the sea and are obstructed by many falls and rapids. Only the Zambezi is of some use below the Kebrabasa Rapids and above the Victoria Falls. Southern Africa is also badly served with natural harbours because of its regularity of outline, lack of major bays and siltation of river mouths.

CLIMATE

Southern Africa has a great diversity of climate, ranging from the sub-Alpine climate of the high Lesotho plateau to the heat and humidity of the Zambezi delta and the aridity of the Namib. The influence of the Great Escarpment on the climate is profound and there is a marked discontinuity between circulations over the plateau and the marginal areas of Southern Africa. These circulations have a significant influence on the Tropical Continental, Superior and Sub-Tropical, Polar and Sub-Polar, and Equatorial air masses which affect Southern Africa. Temperatures are modified by the altitude of the plateau which may be 10°–20° F. (6°–11° C.) cooler than along the east coast. The dry air, maximum insolation and radiation and high elevation of the plateau cause a considerable temperature range. After a warm day in the winter, with a maximum temperature of over 80° F. (27° C.) the chill of night descends swiftly and frosts occur frequently. Frost may indeed occur for 100–150 days over much of the interior of South Africa.

Ocean currents play an important part along the east and west coasts. The warm Mozambique current along the east coast causes high temperatures and a rainfall of over 50 inches in places, while along the west coast the cold mass of the Benguela current swinging equatorwards causes arid, yet cool conditions.

Rainfall is mostly summer maximum and is brought by the Tropical Maritime air mass, blowing in from the Indian Ocean from November to April. Winter, except for the cyclonic rainfall of the south-western Cape, is a dry period when the sub-continent is dominated by high pressure and outward-flowing winds. Snow may fall on the higher parts of the plateau and the Cape mountains when invasions of sub-polar air surge far north. Rainfall is characterized by its great variability and intensity, the bulk of

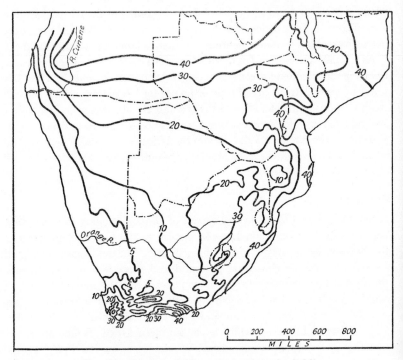

Fig. 68—Southern Africa: average annual rainfall

the annual rainfall occurring in a few heavy showers. About half of Southern Africa has less than 25 inches of rain, while in the Republic of South Africa two-thirds of the country has less than 15 inches (Fig. 68). The eastern and northern sections of the region receive most rainfall, Zambia, Rhodesia, Angola and Malawi having a more tropical régime than areas further south. Rainfall is also high along the edge of the Great Escarpment, but diminishes gradually to the west, the lower Orange River valley and Namib receiving less than 5 inches. Evaporation is considerable and is from 60 to 120 inches a year compared to 20 inches in south-east England.

As in Australia, lack of water is probably the greatest limitation to agricultural and industrial development. It is calculated that less than 10% of the precipitation forms ground-water and much of this is being jeopardized by the ravages of soil erosion.[1] By 1975 the water resources of the Vaal

[1] Despite this fact underground water supplies are very important and it is estimated that half the farming population depends on boreholes for water. Problems are the high salinity of underground water and the falling water table in many areas.

354

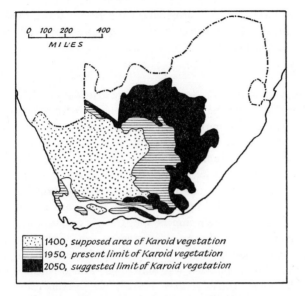

FIG. 69—Southern Africa: the spread of Karoid vegetation

may be used up because of the rapid mining, industrial and urban growth in the Southern Transvaal. Further industrial expansion will have to take place on the better watered eastern plateau slopes and river basins. The need for an integrated approach to problems of controlling and developing water resources and the importance of utilizing river valleys and basins as regional units for social and economic development has been stressed in recent planning.

VEGETATION

Of great importance to pastoralism in Southern Africa is the distribution of vegetation, as livestock rely so heavily on natural grazing. In areas of less than 15 inches of rainfall the Karoid vegetation of low succulent bushes and sparse grass is valuable for small stock (Fig. 69). Grasses are predominant in areas of over 4,000 feet in altitude and over 15 inches of rain, and in the higher wetter areas sourveld with *Themeda triandra* or red grass is character- istic and is very similar to the Kangaroo grass of Australia; in the lower and drier parts sweetveld, in contrast to the sourveld, can be grazed for most of the year. To the north and west of the grassland is bushveld, important as a ranching area with trees of a deciduous variety, larger and more closely spaced in the wetter areas, but in the Kalahari degrading to stunted acacias and thorny bushes. The sclerophyllous vegetation of the south-western Cape consists of small evergreen trees and shrubs not unlike the maquis of the Mediterranean coast of North Africa.

HISTORY

European contact with Southern Africa dates from the end of the fifteenth century with the rounding of the Cape by the Portuguese. The Cape Peninsula was the most suitable watering point south of the Portuguese domains and it was used regularly by East Indiamen on the long voyages to and from Europe and India. Not until 1652 was it occupied permanently by the Dutch, although its strategic value was also appreciated by the French and English. It was taken over by the English during the Napoleonic Wars and finally occupied by them in 1806. European powers, however, saw the Cape more as a refreshment and strategic outpost than as a colony of settlement. As a result economic progress was very slow away from Cape Town and its immediate environs. By the end of the eighteenth century the white population was only about 20,000, and even by 1820 when some 4,000 British settlers arrived, the forerunners of the substantial English-speaking population of South Africa, it was only about 40,000.

Apart from the small town of Cape Town and the settled farming community growing fruit, vegetables, wheat and vines, the rest of the white population consisted of wandering pastoralists who had moved slowly eastwards displacing the Bushmen and forcing the Hottentots to labour for them. It was in the eastern Cape that the Boers clashed with the vanguard of the southward-moving Bantu pastoralists who were stronger and better organized than the Hottentots and Bushmen. In the 1830s the Boers moved away from the hinge-line of Border conflict on to the high veld of the Orange Free State and Transvaal and established their republics on these wide expanses of treeless grassland. The British advanced spasmodically along the coast and on occasions extended their authority over the interior plateau. The Cape and Natal became British colonies, and the Bantu, defeated in a hundred years of warfare, were confined to a scattered horseshoe formation of reserves half surrounding the high veld.

Although South Africa can be regarded as a white man's country, yet with its majority of Bantu inhabitants it is equally an African country. The *Bantu*, superior in numbers and organization and with far greater receptivity to new ideas and crops, did not suffer the fate of other indigenous groups, such as the Bushmen and Hottentots in Africa, Red Indians and Australian Aborigines. Their numbers have increased greatly and they now outnumber the whites by more than three to one in South Africa and by nearly three hundred to one in Malawi. The minority of whites in South Africa, Rhodesia, Angola and Mozambique is, however, determined to retain political control. In South Africa the Afrikaners are attempting to achieve this by separate development of black and white areas. In Angola and Mozambique the policy of identity allows only a trickle of assimilated Africans to become Portuguese citizens and so effectively cuts down the number of aspirants.

The discovery of gold and diamonds in the 1870s caused a *mineral* and later urban and industrial revolution that transformed the economy and society of South Africa overnight. There is also a wealth of other minerals in South Africa, including great reserves of coal and iron ore. Rhodesia was thought to be a second Rand, but expectations were disappointed and the economy of the country has come to rest on a basis of mining, agriculture and secondary industry.

Productive sector	Percentage contribution of individual sectors	
	1952–3	1962–3
Agriculture, forestry and fishing	16·2	10·4
Mining	12·2	13·1
Private manufacturing	24·3	25·5

Nevertheless, gold, asbestos, coal, chrome, copper and iron ore are mined in significant quantities in Rhodesia, and in Zambia copper, exploited after 1930, dominates the economy in much the same way as gold has dominated the economy of South Africa.

ECONOMY

The accompanying map shows the distribution of *economic minerals* and their association with Southern Africa's pre-Cambrian foundation (Fig. 70). They stretch in a long curving line from the Katanga to the northern Cape, sandwiched in between Karoo and coastal sedimentaries in the east and Kalahari sands in the west. There is an offshoot of this mineralized belt in South-West Africa. This great mineralized zone with its treasure house of wealth is probably the richest in the world, and to date over £7,000 million worth of minerals have been extracted from it. Distribution of exploitable minerals is uneven and this factor in particular has given rise to the patchy growth of urban, mining and industrial communities in Southern Africa.

Among the various countries of Southern Africa the Republic of South Africa is an economic giant, with an income *per capita* twice the average for African countries. South Africa produces over 40% of Africa's minerals by value and 20% of the food in Africa south of the Sahara, and is far more developed industrially than any other of the African countries. The economy of Southern Africa is the most complex and developed in Africa and most dominated by Europeans. The resultant pattern is of a series of islands of intense European activity set in a sea of backward rural areas inhabited predominantly by Africans. In 1965 South Africa had an estimated population of 17·8 million, nearly half the population of Southern Africa, and consisting of 12½ million Africans, 3⅓ million Europeans, 1¾ million Coloured and ½ million Asians. The rural density of population may exceed 200–300 per square mile in parts of the Ciskei, Transkei and Natal in South Africa, and in Mozambique and Malawi. These densely populated areas with a backward economy export labour in large numbers to the nodal areas of great economic activity, often located at considerable distances from the labour reserves.

LABOUR SUPPLIES

Migration of labour occurs on a scale unparalleled in Africa and is carefully controlled and organized. Over half of Rhodesia's 600,000 African workers come from beyond her borders, nearly 40% of them from Malawi. The biggest movement of migrant Africans in South Africa is to the gold and coal mines of the Transvaal, Orange Free State and Natal which employ nearly 400,000 Africans, most of whom come from outside the Republic. This massive movement of labour is in response to the scattered distribution of the densely populated labour reservoirs, such as Rwanda and Burundi, Mozambique and Malawi and their great distance from the economically developed areas. Migration of labour may be viewed on the one hand as an attempt to bridge the gap between subsistence and exchange economies, but on the other hand as a social and economic problem of great magnitude because of the constant drain of so many of the African males from the rural areas.

URBAN SETTLEMENT

Rapid urbanization is now the major economic, social and political factor in Africa and the drift to the towns is most marked in Southern Africa. From 1936 to 1960 the population of Salisbury expanded sevenfold and that of Johannesburg doubled in size. From 1936 to 1951 the European population of rural areas and of small towns with less than 1,500 whites

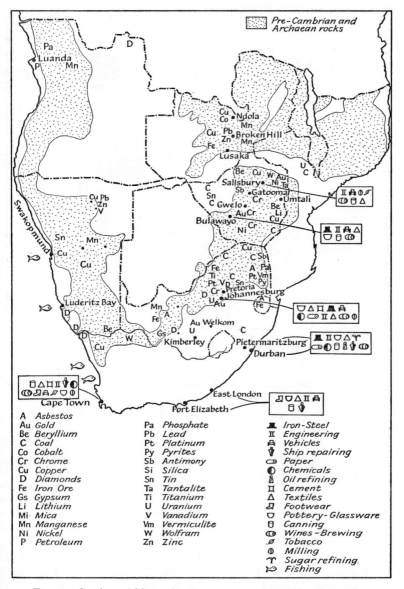

Fig. 70—Southern Africa: mineral resources and secondary industries

A   Asbestos
Au   Gold
Be   Beryllium
C   Coal
Co   Cobalt
Cr   Chrome
Cu   Copper
D   Diamonds
Fe   Iron Ore
Gs   Gypsum
Li   Lithium
Mi   Mica
Mn   Manganese
Ni   Nickel
P   Petroleum

Pa   Phosphate
Pb   Lead
Pt   Platinum
Py   Pyrites
Sb   Antimony
Si   Silica
Sn   Tin
Ta   Tantalite
Ti   Titanium
U   Uranium
V   Vanadium
Vm   Vermiculite
W   Wolfram
Zn   Zinc

Iron-Steel
Engineering
Vehicles
Ship repairing
Paper
Chemicals
Oil refining
Cement
Textiles
Footwear
Pottery-Glassware
Canning
Wines-Brewing
Tobacco
Milling
Sugar refining
Fishing

359

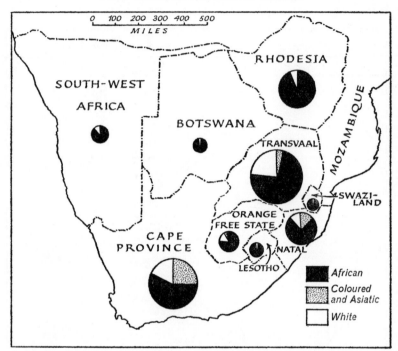

Fig. 71—Southern Africa: the racial composition of the population, 1960

declined in South Africa by over 30,000. In Rhodesia two-thirds of the increase of European population from 1936 to 1951 occurred in Salisbury and Bulawayo. The African urban percentage in South Africa has trebled in the last 25 years and the Africans are increasing at a faster rate than the other urban groups of Europeans, Asiatics and Coloureds. The urban nodes of the southern Transvaal, Rhodesia and the Copper Belt of Zambia and their associated ports of Cape Town, Port Elizabeth, East London, Durban, Lourenço Marques and Beira contain a population of about 5 million which is growing rapidly and is about one-eighth of the total population of Southern Africa.

The contrast between the small areas of marked economic growth and the rest of Southern Africa is borne out even more strongly by the vast areas of very sparse population such as the Kalahari and the Limpopo-Zambezi valleys, where controls such as aridity, tsetse fly, disease, soil poverty, heat and low altitude restrict population to less than 10 per square mile. There is a somewhat fortunate coincidence between the distribution

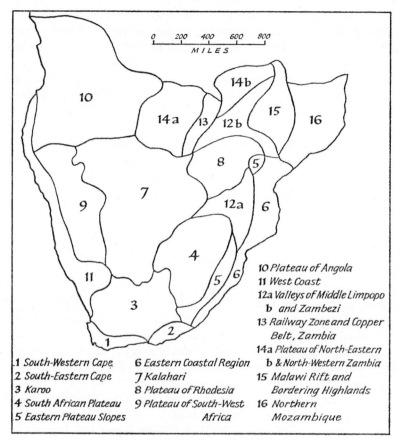

0   200   400   600   800
MILES

10 Plateau of Angola
11 West Coast
12a Valleys of Middle Limpopo
  b and Zambezi
13 Railway Zone and Copper
   Belt, Zambia
14a Plateau of North-Eastern
  b & North-Western Zambia
15 Malawi Rift and
   Bordering Highlands
16 Northern
   Mozambique

1 South-Western Cape   6 Eastern Coastal Region
2 South-Eastern Cape   7 Kalahari
3 Karoo   8 Plateau of Rhodesia
4 South African Plateau   9 Plateau of South-West
5 Eastern Plateau Slopes   Africa

FIG. 72—Southern Africa: geographical regions

of economic minerals and high healthy uplands in the Transvaal, Zambia and Rhodesia.

The pattern of population distribution and economic activity in Southern Africa is more akin to that of Australia and South America than to Europe where settlement and economic activity are more evenly spread. Nevertheless, the stratification of society and economy in Southern Africa, where Africans perform most of the unskilled work at low wages and Europeans the skilled and professional work, is quite unlike that of New Zealand or Canada where manual labour is given a high reward and there are no social and political barriers to advancement. Race relations become increasingly strained in a society where, besides distinctions based on class, there are also other and more severe limitations based on colour (Fig. 71). The social, economic and political supremacy of the white man is being challenged successfully in the northernmost parts of Southern Africa where African majorities control Malawi and Zambia. The remaining bastions of white domination are being assailed: the Federation of the Rhodesias and Nyasaland has not survived and it may not be long before the Portuguese and Afrikaans forms of colonialism succumb to the lever of African nationalism.

The map outlining the minor regions of Southern Africa is the most detailed for any major African region and reflects the greater diversity of the physical background and the greater complexity of man's reaction to the physical forces (Fig. 72).

REGION I—SOUTH-WESTERN CAPE

CLIMATE

The first subregion is the south-western Cape, the main feature distinguishing it from the rest of the region being its unique Mediterranean climate. Rainfall, brought by the north-westerly frontal depressions from the South Atlantic, has a winter maximum. In contrast to the rest of Southern Africa, much of the rain falls in steady showers. As evaporation and transpiration are at a minimum during this season this type of rainfall is of greater value to agriculture than elsewhere, and for cultivation without irrigation the minimum is about 15 inches. The effect of the Cape folded mountains causes sharp variations in rainfall, with a marked rainshadow in the longitudinal depressions such as the Little Karoo and Breede river valley, and amounts of up to 100 inches on some of the Cape mountains.

Summer is a warm dry season with little rainfall and higher temperatures, mitigated to some extent by the boisterous south-easterly wind known as the Cape Doctor. Temperatures vary according to relief, aspect and season, but are about 55°F. (13° C.) in the lowlands in winter and 70° F. (21° C.) in summer. There is a higher range of temperature in the interior basins with frost in the winter and maxima of over 100° F. (38° C.) in the summer.

RELIEF AND STRUCTURE

Another major distinguishing feature of the south-western Cape is the belt of *folded mountains* trending approximately north and south in ranges such as the Olifants river mountains, the Cold Bokkeveld mountains and

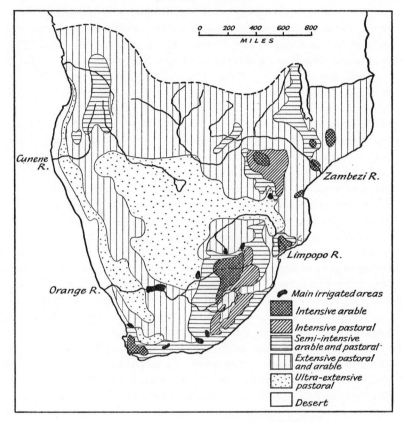

FIG. 73—Southern Africa: land use (after Wellington)

the Great Winterhoek mountains which meet the east-west trending ranges of the Langeberg and Zwarteberg in the Hex river valley. Pressure from the south caused folding in Permo-Triassic times which was followed by later marine deposition, Cretaceous faulting and finally river erosion that exhumed the earlier folds. The main ranges, generally anti-clinal, are composed of Table Mountain Sandstone and rise gauntly and spectacularly from the valleys and basins eroded from the soft Bokkeveld shales.

AGRICULTURE

The south-western Cape has been farmed continuously for over 300 years. It was in the Cape Peninsula and on the slopes of the Hottentot

Holland mountains that the first free burger farms were established. Vineyards, orchards and wheatfields proclaim the Mediterranean climate and the graceful Cape Dutch farm-houses with gabled roofs the longevity of vine, wheat and fruit farming at the Cape. The western and southern lowlands, lacking permanent water for irrigation, grow wheat, while the well drained lower mountain slopes with loamy soils derived from shales, which can be irrigated in the summer, are used for fruit growing.

The main *wheat* areas are in the Swartland and the Ruêns. The Swartland stretches north of Cape Town for some 60 miles between the Olifants river mountains and the sea. Soils are shallow and consist of gravelly and sandy loams lacking nitrogen and phosphate and liable to erosion, which has become serious through long years of overcultivation. The Ruêns, or ridges, consist of a more undulating wedge of country between the small towns of Caledon, Bredasdorp and Swellendam and to the west of Cape Town. Soils are more fertile than in the Swartland. Rainfall in both areas is from 11 to 15 inches which is sufficient for the cultivation of wheat. Owing to government subsidies the production of wheat has trebled since 1930 and is now between 7 and 9 million bags a year, but the average yield is very low and in the south-western Cape is only about 3 bags to the acre. Were it not for the government subsidy many areas here would have to concentrate on other crops and livestock rather than wheat.

The south-western Cape is the main *deciduous fruit* growing region in Southern Africa. Winter temperatures, although generally suitable, are often too high for the trees to have the necessary period of dormancy and, therefore, cause excess growth and insufficient fruit. The various types of fruit trees are grown under different climatic conditions; for example the mean summer temperatures for peaches and pears should be more than 75° F. (24° C.) and for apples and plums below 68°–70° F. (20°–21° C.). About one-third of deciduous fruit trees are peaches, and they are more widely distributed than pears, which are grown in the Ceres area and Hex river valley, and apples, which come from the Langkloof or Elgin.

Viticulture is important and vineyards, covering some 160,000 acres, are widely distributed, about half the production coming from irrigated areas in the Hex and Breede river valleys, the rest from the coastal belt where the rainfall is higher. Areas near the railway concentrate on table grapes, but those further away, with higher transport costs, produce wine. Climatic criteria are also important for viticulture and those parts with mean summer temperatures of over 85° F. (30° C.) can only produce sweet wines with a high alcoholic content, while light dry wines come from areas with temperatures below 75° F. (24° C.). A great variety of wines is produced, while the quality and quantity of wine has fluctuated widely for a long period. About 70 million gallons of wine are now produced. Exports were considerable in the early nineteenth century to a protected market in Britain, but fell drastically when protection was removed. Quality has been improved in recent years and exports of wine are now worth over £2 million a year (1964). Table grapes, raisins, sultanas and currants are also exported.

The deciduous fruit industry is well organized and exports of fresh, dried and canned fruits and jam have increased in the last decade. During the 1964–5 season 155,000 tons of fruit were exported, over half consisting of apples, with grapes and pears nearly a quarter each.[2] Britain obtains one-third of her imports of canned fruit from South Africa. Exports of preserved fruit were worth £18 million in 1964. The south-western Cape is favourably situated for export and has the advantage of being able to market its fruit during the European winter.

The main Turkish *tobacco* growing region in South Africa is in the Stellenbosch district and leaf of a high quality is produced.

Dairying has developed near Cape Town and the larger towns of the south-western Cape. Lambs are reared in the Swartland, while in the drier areas further east the main concentration is on merino wool.

CAPE TOWN

Cape Town (Plate XXIV) with a population of ¾ million is Southern Africa's second largest city. It is the oldest urban settlement in South Africa dating back to the Dutch occupation in 1652. It grew up around the fort, stream and vegetable gardens on the northern slopes of Table Mountain and at the head of the Cape Peninsula. Its growth during the eighteenth century was slow and at the beginning of the nineteenth century its population was about 15,000. Urban development was fairly rapid after the mineral discoveries in 1870, but restricted by the steep slopes of Table Mountain and the outliers of Signal Hill, Lion's Head, Devil's Peak and the sandy flats, vleis and marshes of the Cape Flats to the east, a tombola joining the Cape Peninsula to the mainland. Cape Town has, therefore, tended to develop in linear form along the eastern slopes of the mountains comprising the backbone of the Cape Peninsula. The Cape Flats have allowed an easterly urban and industrial sprawl. The municipalities of Bellville, Parow and Goodwood are situated on the Cape Flats and have a combined population of over 160,000. Greater Cape Town, therefore, has a population of about 1 million. The construction of the new harbour, the Duncan Dock, on reclaimed land and the moving of railway station

[2] In 1964 exports of apples and grapes were worth over £10 million.

and goods yard on to this land has enabled the crowded commercial core to expand seawards. The majority of Cape Town's population of $\frac{3}{4}$ million consists of non-whites, mostly Coloured who have invaded former deteriorating European areas such as Woodstock and Salt River. Rapid urban expansion has caused a serious lack of housing. Temporary huts known as pondokkies were put up on the badly drained Cape Flats by Coloureds and Africans who have drifted into Cape Town in large numbers since the war. Most of these appalling slums have been eliminated and new sub-economic housing estates built for the Coloured people on the Cape Flats.

Cape Town is the legislative capital of South Africa and the main cultural centre. It is the largest passenger port in Southern Africa with over half the passenger traffic of the Republic. In the period 1963–4 Cape Town handled $5\frac{1}{2}$ million tons of cargo. It can accommodate 40 ocean-going ships and has the largest dry dock in the Southern Hemisphere. It is also an important railway centre with main lines (although single track) to the Eastern Cape, Orange Free State, Transvaal, Zambia and Rhodesia. It is one of the premier tourist resorts in Southern Africa, catering for many thousands of visitors each summer who enjoy the hot dry weather and splendid beaches.

Despite its location a thousand miles from the Witwatersrand Cape Town is second in importance to the Southern Transvaal as an industrial centre. It has a good location for imported raw materials and industries catering for overseas markets. There is a wide range of industries reflecting the local hinterland with its Mediterranean climate and port activities with textile, clothing and footwear factories and food-processing, engineering and other industries.

Stellenbosch was founded soon after Cape Town and now has a large Afrikaans-speaking university, the counterpart of Cape Town's English-speaking university. Paarl, Worcester, Wellington, Ceres, Robertson, Bredasdorp and Swellendam are agricultural centres of some importance. Paarl, Worcester and Wellington are also minor industrial centres.

REGION 2—SOUTH-EASTERN CAPE

The south-eastern Cape is a transitional area five hundred miles to the east that has acquired a distinctive regional personality largely for historical reasons. It was here that the Boer pastoralists first met the Xhosa in the van of the Bantu invaders from the north. A hundred years of warfare has caused this region to be known also as the Border. It was here, too, that the 1820 settlers of British stock settled, and from here that most of the Voortrekkers left on the Great Trek in the 1830s. Later German settlers arrived. Today the region bears the stamp of English, German and Bantu settlement and warfare, with European farms interspersed among African reserves in the Ciskei.

RELIEF AND CLIMATE

The region is physiographically a mixture of the Cape folds and eastern plateau slopes. Cape folds die out near Port Elizabeth, while further north the line of the Winterberg-Amatola mountains, forming part of the Great Escarpment, overlooks a low plateau to the south and broken country to the north, which forms the basin of the Great Kei. Drainage is superimposed and deeply incised. The region is also transitional in climate with elements of both winter and summer rainfall, amounts varying from less than 10 inches inland in the west and the Sundays river valley to 50 inches in the Amatola and Kologha mountains. The temperature range increases inland with snow on the Amatolas in the winter and maximum temperatures of well over 100° F. (38° C.) in the deep valleys in summer. The coastal strip has no frost, and lower summer temperatures than inland.

AGRICULTURE

Agriculture is the most important activity and varies greatly from the backward African farming of the Ciskei reserve to extensive pastoralism on Karoo farms and specialized fruit production in the Sundays, Kat and other valleys. In parts of the Ciskei and on the slopes of the Amatolas the density of African population is too high and the carrying capacity of the land for stock greatly exceeded. A weed Helichrysum has invaded part of the Amatolas, much of the grass cover has been destroyed and gullying is serious.

Among the Nguni tribes the household or *kraal* is the basic social unit. With increase of population in the Ciskei the former scattered kraals are replaced by a dense pattern of settlement with straggling groups of huts not unlike villages. An increasing number of Africans are landless and in some areas land is held under individual tenure. Small patches of arable land, usually less than 10 acres per family, and a few livestock are insufficient to feed the family, so that the majority of adult males have to seek work in the European areas. Maize is the main crop, followed by kaffircorn (sorghum), beans and peas, but yields are very low with 1-2 bags to the acre in a good year and nothing in a bad year.

With its multiplicity of small plots forming a patchwork quilt, and large unenclosed areas of land reminiscent of medieval Europe, the African reserves, scattered among the more orderly European farms, illustrate clearly the wide difference in agricultural technique between European and African farmers that is prevalent elsewhere in Africa.

The production of *wool* from merino sheep was developed by the 1820 settlers. Today woolled sheep are predominant in the arid area of the west and also on the better watered areas south of the Great Escarpment, where beef cattle are also important. *Dairying* is carried on in the coastal zone near the urban markets of Port Elizabeth and East London. Cattle graze on the sour grasses and their feed is supplemented by fodder crops. Dairying has also developed in the irrigated areas of the lower Sundays valley where much lucerne is grown. Besides lucerne the growing of citrus fruit is important in the lower Sundays valley. Pineapples are produced on a considerable scale in the coastal area near East London and Peddie. The rapid siltation of Grassridge dam and Lake Arthur has reduced greatly the amount of water available for irrigation, but the implementation of the Orange River Scheme, whereby water is to be diverted from the upper Orange into the headwaters of the Great Fish river by means of a long tunnel and canals, would increase greatly the amount of water available for irrigation in the south-eastern Cape.

TOWNS

The region is served by the two major ports of East London and Port Elizabeth which has helped further to focus regional consciousness. Railways go from these ports to the Orange Free State, Transvaal, Zambia and Rhodesia. They are the outlet for agricultural exports of fruit and wool and have developed food-processing, tanning, boot and shoe, clothing and furniture and motor-car assembly industries.

Port Elizabeth is the larger of the two cities with a population of about 200,000, and it is the third port of the Republic, handling a record total of 4 million tons of cargo in 1963-4. It developed after the arrival of the 1820 settlers and displaced Uitenhage and Graaf Reinet as the chief centre for the South eastern Cape. *East London* has only half the population of Port Elizabeth, but is South Africa's only river port at the mouth of the Buffalo river. Port facilities have been improved by the provision of a turning basin at the mouth of the river. *Grahamstown* is a university city with somewhat the same relationship to Port Elizabeth that Pietermaritzburg bears to

[South Africa House

XXV. Western Deep Levels Mine, South Africa.

XXVI. Johannesburg from Hillbrow, with mine dumps in the background.

[South African Information Service, Pretoria

XXVII. The Drakensberg near Bergville, Natal.

XXVIII. Limpopo Valley Settlement Scheme, Mozambique.

Durban. During the nineteenth century it was the main centre for the 1820 settlers. Smaller towns such as Alice and Fort Beaufort have developed round forts established by the British government during the Border wars. They have not grown greatly in size and serve as local route and market centres with shops grouped round the large main square where ox wagons used to outspan. *King Williamstown* and *Queenstown*, are larger in size and serve a wider agricultural area.

REGION 3–KAROO

Resting on a long coastal base consisting of the south-western and south-eastern Cape is the large arid region of the Karoo which stretches north to include the middle Orange river valley. Its eastern and western boundaries are transitional, the eastern boundary being slightly west of the Sundays river and Graaf Reinet. Bisected by the Great Escarpment it consists of two major parts with the Great Karoo below the Escarpment and the Upper Karoo on the plateau. The main regional distinguishing features are the *predominance of Karoo rocks* and the general aridity which precludes cultivation without irrigation.

THE GREAT KAROO

The Great Karoo consists of a series of dissected basins of rivers such as the Gouritz which rise in the Great Escarpment and cut through the Cape mountains by means of spectacular poorts. River beds are dry for most of the year and the bulk of the water is from underground sources. There is no artesian basin in the Great Karoo as the Beaufort and Stormberg sandstones are not porous enough and doleritic intrusions impede the flow of underground water. *Rainfall* is from 5 to 15 inches a year, mostly occurring in winter. The vegetation consists of scattered Karoo bushes which afford year round grazing for sheep, but there is little grass and much of the ground is bare.

Extensive pastoralism with merino sheep is the main activity. The Cape Province has $\frac{2}{3}$ of South Africa's sheep of which over 80% are woolled. The percentage of merino sheep increases further east where higher rainfall encourages a meagre growth of sweetveld as well as Karoo bushes. The carrying capacity of the western Karoo is low and 10 acres will only support 1 sheep compared with 3 acres in the east. Farms in the west are usually very large—over 5,000 acres in size. Wool production and the numbers of woolled sheep have fluctuated widely in South Africa as the result of very low prices for wool during the 1930s and the recurrent effects of drought. Numbers reached a peak of 44 million in 1931 but now only number 37 million. The average yield of wool increased from about 6 pounds per sheep in the 1930s to nearly 9 pounds in the 1950s. The wool clip is second only to gold in export value and in 1964–5 was worth £50 million.

Throughout the Karoo windmills tapping underground sources of water are a familiar sight. High wool prices since the Second World War have encouraged farmers to build a large number of small dams to provide water for stock. A limited amount of irrigation is possible, using water from streams that flow from the Great Escarpment, and crops such as lucerne are grown.

THE UPPER KAROO

The Upper Karoo is somewhat similar structurally to the Great Karoo most of the area being occupied by Karoo sedimentaries. It lies at a higher level than the Great Karoo at an altitude of 3,000 to 4,000 feet and comprises the dry southern tip of the great plateau. Rainfall is from 5 to 15 inches each year, but is more unreliable and variable than in the Great Karoo.

The main activity is *extensive pastoralism* with sheep and goats. There are fewer merinos than in the eastern part of the Great Karoo and Afrikander and Blackhead Persian sheep are reared for mutton. In areas with less than 10 inches of rainfall Karakuls are reared for their pelts, production being about one-fifth of that in South-West Africa. The number of goats has declined greatly and with it the production of mohair. Farms are very large, up to 20,000 acres in size, and only the windmills and groves of eucalyptus trees break the monotony of the wide dusty plains.

Irrigation farming is important along the middle Orange between Upington and Kakamas where some 40,000 acres are under irrigation and crops such as lucerne, cotton, wheat and dried fruits are produced. For most of its course the Orange river is deeply incised and little deposition occurs. It is only near Upington that the river splays out and conditions are suitable for irrigation. The government has done much to encourage irrigation cultivation near Upington as part of a general plan to assist indigent white farmers. Saaidam or flood irrigation is practised along intermittent tributaries of the Orange such as the Zak and Hartebeest and wheat is produced, when rainfall is sufficient.

The total irrigable area in South Africa is now about 1½ million acres. The possibility of increasing the irrigable area is fairly limited because of the lack of surface water, infertile soil, deeply incised valleys and high rate of evaporation and siltation. The Orange River irrigation and hydro-electric scheme, the largest yet undertaken by the government, should aid greatly in the development of the Upper Karoo. The whole scheme will take 30 years to complete and cost over £200 million. Three dams are to be built on the middle Orange and the largest will have a capacity of 133 million acre feet. Most of the water will be led by means of tunnels, one 51 miles and the other 32 miles long, and canals into the Great Fish and Sundays rivers. It is hoped that the project will bring about ¾ million acres under irrigation, and the main crops to be grown are lucerne, cotton, wheat and fruit. An increase in the production of lucerne will eke out the poor fodder supplies and increase wool production and dairy products. In addition to providing irrigation the Orange river's water will be harnessed to provide hydro-electric power, and 20 hydro-electric stations are planned along the course of the river with a potential of 177,000 kw.

REGION 4–SOUTH AFRICAN PLATEAU

The South African Plateau is the most important region in Southern Africa as it contains the most valuable resources, the largest urban settlement and the most developed agriculture and secondary industry. The western boundary is transitional, but is approximately along the 15-inch isohyet and therefore includes Kimberley and the Kaap plateau. To the north the region stretches as far as the Zoutpansberg and overlooks the Limpopo trough, while to the south it has a narrow base along the northern border of the Eastern Cape. The eastern boundary is approximately along the line of the Great Escarpment and includes Lesotho (Basutoland).

RELIEF, STRUCTURE AND CLIMATE

The core of this region is the *high veld* whose extensive even surface is a legacy of the Gondwana peneplane and horizontal bedding of Karoo rocks. Variety of surface relief occurs only where the Karoo rocks have been stripped off, as in the northern Transvaal, or overlain by vast deposits of Stormberg lava, as in Lesotho. Tertiary updoming tilted the plateau to the west so that the altitude gradually declines to below 4,000 feet along the western border and the main drainage flows to the west via the Orange and Vaal. Because of the altitude of 4,000 to more than 6,000 feet in places the climate is cool for the latitude, frost occurring for a third of the year over most of the area and in the Basuto Highlands at any time of the year. Maximum temperatures are usually less than 90° F. (32° C.) except in the lowest parts of the northern Transvaal. Rainfall varies from 15 inches along the western border to nearly 80 inches in eastern Lesotho, the south-western Transvaal usually having over 30 inches. Rainfall is, however, unreliable, particularly in the west, and over 80% occurs in the summer.

ECONOMY

The *Witwatersrand* dominates the economy of the whole region. *Gold* was first discovered here in 1886, having been mined in the Barberton area and elsewhere in the Transvaal prior to this date. The vast gold-bearing reefs are the largest in the world and occur over a distance of about 80 miles to the east and west of Johannesburg including the Far West Rand, and again some 60 miles to the south in the northern Orange Free State (Fig 74). The auriferous deposits are found in a series of reefs which curve deeply into the earth, the angle of inclination at depth being about 30°. Mining is carried on at great depths, the deepest mines being up to 10,000 feet down, in order to tap the gold-bearing reefs, of which the most import-ant is the Main Reef. The gold occurs as minute specks in the quartz conglomerate of the Witwatersrand series. Due to the tremendous extent of the Rand gold resources, the pennyweight content of gold per ton ore milled is very low, although fairly uniform, and on an average below 6 cwt. but higher in the new mines (Plate XXV).

The nature and distribution of the vast mass of gold-bearing quartz with its uniformity of gold content has given rise to a great and stable industry, unlike many other temporary gold-mining areas of the world, where boom towns soon became ghost settlements and alluvial deposits were exhausted rapidly.

Supplies of water were available at first from underground sources in the Dolomite near Johannesburg, and later in much greater quantities from the Vaal after construction of the Vaaldam.[3]

There are very large proven reserves of *coal* estimated at some 70,000 million tons, sufficient to last the country for nearly 2,000 years at the present rate of exploitation. However, the quality is medium to poor, and less than 300 million tons are of coking quality. About three-quarters of the Republic's annual coal production and over 40 million tons comes from the southern Transvaal and much of this from Witbank only 80 miles east of Johannesburg. Coal is found in the lower Karoo beds, particularly the Ecca. Much of it is good quality and easily worked as seams are horizontal,

[3] Although rapidly increasing economic and urban growth in the Southern Transvaal and northern Free State is demanding so much water that supplies may well be exhausted by 1975.

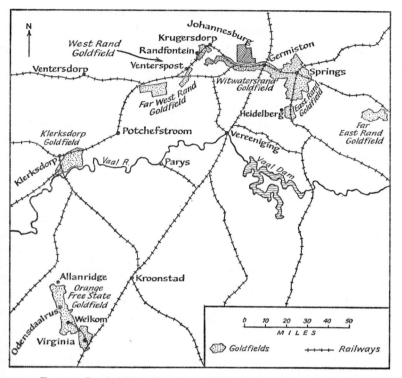

FIG. 74—South Africa: Transvaal and Orange Free State goldfields

wide and near the surface. The price of coal is among the lowest in the world and it is the main source of power in South Africa, hydro-electricity being as yet of little importance.

Although the provision of adequate labour has been a recurrent problem, met to some extent after the Boer War by the temporary employment of Chinese, nevertheless over a long period careful organization has built up a large labour force of migrant Africans. This cheap labour, compared to the high cost of white mining labour, has added greatly to the profitability of exploiting the Rand gold ores. Capital, mining techniques and skill acquired during the operation of the diamond mines at Kimberley were also to prove of considerable assistance in the early days of the gold-mining industry on the Rand. Another important factor is the centralizing organization of the gold-mining industry; some 57 gold-mining companies, 26 coal-mining companies and 13 financial corporations were incorporated in the Chamber of Mines at the end of 1963.

By 1894 gold production was valued at about £15 million, in 1932 £32 million and in 1965 £383 million. The Central Rand has declined in importance and the centre of mining has now shifted to the Far West Rand, the Klerksdorp field and the Orange Free State fields, which now produce about a third of the gold. Gold mining is of vital importance to the economy of South Africa. It is estimated to support nearly half the population directly and indirectly. Its stabilizing effects are particularly noticeable during periods of economic depression or political stress. It provides a large and assured market for agricultural produce and manufactured goods. Uranium is a valuable by-product of gold mining and in 1965 production was worth nearly £16 million.

INDUSTRY AND URBAN GROWTH

The spectacular growth of the string of towns along the Witwatersrand is in itself an epitome of the economic success of the Rand gold mines. A metropolitan area with over 2 million people has developed where less than eighty years ago there was bare veld and a tiny pastoral population of Boers. The Rand is flanked to north and south by the minor urban and industrial complexes of Pretoria, Vereeniging and Vanderbijl Park. *Pretoria*, 20 miles north of Johannesburg, was the capital of republican Transvaal. It is now the administrative capital of South Africa and a centre of heavy industry with nearly half the industrial workers employed by the Iron and Steel Corporation. Pretoria and Vanderbijl Park have the two largest steel plants in South Africa, with a combined capacity of about $2\frac{1}{2}$ million tons of pig iron and steel produced at very low cost. Both centres are favourably situated in relation to South Africa's major economic region of the Southern Transvaal. Since 1921 the population has expanded greatly from 64,000 to over 400,000.

The area of the Witwatersrand metropolitan region is over 1,600 square miles and the population about $2\frac{1}{4}$ million people. *Johannesburg* (Plate XXVI) is the centre of this conurbation, stretching some 60 miles in an easterly direction to Springs and including the towns of Randfontein, Boksburg, Benoni and Springs. The origin of this conurbation lay in the separate growth and gradual coalescence of separate mining camps which became towns. The rise of secondary industry and commerce, particularly during the last 30 years, has accelerated urban growth.

Manufacturing accounted for over a quarter of the national income in 1964–5 and has doubled in value in the last 10 years. The number of establishments has grown from about 12,300 in 1954–5 to nearly 15,000 in 1960–1. However, despite these considerable advances, an impressive

374

array of minerals and low costs of production, South African industry is held back by a maze of regulations controlling labour and by the low purchasing power of the majority of the population.

Industry has concentrated more in Johannesburg and the East Rand towns than the West Rand because of the availability of flat land, Witbank coal supplies and the convergence here of railways from Durban and Lourenço Marques. The Southern Transvaal is South Africa's major industrial region employing about half the workers and it has a great variety of industries such as iron and steel engineering, explosives, cement, hardware, food-processing, furniture, textiles and clothing. The largest oil-from-coal plant in the world was established at Sasolburg, about 50 miles south of Johannesburg, in 1955, based on cheap local supplies of coal. It was hoped that it would help to reduce dependence on supplies of oil from the Middle East.

The broad pattern of urban growth along the Witwatersrand is now of a central east-west trending belt with mining, manufacturing and commerce surrounded by residential areas, Europeans mainly to the north and non-Europeans to the south. Johannesburg's rapid industrial growth has caused grave social problems with a great post-war influx of Africans seeking to stay permanently rather than on a temporary basis as on the mines. Overcrowded shanty towns grew up rapidly next to the old native townships. A rapidly changing population, instability, lawlessness, high crime rate, and a high incidence of disease are characteristic of the shanty town, with its crowded shacks and huts and haphazard street pattern, somewhat like the slums of a West African town. There has been a vast scheme of slum reclamation in the last ten years and the great African urban complex of Soweto with an area of 26 square miles has a population of over 700,000 in 23 townships. *Bloemfontein* is an important route centre and capital of the Free State.

AGRICULTURE

Agriculture has undergone major changes in the last 30 years. Since 1950 even the value of agricultural and pastoral production has increased by over 100%. Mechanization has increased greatly, tractors have replaced oxen, the numbers of beef and dairy cattle have increased, pastures have improved, more fertilizers are used and there is greater local specialization.

The high veld is South Africa's main agricultural region, producing over half the country's maize, nearly a third of the wheat and a high percentage of the groundnuts, kaffircorn and potatoes and also considerable quantities of dairy produce, meat and wool (Fig. 75).

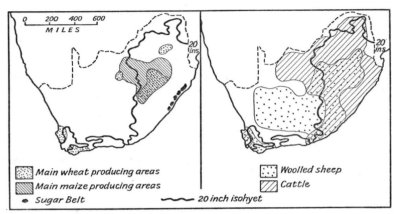

FIG. 75—Southern Africa: livestock and crops

The western Free State and Transvaal are marginal for crop production although maize still supplies the bulk of the farm income. Wheat, ground-nuts and livestock are becoming more important in the western Free State. The yield of maize is low but quality is high partly because of the low humidity. Yields could be improved by increased cultivation and weed control, greater application of fertilizers and use of hybrid maize strains. Maize is a staple food for the African and is used increasingly as a fodder crop for cattle. It is the most important crop grown, accounting for 40% of the European cultivable area. The 1962–3 crop was a record with over 67 million bags, much of which was available for export. There is a problem of surplus production as the Mealie Industry Control Board often has to export maize at a loss.

North of the Witwatersrand the country is 1,000 to 2,000 feet lower with higher temperatures and vegetation of thorn bushes and savanna grasses. This is the great crescent of the northern Transvaal, sometimes called the Bushveld, which consists of the flat expanses and Karoo sediments of the Springbok Flats, the granite Pietersburg Plain, the great laccolith of the Bushveld Igneous Complex, the cuestas of the Bankeveld, and, rising boldly above the level surface of the plateau, the dissected masses of Archaean rocks forming the Waterberg and Zoutpansberg. Because of the low seasonal *rainfall* over much of the area irrigation is necessary for crops such as tobacco, cotton, fruit and vegetables. Warmer temperatures throughout the year also encourage the *cultivation* of these subtropical crops.

Tobacco is grown mainly in the Rustenburg area and *cotton* further east near Barberton. *Citrus fruits* are important and expanding in acreage in the irrigable valleys of the Limpopo and Crocodile-Komati in the northern and eastern Transvaal. Much of the crop is produced by large companies such as Zebediela and Letaba Estates. The estates also grow a variety of subtropical fruits and vegetables. Before the war the low altitude of much of the area and the prevalence of malaria caused it to be neglected by white farmers. In the nineteenth century the Boers had been driven back by disease from settlements such as Schoemansdal on to the healthier high veld. However, with the control of malaria, the extension of irrigation facilities and the growth of urban markets since 1940 this region has developed greatly. The Bushveld is an important *mineralized* region with limestone and marble obtained from the Dolomite, platinum and chrome occurring in the norite and tin in the Rooiberg series. Asbestos is obtained from the districts of Pietersburg, Carolina and Barberton; chrome and platinum from the Lydenburg and Rustenburg districts, and iron ore chiefly from Thabazimbi, although reserves are scattered widely over South Africa.

DIAMONDS

To the west the South African plateau region includes part of *Griqualand West* and *Kimberley*, whose diamond fields sparked off South Africa's economic revolution. *Diamonds* were discovered first in the gravel terraces of the Vaal and from 1870 on in the dry diggings and volcanic pipes of the Kimberley area. Mine diamonds are obtained from three levels in the volcanic pipes, firstly from the yellow ground on the surface, then from the blue ground beneath and lastly the harder rock at depth. The blue ground and underlying rock have to be crushed and washed before the diamonds can be obtained. These Cretaceous volcanic pipes, some bearing diamonds, are found sporadically from the Cape to Tanzania.

Kimberley grew little in size after the first diamond boom and the population now is only about 70,000. Kimberley is a route and market centre for the northern Cape and organizational headquarters for the diamond industry.

Since 1926 much of South Africa's diamond production has been alluvial, the middle Orange, Vaal and Hartz valleys, the Lichtenburg fields in the south-west Transvaal and later near the mouth of the Orange River.[4] In 1963 the value of South African diamonds sold was about £18·6 million.

RURAL SETTLEMENT

Little mention has been made of the distribution of African rural popula-

[4] See p. 406.

tion on the plateau, much of the Bushveld and the Zoutpansberg being demarcated as native reserves with a subsistence economy based on cattle and kaffircorn. Usually the areas of African occupation are on poor soil and steep slopes, while the more level areas with more fertile soil have been taken over for European farms.

LESOTHO

The most important area of African-held land is farther south in Lesotho (Basutoland, prior to independence in October 1966) which is politically distinct from the reserves having been for long under British control as one of the three Protectorates, the others being Swaziland and Bechuanaland (renamed Botswana on becoming independent in September 1966). Lesotho has an area of only 11,700 square miles and an estimated population in 1964, including migrants in South Africa, of over 1 million. As half the region is virtually uninhabitable owing to its high altitude of 8,000 to 11,000 feet and bleak climate, there is much overcrowding of population and serious soil erosion. Lesotho can be divided into *two main regions*, the so-called western lowlands and the basaltic plateau. The *narrow lowland strip* of land between the Malutis and the Caledon river is only about a quarter of the area of Lesotho, but contains about two-thirds of the population. Structurally the lowlands are part of the high veld and consist of Karoo sedimentaries, chiefly sandstone, although more dissected than in the Free State and with many flat topped koppies, some of wide extent, which afforded natural protection to the Basuto during the wars of the nineteenth century and assured national survival. Settlements concentrate on the lower slopes of these plateaux, their site being influenced by defensive land use and climatic factors. Huts, many of them well built of stone, face east to gain maximum sunshine and avoid the cold air of the valley bottoms.

Soils are light, sandy and overworked. Soil erosion had reached such serious proportions by the mid 1930s that the future of Basutoland was imperilled. Great dongas (gullies) cut deeply into the soft Karoo rocks and the limited area of arable soil was still further reduced. A vigorous campaign of soil conservation over the last 25 years has largely stabilized lowland soils and improved mountain pastures, but the problem of soil erosion is still serious, and now more social and economic than physical in nature. Only about an eighth of Lesotho is arable and maize, wheat and sorghum are the chief grain crops; these are valued at about £4 million a year. Yields are low and could be increased greatly by application of superphosphates. Cattle dung is not used on the land, but as fuel as the country is practically treeless. Land tenure is still largely communal, but groups of individual farmers have pooled their land and are attempting to carry out a cooperative form of mechanized farming which may cause a revolution in farming techniques. Cooperatives have nearly 3,000 members and there is also a Progressive Farmers' Scheme.

The *basaltic highlands* lie to the east of the lowlands and consist of a great dissected dome of Stormberg basalt overlying Karoo sedimentaries. They are South Africa's main watershed and the Orange river and its tributaries have cut spectacular gorges thousands of feet deep into the plateau. The volcanic soils are more fertile than the sandy lowland soils and rainfall is higher. Wheat, barley, beans and peas are the main crops grown in the highlands below 8,000 feet. However, the major activity is pastoralism and wool and mohair are Lesotho's major exports with a value of about £1¼ million in 1962.

Sheep usually remain at high altitudes throughout the year. Cattle stay for most of the year at the cattle posts, going down to the villages for a short while only after the harvest has been gathered and returning to the mountain pastures with the first flush of spring grass. In 1961 there were about 400,000 cattle, 600,000 goats and 1¼ million sheep. Yields of wool and mohair are low and quality poor. Improvement can only come about through better breeding stock, control of communal grazing and provision of winter fodder supplies. Severe population pressure on the lowlands has forced people to move into the highland areas, a process that has been going on for fifty years. The upper limit of settlement is now 8,000 feet.

Lesotho has practically no minerals and few prospects for industrial development.[5] Communications are poor with gravel roads in the lowlands while over most of the high plateau Basuto ponies and bridle paths are the only means of transport, although a new mountain road has been built into the heart of the plateau. There is, however, a local airline and charter flights can be made over most of the country. Lesotho's basic problem is thus of an impoverished bleak highland area unable to support its population and suffering a continuous drain of people, both temporary and permanent, to the Republic of South Africa, despite the Basuto dislike of that country's policy of apartheid. The movement of migrant labour is most marked from the impoverished parts of the lowlands. Remittances and money brought back by migrant workers play a considerable part in helping to reduce the deficit between imports and exports. In 1960 almost £1 million was brought into Lesotho in this way.

The budget has been only about £3 million a year, of which less than

[5] Diamonds are now being mined in the Mokhotlong area on a small scale; the value of diamonds sold in 1965 was nearly £260,000.

one-third comes from local sources, the rest accruing from customs collected on Lesotho's behalf by South Africa, and from British loans and grants. With the rapid approach of political independence, the power of the chiefs has waned, but the country is faced with the difficult task of combining economic dependence on South Africa, whose territory completely surrounds Lesotho, with independence of political action.

REGION 5–EASTERN PLATEAU SLOPES

RELIEF AND STRUCTURE

Below the plateau is the long narrow region of the eastern plateau slopes or eastern terraced foreland. This has originated through retreat of the Great Escarpment (Plate XXVII) to the west. It is situated between the escarpment and the coastal lowlands in South Africa and Mozambique. Altitudes vary from about 2,000 feet to over 6,000 feet in places. Tiers of terraces marking four successive *cycles of erosion* rise in succession from the coast to the plateau edge. The region is traversed by the deep gorges of rejuvenated and incised rivers such as the Tugela, Usutu, Komati and Sabi. Most of the rocks in the Eastern Cape and Natal are Karoo sedimentaries with a belt of Table Mountain Sandstone and Archaean rocks from the Umzimvubu to the Umfolosi river. The Table Mountain Sandstone forms bold mesas or plateaux, while the Old Granite has been eroded into a bewildering series of steep hills and valleys, perhaps best illustrated by the Valley of a Thousand Hills near Durban.

CLIMATE AND NATURAL VEGETATION

The climate varies greatly with little frost and warm temperatures at about 2,000 feet, but cold winters and severe frost in the enclosed basins and valleys of the higher plateau slopes. Rainfall is comparatively high over most of the region varying from less than 30 inches to over 50 inches a year. Some occurs in each month of the year in the south, but with an increasing summer maximum to the north.

Grass is the dominant vegetation type varying from tall grasses of the *Hyparrhenia* species in the areas below 4,000 feet and with 30 to 40 inches of rain to *Themeda triandra* in the drier valleys.

SETTLEMENT

Because of the dissected terrain, the steep slopes and the social traditions of the African, *pastoralism* is more important than arable farming. This region was occupied and densely settled by Bantu pastoralists before the eighteenth century. It was subsequently invaded by the Boers who retreated back to the high veld after Natal became British. Much of Natal was settled by British immigrants in the second half of the nineteenth century and the whole region of the eastern plateau slopes is now interpenetrated by white farmers and African peasants.

THE NATAL MIDLANDS

The Natal Midlands, from 2,000 to 5,000 feet high, is the most favourable agricultural area of this region. In those parts with higher rainfall and frequent mists wattle plantations are numerous. Farmers combine wattle growing with pastoral farming. Dairying is important in areas accessible to Durban and Pietermaritzburg, while stock rearing is carried on in more isolated areas.

Areas above 5,000 feet in altitude have sour grasses which become tough and unpalatable in winter and can only be used for summer grazing. The Drakensberg foothills are a zone of transhumant pastoralism. In the lower areas of thornveld in the middle Tugela basin much of the land is in native reserves and badly eroded, and cattle kraals with round bee hive-shaped huts occupy the hilltops. *Pietermaritzburg*, named after a Voortrekker leader, is situated at an altitude of 2,218 feet on the eastern edge of the region and about 50 miles from Durban. It is the capital of the Province of Natal and is an important administrative and educational centre.

THE TRANSKEI

The Transkei is the most extensive area of African settlement in South Africa (Fig 76). It has an area of 16,554 square miles and a population of about 1·4 million Africans. It is not unlike the Midlands of Natal and consists of rolling country with short grass cut into by deep valleys. Most of it lies below 3,000 feet so that the climate is warmer and rainfall is adequate and well distributed. Some of the soils are derived from dolerite and are fertile. The region is densely populated and considerably overstocked. In common with other African rural areas there is a large migration of males each year to seek work in the European areas.

In contrast to Zululand, huts are of the rondavel type with round mud-brick walls and steep conical roofs thatched with grass. Homesteads are perched on hill slopes and ridge tops with small patches of arable land nearby growing *crops* of maize, kaffircorn, pumpkins and beans. Livestock are less important than in Zululand where the cattle kraal is the core of the settlement. Ploughing of steep slopes against the contour, monoculture of

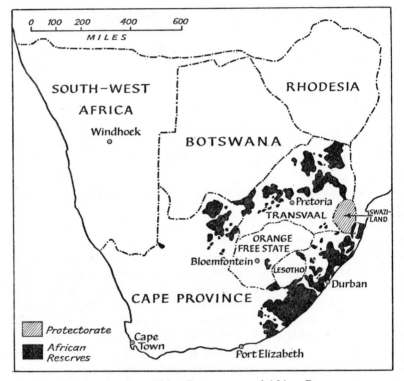

FIG. 76—Southern Africa: Protectorate and African Reserves

crops like maize for a long period, overgrazing and veld-burning, have caused serious soil erosion.

The South African government have attempted to improve farming techniques with scant success until recently, but now measures to arrest erosion are noticeable in many areas. As part of the general plan to develop the African rural areas, styled 'the Bantu homelands', the government envisages the removal of some 300,000 African families from the land throughout the reserves, the establishment of towns, the development of secondary industry and the growth of mixed farming and other agricultural ventures. As yet there has been little agricultural development, so that the growth of secondary industry based on agricultural raw materials has been very slight. Greater industrial development is taking place on the fringes of the African areas in towns such as King Williamstown, East London and Pietermaritzburg. The establishment of new towns is also very slow as the economic basis for urban life in a backward rural area such as the Transkei is scanty and uncertain. The inhabitants of the Transkei have limited political control with a legislative assembly of 4 paramount chiefs, 60 chiefs and 45 elected members, a representation which reflects the desire of the Bantu Affairs Department to revive the powers of the chiefs in South Africa. Subjects such as defence, external affairs, posts and telegraphs, customs, national communications, internal security and constitutional change are reserved to the South African government.

SWAZILAND

On the northern edge of the South African section of the eastern plateau slopes is the tiny Protectorate of Swaziland with an area of only 6,700 square miles and an estimated population in 1966 of nearly 390,000 including over 8,000 whites. In contrast to Botswana, and more particularly Lesotho, where no European may own land, nearly half the area of Swaziland is in European ownership, causing a chequerboard pattern of contrasting land use. Partly because of this, economic development has been and is greater in Swaziland and there is more economic potential. During the period 1960–3 exports have more than doubled in value and are now worth over £10 million, of which sugar counted for over £4 million, asbestos £2½ million, wood pulp and other products £2 million, with cattle, cotton, rice, citrus, pineapples and other products completing the list.

Swaziland consists of three north-south trending belts of high, middle and low veld, flanked by the Lebombo plateau on the eastern border. The *high veld*, covering nearly half the country, is a much dissected strip of country above 3,500 feet in height, most of it with over 50 inches rainfall a year and therefore very suitable for plantations of eucalyptus and conifers. The Usutu *coniferous forests* of nearly 100,000 acres were developed by the Colonial Development Corporation and Courtaulds, while Peak Timbers and Swaziland Plantations have 75,000 acres in the Piggs Peak area. Sawmills and box factories have been established and a pulp mill with an eventual output of 100,000 tons of wood pulp a year. The Havelock *asbestos* mine produces chrysotile asbestos worth over £2 million a year which is exported by aerial ropeway to the railhead at Barberton. A little gold and tin are also mined. There are coal reserves and large deposits of high-grade iron ore at Bomvu Ridge, north-west of Mbabane, to which a railway was completed from Goba in Mozambique in September 1964. This is Swaziland's first railway and should be of great economic significance.

The next zone of the *middle veld* is below 3,000 feet and has a rainfall of 30 to 40 inches per annum. The area between Mbabane and Bremersdorp is the main centre of European farming in Swaziland. The Malkerns irrigation scheme, with some 10,000 acres under irrigation and drawing water from the Great Usutu river, produces crops such as pineapples, citrus, rice and vegetables on about 2,000 acres of land. Sales of citrus should increase greatly in 5 years time when trees come into full bearing.

The *low veld* is below 1,000 feet and drier with less than 30 inches of rainfall per annum. Swaziland is well watered with perennial rivers such as the Usutu and Komati and the low veld has considerable potential for the irrigation of tropical crops. There are two major schemes, the first in the north-east drawing water from the Komati and producing sugar, rice and citrus, the other the Big Bend irrigation scheme on the Great Usutu river in the south-east with sugar as the main crop. Exports of sugar, which were nil in 1955, were worth over £1½ million in 1961, and production was about 170,000 tons in 1966.

Less than one-tenth of Swaziland is under cultivation so that there are great possibilities for extension of the cultivable area. There are about 600,000 head of cattle and ¼ million sheep and goats, the majority owned by Swazi, who characteristically value cattle more for social and ritual than economic reasons.

Although Swaziland has been developed economically far more than Lesotho and Botswana there is still an imbalance of wealth and revenue in the territory between the European and Swazi sectors. There has been a marked drop in the ratio of African tax receipts to total revenue from nearly 50% in the 1920s to only 5% in 1956. The Swazi areas are backward compared to the European areas. A rapid agricultural advance is needed to bring about a change from the present subsistence economy to a wider participation in commercial agriculture.

THE MANICA HIGHLANDS

Between the Sabi and Zambezi rivers in Mozambique there is a small strip of plateau slope in the Manica highlands. It receives 40–60 inches of of rain per annum which is sufficient for the production of tea. A small colony of Portuguese farmers grow maize, tobacco, fruit, vegetables and sisal. There is a scheme to settle 200 Portuguese families near Vila Pery, each settler to receive 1,250 acres of land. Sisal will be the main cash crop. A large hydroelectric plant has been constructed on the Revue river and power supplied to Umtali in Rhodesia and Vila Pery where a cotton textile factory and jute mill have been established. Vila de Manica, near the Rhodesian border, is an important agricultural centre and the main town of Manicaland. The region is well served by the Beira-Salisbury railway.

XXIX. Victoria Falls, looking across the chasm from above the Devil's Cataract.

[Rhodesia House

XXX. Salisbury and the high veld of Rhodesia.

XXXI. Kariba, February 1961.

[Rhodesia House

REGION 6–EASTERN COASTAL REGION

Between the eastern plateau slopes and the sea is the Eastern Coastal Region, only a few miles wide in the south but bulging out to a width of 150 miles in Mozambique, which has no true coastal plain in the south. The region stretches from the Great Kei river in the south to the Zambezi in the north. The 2,000-feet contour forms the western boundary and the coastline the eastern boundary.

RELIEF AND STRUCTURE

The area below 2,000 feet in the Transkeian Territories is similar structurally to the uplands above 2,000 feet with Karoo rocks exposed over most of the area. Further north Cape and Archaean rocks are found parallel with the coast. Rivers have cut deeply into the Karoo sediments and the gorge sections are often continued right to the sea, as, for example, the Umzimvubu river which cuts through Table Mountain Sandstone at Port St Johns. Sand bars have been built up across the mouths of rivers, direct contact with the sea only being possible where rivers are strong enough to break through the sand barrier. The coastline in the south is very smooth.

In northern Zululand a coastal plain of uniform surface develops, which widens out in Mozambique and is composed of sediments of Cretaceous to Recent age consisting of conglomerates, sandstones and limestones. Sand dunes in northern Zululand rise to hundreds of feet in height and force rivers to flow parallel with the coast. Further north in Mozambique the plateau edge trends north-west, cut into by the Limpopo embayment, while the coastline turns north-east and north. The even surface of the plain rises slowly to the granites of the Manica platform in the north. This wide swathe of coastal lowland is crossed by a number of large rivers such as the Limpopo, Sabi, Buzi and Pungwe.

CLIMATE AND VEGETATION

One of the main distinguishing features of the coastal region is the warm climate. The coast is washed by the warm Mozambique current which causes high diurnal temperatures with little diurnal range. Mean monthly temperatures vary from 60°–75° F. (15°–24° C.) in the south and 75°–85° F. (24°–29° C.) in the north. North of Beira rainfall along the coast is over 60 inches, but to the south there is less than 40 inches a year, and inland much of Gazaland is semi-arid with less than 25 inches. The incidence of summer rainfall increases rapidly to the north and winter from May to October is a markedly dry period.

Vegetation varies from mangrove swamp near the lagoons and river mouths to raphia and other varieties of palm in Natal, and deciduous thorn forest over much of the interior of Mozambique.

The eastern coastal region can be subdivided into minor regions, consisting of the palm belt of Natal and the Eastern Cape and the coastal plain of Southern Mozambique.

NATAL COASTAL BELT

In the coastal strip of Natal *warm temperatures* throughout the year encourage the growing of subtropical crops such as sugar and fruit. Bananas, mangoes and pawpaws are grown on the hill slopes and vegetables on the river alluvium. Near Durban there is a zone of Indian market gardening.

Sugar-cane is the outstanding cash crop of the Natal coastal belt north of Durban. Cultivation expanded after the introduction of Indian indentured labour in 1860–6. Production which was about $\frac{1}{2}$ million tons in the 1930s has now more than doubled and for the 1964–5 season was 1·4 million tons, of which about 600,000 tons was available for export.[6] The bulk of the cane is grown in rolling country in a belt about seven miles wide from Durban to Stanger, with extensions south to Port Shepstone and north to Mtubatuba. The area under cultivation is about 640,000 acres. Despite marginal climatic conditions of low temperatures and rainfall of 40 inches per annum yields are fairly high. Planting begins after the August–September rains and again in March and cutting of the cane takes place about two years later. Most of the field work is done by Africans, and not by Indians as before. Production is mainly from large estates run by Europeans; there are about 1,200 European cane growers, but over 5,000 non-European growers.

The main town of the southern coastal belt is *Durban* founded in 1824 on the shores of a lagoon and providing one of the few harbours along this almost harbourless stretch of the coastline. Less than 500 miles from the southern Transvaal, Durban has developed as the main outlet for a large and wealthy hinterland and is the major port on the east coast of Africa. Population has expanded from less than 70,000 in 1904 to 655,000 now, with almost equal numbers of Europeans, Asiatics and Africans, and with a small Coloured population. Durban is the most cosmopolitan city in South Africa.

Urban growth has been influenced greatly by relief. The only flat land composed of alluvium is in a narrow strip between steep Pleistocene ridges and the bay. It is now occupied largely by industry, commerce, port facilities,

---

[6] South Africa lost its share of the British market after leaving the Commonwealth, but an agreement with Britain assured South Africa of a market until 1966.

hotels and flats. Residential areas have made use of the steep hills and ridges composed of Pleistocene and Karoo rocks north of the alluvial strip. Sea breezes, cooler temperatures and a view across the Bay have attracted Europeans to the Berea and the hills overlooking the sea. Africans and Asians occupy less congenial climatic areas behind the coastal hills and inland. The Africans used to live under bad slum conditions in areas such as Cato Manor, but they have now been rehoused in the African townships of Umlazi and Kwa Mashu. The warm climate, particularly in winter, and the attractive coastline have attracted large numbers of visitors. A number of small resorts stretch for 100 miles along the coast from Port Edward to north of Durban. Durban is South Africa's main cargo port with nearly half South Africa's imports and over half the exports; it handled $14\frac{1}{2}$ million tons in 1963-4, an increase of over 3 million tons compared with 1962-3. It is also an important industrial centre with an oil refinery, sugar refinery, and manufacture of soap, margarine, paint, fertilizers, explosives, the preparation of wattle extract, etc. Durban is the outlet for the main railway line from Natal to the Witwatersrand.

MOZAMBIQUE COASTAL PLAIN

Economy

Mozambique has a total area of about 300,000 square miles and a population of about $6\frac{3}{4}$ million, including 100,000 Europeans in 1961. The fortunes of the coastal belt south of the Zambezi are closely bound up with the transit trade in copper, coal, asbestos, chrome, oil, maize, tobacco, sugar and manufactured products that has grown up with the economically developed plateaux of the Transvaal, Zambia and Rhodesia; railway lines from the Copper Belt, Bulawayo, Salisbury and Johannesburg converge on the ports of Lourenço Marques and Beira. Although at the southern tip of Mozambique *Lourenço Marques*, with a population of about 185,000 in 1961, is the capital and nearest port to the southern Transvaal, a factor of great economic importance (Fig. 77). The harbour is one of the best in Southern Africa and consists of an outer harbour, Delagoa Bay and an inner harbour leading off the Bay and which can handle sixteen ocean-going steamers. The port handles a greater volume of trade than Beira, with a total of nearly 9 million tons in 1964 compared with 4 million tons for Beira. The Mozambique Convention of 1909 regulates traffic with South Africa, guaranteeing a minimum of $47\frac{1}{2}\%$ of the lucrative Transvaal traffic to Lourenço Marques, which handles much of the Transvaal's exports of coal and imports of oil. The recently completed line from Guija to Bannockburn has enlarged Lourenço Marques's hinterland greatly and much of Zambia's exports of copper now go via Lourenço Marques rather than Beira. Traffic should increase further with the exploitation of iron ore deposits in Swaziland and the building of the Swaziland railway link to Goba which is connected by rail with Lourenço Marques. The harbour has been dredged to facilitate the docking of 50,000-ton ore ships. Lourenço Marques is the major commercial and industrial centre in Mozambique, with an oil refinery, flour mill and manufacture of cement, furniture, textiles, footwear, soap, etc. It is also an important centre for the tourist industry.

Beira is the terminal of the central railway to Rhodesia and also the railway to Malawi and is situated at the mouths of the Buzi and Pungwe rivers. The port has a poor site with shallow water, silting and a considerable tidal range. There was great congestion during the early 1950s, but with the improvement of harbour facilities and completion of the Limpopo railway further south the situation has been greatly eased. Exports consist of minerals such as copper and chrome and agricultural products such as tea, cotton and sugar.

Apart from the transit trade with the plateau, economic development elsewhere has been very slow and until recently almost stagnant. In the past Portugal has neglected her major colonial territories of Angola and Mozambique, partly because of her own lack of capital, and partly because of colonial policy which restricted trade to her own markets. But restrictions on trade with other countries have broken down and now Portugal handles only about 30% of Mozambique's imports and exports. Britain, the United States and South Africa also have important shares of the trade. A major development plan was initiated in 1953. An important feature was the Limpopo Valley Scheme (Plate XXVIII) which aims at the settlement of nearly 100,000 Europeans and Africans on ½ million acres of land much of it under irrigation. However, by 1962 only about 5,000 people, two-thirds of them Portuguese, had been settled and many had left for Lourenço Marques. Disease, an enervating climate, small holdings, poor farming techniques, high costs of establishing the settlers have all contributed to the scheme's lack of success.

Commercial development is controlled mainly by large companies on plantation lines. A wide variety of tropical crops can be grown, but much of the south has poor sandy soils on which maize is the main crop. Fertile alluvial soils are found in the major river valleys and the Zambezi delta, and profitable crops of bananas, citrus and sugar are produced in these areas. Dairying and market gardening are carried on near Lourenço Marques. There are large sugar plantations in the Lower Zambezi valley and the Buzi

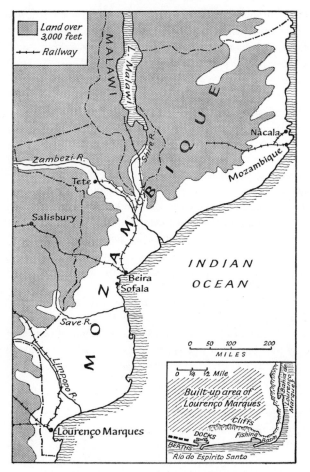

FIG. 77—Mozambique and Lourenço Marques

and Incomati valleys. Sugar refineries are being built near Lourenço Marques and Beira. Production of sugar in 1964 was about 150,000 tons.

The deficit in the trade gap is largely made up through port and railway receipts and the export of nearly ½ million African labourers to South Africa and Rhodesia. The Mozambique Convention permits the recruitment of some 80,000 Africans a year to work on the Rand mines. From 1942 to 1951 two-thirds of the value of traffic passing through Mozambique was transit trade. The invisible items in the balance of trade (transport charges, migrant labour and tourism) are probably worth more than half the

389

value of visible items. Tourism has become a useful source of revenue and because of the mild winter climate large numbers of people escape the cold winter of the high plateau, those from the Transvaal going to Lourenço Marques, those from Rhodesia going to Beira.

REGION 7–KALAHARI AND BOTSWANA

Far from Mozambique and occupying geographically the heart of Southern Africa is the semi-desert of the Kalahari. This is one of the regions that is poorest in natural resources in Southern Africa. The Kalahari region stretches northwards approximately from the line of the Molopo river to the highlands of Angola and the Zambezi river and from the plateau of South-West Africa in the west to the Transvaal middle veld in the east. Politically most of the area is occupied by Botswana which has an area of 222,000 square miles.

RELIEF, STRUCTURE AND DRAINAGE

It lies at an altitude of 2,000 to 4,000 feet and consists of a vast area of *Kalahari sands* overlying an Archaean platform. There is no surface drainage in the south and in the north the Okovango drains into an interior basin occupied by the ephemeral waters of Lake Ngami and the Makarikari Salt Pan. On the north-western fringe of the Kalahari drainage is to the Zambezi.

CLIMATE AND VEGETATION

The Kalahari is incorrectly styled a desert and is rather semi-desert with rainfall varying from 5 inches in the south to over 20 inches in the north. Vegetation for the most part consists of thorn veld with tall and widely spaced grasses and thorny acacia trees and bushes. The scenery is monotonous in the extreme with hundreds of miles of sand-covered, bush-clad wastes and ephemeral sand-choked streams.

POPULATION AND SETTLEMENT

The population is very sparse and over much of the central Kalahari the only inhabitants are nomadic Bushmen, hunters and collectors, driven into the wilds of the Kalarhari by Bantu and Boers. There are a few European farmers at Ghanzi near the South-West African border. To the east in a strip of country along the Transvaal border with a rainfall of 15–20 inches, crossed by the main railway from the Cape to Rhodesia, live the great majority of the population of 560,000 including over 3,000 Europeans.

There are eight different Tswana tribes, the N'gwato being the largest. In contrast to the scattered Bantu settlement pattern of the rest of Southern Africa the Tswana live in nucleated villages, some of which like Serowe and Kanye have populations of up to 20,000. Scarcity of water is the main factor dictating concentration of settlement. Only about 5% of Botswana is cultivated. Scattered plots of land near the villages grow sorghum, millet and maize, but yields fluctuate widely because of the highly variable rainfall and grain often has to be imported. Sorghum, millet, peas and beans are exported in small quantities in an average year. Nearly all Botswana consists of African-owned land or former Crown land and there are only four small blocks of land in European ownership.

AGRICULTURE

Extensive *pastoralism* is all that the meagre environment permits and Botswana has an estimated cattle population of 1,350,000. Cattle and hides and skins comprise nearly three-quarters of the exports. Until recently South Africa, Zambia and Rhodesia were the main markets, but in 1961 Britain was the largest buyer of meat from Botswana. The annual average of sales from African areas is only 5% of stock, compared to 10–15% from European farms. Grazing round existing water points is being exhausted, so that to increase cattle exports greater control of grazing must be exercised and there must be a more rapid turnover of stock. Cattle posts are often situated at a considerable distance from the villages as in Lesotho. In contrast, however, to Lesotho, the major problem is not lack of land, but lack of water, which is obtained from bore-holes and small earth dams or by digging in the sandy river beds.

IRRIGATION

The Kalahari unfortunately lacks the great artesian water supplies of Australia. Even if water were available the carrying capacity of much of the thorn veld is only about 1 beast to 50 acres. In the north where water is more readily available tsetse infestation precludes cattle keeping. Here the Okovango is a perennial stream flowing south-east from the highlands of Angola, its reed-choked waters dissipating in the Okovango delta, an important source of water for irrigation in the future.

MINERALS

Mining on a small scale is carried out and gold is obtained in the Tati concession, asbestos near Kanye and manganese near Lobatsi, the last two being the most important minerals mined at the moment, with exports worth about £180,000 in 1963. Recent prospecting has revealed the existence of a large deposit of medium-grade bituminous coal estimated at over 400 million tons near Gaberones. The Vacuum Oil Company has been granted a concession to prospect for oil.

POLITICAL STRUCTURE

As in Lesotho and Swaziland there was the usual panoply of indirect rule until recently, with tribal authorities, councils and treasuries. In June 1961 a new constitution provided for a legislative council in which about a third of the seats were held by Africans. On achieving independence in September 1966 the country changed its name from Bechuanaland to Botswana. The anomaly of having a capital situated beyond the territorial boundaries has been rectified by the transference of administration from Mafeking in the Cape to Gaberones, a village about 50 miles north of Lobatsi.

REGION 8–PLATEAU OF RHODESIA

RELIEF, STRUCTURE AND DRAINAGE

The high veld of Rhodesia, oriented from south-west to north-east, lies at an altitude of about 4,000 feet and is cut off from the South African plateau by the Limpopo Depression. Along the eastern rim overlooking Mozambique the plateau reaches an altitude of 8,500 feet in the Inyanga district. As in South Africa the core of high veld is almost surrounded by the middle veld at an altitude of 2,000–4,000 feet. These two zones of high and middle veld constitute the plateau region of Rhodesia. Karoo beds have been largely stripped off the high veld to expose the *Archaean platform* of schists, gneisses and granites which cover nearly three-quarters of the country and are the source of its mineral wealth. Karoo rocks are most extensive in the low veld and Kalahari sands in the west. Drainage from the watershed of the high veld is north to the Zambezi and south to the Sabi and Limpopo.

CLIMATE

Rainfall is higher than in South Africa, most of the country receiving more than 20 inches a year and a narrow belt along the eastern border more than 40 inches. This is offset to some extent by the more tropical location and greater seasonality. On the high veld temperatures vary from about 55° F. (13° C.) to over 70° F. (21° C.) thus creating an island of

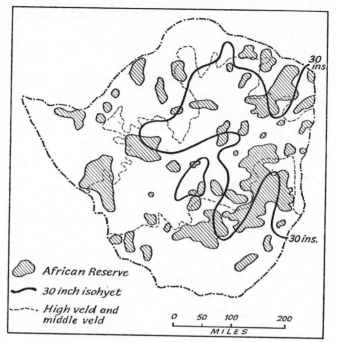

Fig. 78—Rhodesia: African Reserves, relief and rainfall

almost temperate climate which is partly reflected in the comparatively large white population of nearly ¼ million.

POPULATION AND SETTLEMENT

Most of the Europeans in Rhodesia live on the high veld which covers nearly a quarter of the country. The majority of African reserves are situated below 4,000 feet, particularly in the eastern half of Rhodesia. The African population, which was estimated at about ½ million in 1898, has been growing at a very rapid rate and by 1964 had reached 4 million in number. Some of this great increase of population may represent, as in Malawi, permanent migration of Africans from Mozambique into Rhodesia.

The land of Rhodesia has been allocated between Europeans and Africans in nearly equal proportions by the Land Apportionment Act of 1930 and various amendments (Fig. 78). The European area consists of about 35 million acres, nearly 5 million acres are Unreserved Land, nearly 11 million acres consist of Forest Area and National Parks, and the African area about 40 million acres. About 60% of the African population still lives in the African areas, the rest on European farms, in the mining areas and the towns.

AGRICULTURE

The following figures for gross value of agricultural production of European and African agriculture and land use in 1960 give some indication as to the relative value of the different sectors of the agricultural economy and varying types of land use.

Gross Value of Agricultural Production 1960 in £ million

	European	African Subsistence	African Sales
Crops	40·8	6·1	1
Livestock	8·9	2·4	3
TOTAL	49·7	8·5	4

European Farming 1960		African Farming 1960	
Total area of farms under crops	34·48 m.acres	Usable area	27·6 m.acres
Number of farms	7,064	Area being cultivated by present cultivators	2·9 m.acres
Under 1,000 acres	2,100		
1,000–5,000 acres	3,564	Number of cultivators	346,136
5,000–20,000 acres	1,156	Master farmers	8,966
20,000 acres and over	244	Demonstration	
Male European workers	8,632	plot-holders	5,327
Male African workers	208,971	Cooperators[7]	90,499
		Others	241,344

[7] Cooperators are farmers who use manure or fertilizer, carry out some rotation and plant their crops in rows.

There are nearly 100 times more African than European farmers and the average holding is about 40 acres compared with the average European holding of about 5,000 acres.

Soils vary from a sandy light soil which overlies the granite to dark red loams which have developed from schists and produce much of the maize.

The great majority of the crops are grown in the summer, as winter crops have to be grown under irrigation. It is estimated that the present area of nearly 600,000 acres under *irrigation* could be expanded greatly to nearly 1 million acres. The Sabi-Limpopo Authority has been set up to encourage development in the low veld and three major dams have been built. High yields of sugar, wheat and cotton are being obtained, and about 70,000 acres were under irrigation in 1966.

Crops such as maize, kaffircorn, tobacco and groundnuts are therefore planted in October and November and those requiring a shorter season such as oats, beans and linseed are planted up to January.

Maize is the chief crop and the main area of production is in the higher rainfall area of the east. The European-grown crop in 1964 was nearly 4½ million bags with the comparatively high yield of 11 bags to the acre, yields having nearly trebled in the last few years, while the African-grown crop was 3 million bags. Maize is a staple diet for the Africans.

Tobacco is by far the most important cash crop, production having expanded greatly since the war and in the 1964–5 season was worth nearly £34 million. Britain takes about half the crop on an average. There are nearly 3,000 tobacco growers and the acreage planted in 1965 was about 226,000. The average tobacco farm is about 2,500 acres in size, tobacco being grown on only about 70 acres with maize and grasses such as star grass in rotation, the remainder of the farm being devoted to livestock. The tobacco belt is located in the east and the main tobacco-growing areas are Lomagundi, Salisbury, Mazoe, Marandellas, Makoni, Hartley and Umtali. Nearly all the tobacco is of the light flue-cured Viginian variety which is produced entirely by Europeans, and grows best on light, sandy soils in areas with about 25 inches of rainfall a year.

Minor cash crops are cotton, deciduous and citrus fruit, sugar, tea, ground-nuts, potatoes and other vegetables. Citrus are important in the Mazoe Valley, supplied with water from the Mazoe reservoir, and smaller amounts are produced in the Sinoia and Umtali districts. There are orchards of deciduous fruit in the higher eastern areas, particularly near Melsetter. In 1964 there were over 5,000 acres under tea in the Chipinga and Inyanga districts with a rainfall of 40–45 inches. The acreage under tea is expanding and there is a considerable potential in the small, well watered areas of the eastern border. Sugar is of increasing importance in the low veld, production having risen from 11,000 to 101,000 tons in 1959–63. Forestry is also of growing importance in the east and there are nearly ¼ million acres of plantations, nearly half under conifers, the rest wattles and eucalyptus. Rainfall of over 40 inches and high temperatures cause rapid tree growth and trees reach economic maturity in a third of the time taken in cooler more temperate latitudes.

Pastoralism is second only to tobacco in the agricultural economy of Rhodesia. Dairying is carried on mainly in the east, near the main towns and along the railway belt. It supplies the local demand for liquid milk and cheese, but about a third of the butter needed has to be imported. Ranching, which is more important than dairying, is concentrated in the drier areas of Matabeleland. European-owned cattle number about $1\frac{1}{2}$ million and African-owned cattle about 2 million. There are considerable exports of beef to South Africa and the Copper Belt and of high-grade beef to the British market. European farmers realize the dangers of overconcentration on tobacco and maize and are paying increasing attention to mixed farming. Attempts to improve the standards of African agriculture in Rhodesia by means of the Land Husbandry Act have only been partially successful, mainly because of political tension. By 1962 over 15 million acres had been allocated for individual holdings and improved farming techniques. There are also about 8 million acres available for purchase by improved or master farmers.

MINERALS

Mineral exports were worth over £25 million in 1964 with asbestos valued at £10 million, gold at £7 million, coal £1·6 million, chrome £2½ million and copper £4½ million. There were also exports of pig-iron and ferro-chrome valued at £4·4 million.

Gold is mined principally on the high veld and particularly along the railway belt from Bulawayo to Salisbury. It is distributed widely in small faulted ore bodies, mainly among the metamorphosed rocks. Small mines[8] make a significant contribution to production, a contrast to the southern Transvaal and northern Orange Free State where only very large companies can work the extensive auriferous deposits profitably. However, a number of small operators and companies have had to cease or cut down operations because of rising costs.

Although gold has for long played a significant part in the economy of Rhodesia, *asbestos* recently surpassed it in value. Rhodesia is the world's third most important producer and the quality of the chrysotile asbestos is high. The largest deposits occur at Shabani and Mashaba. A certain amount is used for the asbestos cement industry, but the bulk of the asbestos is exported. Chrome ore is mined chiefly at Selukwe in the Great Dyke, a

[8] About 30% of all mines in Rhodesia employ less than 10 people. However, much of the gold produced in the country comes from 12 main mines.

396

highly mineralized belt of igneous rocks that runs almost north-south through the high veld. An increasing amount of chrome is being used locally with the establishment of a plant at Que-Que in 1962 to produce high-carbon ferro-chrome. Coal reserves of good quality, estimated at 7,000 million tons, are extensive in the northern middle veld and low veld of the south. At present Wankie is the only producer, with an annual production of nearly 4 million tons. There is a long rail haul on single-track railways to the main markets in the Copper Belt of Zambia and towns and industrial centres of the high veld of Rhodesia.

Great congestion on the railways with the post-war economic boom limited production of coal, necessitating imports of American coal via Lobito Bay for the Copper Belt which has also burnt local supplies of wood to provide power for the mines. Lack of power has been one of the main reasons for the construction of the Kariba Gorge hydro-electric scheme. There are also very large deposits of iron ore, estimated at over a 1,000 million tons, some of which is exploited for a growing iron and steel industry at Redcliff near Que-Que while some is now being exported to Japan. Copper mining is carried on near Fort Victoria and Mangula near Sinoia. Rhodesia is fortunate in having a variety of minerals, particularly extensive deposits of good quality coal and iron ore which provide the basis for expanding secondary industries.

MANUFACTURING

With the growth of manufacturing industries since the war Rhodesia's economy has broadened and matured more than that of her neighbours, except for South Africa. Rhodesia produced over 80% of the manufactured goods in the Federation. The number of factories has grown from 299 with a net output value of £2½ million in 1939 to nearly 1,300 in 1961 with a value of about £146 million. Manufacturing has developed particularly in the only two large towns of Salisbury and Bulawayo[9] and is concerned with the processing of some of the main exports, such as minerals, tobacco and wattle bark and supplying the local market with a fairly wide range of consumer and producer goods. The consumer industries include the manufacture of clothing, textiles, footwear, furniture, foodstuffs, beverages, etc. Among the wide range of producer goods are engineering, paper and asbestos-cement products, iron and steel piping and steel sections, windows, door frames, metal containers, fencing, fertilizers, etc. An industry that reflects the growing importance of motor transport is the assembly of motor

[9] Salisbury accounts for over 40% and Bulawayo for over one-third of the manufactured products in Rhodesia.

vehicles, and the manufacture of accessories such as tyres and batteries. During 1961 the Ford Motor Company's factory was opened near Salisbury for the assembly of cars, trucks and tractors. Since 1960 Rhodesia's industrial pattern has grown more complex with the establishment of industries such as the assembly of radio and television sets.

COMMUNICATIONS

As elsewhere in Africa *railways* are the life blood of Rhodesia and Zambia which lack navigable waterways except for a slight amount of traffic on the Upper Zambezi and Lakes Bangweulu, Malawi and Tanganyika. Rhodesia has to rely on railways for the bulk haulage of heavy raw materials, particularly minerals, and manufactured goods.

Because the 3 foot 6 inch gauge single-track railways were laid as expeditiously and inexpensively as possible, making use of water-sheds to avoid heavy bridging costs, the line of route followed is often circuitous. With the economic development of Rhodesia, Zambia and South Africa the embryonic railway pattern has in many areas proved inadequate, causing congestion and delay. Railways are a state concern in both South Africa, Rhodesia and Zambia, competition from road haulage has been restricted and railway rates tapered to encourage the growth of agriculture and export of raw materials. The communications network has been further complicated by the distribution of state boundaries, the Portuguese-controlled middle Zambezi, for example, cutting into the former Federation.[10] Roads in Southern Africa have developed as feeders to the railway in many areas. There is, however, an increasing amount of long-distance road traffic from South Africa to Rhodesia and Zambia and even further afield to East Africa.[11] Although Rhodes's dreams of a Cape-to-Cairo railway did not materialize, the Great North Road is an important feature in the system of communications over much of the eastern side of Africa.

In all, Rhodesia has about 30,000 miles of *road*. The major urban centres of the southern Transvaal, Bulawayo, Salisbury, Lusaka, the Copper Belt and Beira are connected by bitumenized roads. Strip roads played a useful part in the pre-war development of Rhodesia, but cannot stand up to the great post-war increase in motor transport.

URBANIZATION

A combination of high plateau, healthy climate, good communications and the concentration of commercial agriculture, mining and manufactur-

[10] The effect of economic nationalism on railway routing is referred to on p. 118.
[11] Rhodesia Railways operate motor services on about one-sixth of the roads in Rhodesia.

ing along the railway belt of the high veld has led to the growth of the major towns of Salisbury and Bulawayo and minor towns such as Gwelo and Umtali. The towns of Rhodesia, set far apart from each other, have a spacious layout and grid-iron pattern of wide, tree-lined streets. The commercial core of the biggest towns stands out as a collection of miniature skyscrapers, the tall buildings featuring prominently in the clear, dry air of a Rhodesian winter, and their vertical lines affording a welcome contrast to the wide, flat expanses of the plateau. As in South Africa white and African residential areas are segregated.

Towns have expanded greatly in size over the last 25 years and the migration of Europeans into Rhodesia has been directed largely to Salisbury and Bulawayo, which now contain over half the country's white population.

Salisbury (Plate XXX) is the largest of the Rhodesian towns and has a total population of 300,000, including about 90,000 Europeans. It was founded in 1890 by the Pioneer Column and grew from two centres, the one selected by the Pioneer Column at the base of a hill, the other by the Chartered Company on a nearby slope. The town became a municipality in 1897, and a city in 1935. It is now the capital of Rhodesia and was the capital of the abortive Rhodesia-Nyasaland Federation. Salisbury is favourably situated at an altitude of nearly 5,000 feet at the junction of the north-east to south-west and east-west trending lobes of the high veld and near the centre of the tobacco and maize belt and an important mining area. It is Rhodesia's main market for agricultural products such as tobacco, maize and livestock and is an important industrial centre.

Bulawayo (The Place of Killing), about 250 miles south-west of Salisbury, was founded on the site of Lobengula's Great Kraal in 1893. It developed rapidly and in 1904 its white population was twice the size of Salisbury. Its subsequent development has been slower as it lies in more arid country and it is peripheral to the major area of economic development in Rhodesia. The population is now over 200,000 including about 50,000 Europeans. It is an important market town and the major railway centre in Rhodesia. It has a wide variety of industries and was the largest manufacturing area in the Federation.

The smaller towns of Gwelo, Gatooma and Umtali each have a population of less than 50,000 and there is no town of intermediate size. Umtali is attractively situated near the Mozambique border.

REGION 9—THE PLATEAU OF SOUTH-WEST AFRICA

RELIEF AND STRUCTURE

The plateau of South-West Africa, from 4,000 to over 6,000 feet in altitude, forms part of the great plateau of Southern Africa, a western and drier counterpart of the South African high veld. It is flanked to the west by the desert coast of the Namib and to the east by the Kalahari. The Kalahari sands cover much of the ancient platform in the north, while Old Granite, Primitive systems and the Nama system outcrop in the central area of South-West Africa. The plateau slopes gradually to the Kalahari Basin. In the west the escarpment is close to the coast, being far less eroded than in the east.

CLIMATE

The climate is semi-arid with rainfall varying from about 20 inches in the north to 5 inches in the south and Windhoek, approximately midway between the Cunene and Orange rivers, has about 13 inches of rainfall a year. The high altitude of the plateau causes cool temperatures Windhoek having a mean below 70° F. (21° C.). The dry air and great radiation cause an appreciable temperature range.

POPULATION

The total area of the country is about 320,000 square miles and the population, according to the 1960 census, just over ½ million, including 73,000 Europeans. Africans form the majority, the largest tribe being the Ovambos who number nearly 300,000. There are small numbers of Coloureds, Hottentots and Bushmen. The population is, thus, very sparse, and large areas are almost uninhabited.

South-West Africa was a German colony until the First World War when it was occupied by South African forces. Since then it has come increasingly under the aegis of South Africa, despite its status, first as a Mandate of the League of Nations, and later as a Trust Territory of the United Nations. It is now practically a fifth province of the Republic and sends six members to the South African House of Assembly in Cape Town.

Ovamboland in the north is the best watered and most densely populated part of South-West Africa. It is inhabited by the Ovambo who, compared with the rest of the country, have a strong tribal organization. The economy is largely subsistence and maize and kaffircorn are grown on the sandy soils. Cattle are also an important feature of their economy. The Ovambos supply much of the labour needed on the mines and European farms further south.

AGRICULTURE

The semi-arid climate, as in Botswana, precludes arable farming and *pastoralism* is the major occupation. Cattle are predominant in the central and northern areas and sheep and goats in the drier southern areas. Water is obtained from boreholes and the rainfall of 12–15 inches a year supports a fair grass cover in a normal year. Droughts, however, occur frequently in this region of highly variable rainfall and numbers of livestock fluctuate considerably. It is estimated that nearly ½ million head of karakul out of a total of 3½ million were lost through drought in 1960. Exports of 3 million karakul lamb pelts in 1964 were worth nearly £7 million. These sheep thrive on the dry areas of the south. There are some 2½ million head of cattle and in a normal year about 300,000 cattle, valued at £8 million, are exported to South Africa. Despite the aridity *dairying* is an important subsidiary and in 1958 before the last three-year drought production of butter was nearly 9 million pounds compared with half that amount in 1960.

The only large *irrigation* scheme is the Hardap dam, built in 1962, which irrigates 7,500 acres. Plans for hydro-electric power and irrigation projects on the Cunene River may soon be implemented.

MINERALS

Mineral exploitation is of considerable importance on the plateau. Copper mining commenced at the Tsumeb mine in the Grootfontein district in the days of German occupation, but was suspended in 1940 to be resumed on a larger scale in 1947. Most of the production consists of lead and zinc concentrates with some copper, and exports of lead, copper and zinc concentrates for 1964 were worth about £14 million. Copper and lead smelters and a refining plant completed recently now refine these minerals before export.

TOWNS

Windhoek is the capital and is situated in the centre of the main ranching area at an altitude of about 5,600 feet. It has a cool climate with mean monthly temperatures varying from 55° to 75° F. (13°–24° C.), compared with Beit Bridge in the Limpopo Valley which is in much the same latitude but has an altitude of only 1,505 feet and has mean temperatures ranging from 61° F. (16° C.) (July) to 81° F. (27° C.) (January). The total population of Windhoek is 36,000, including some 10,000 whites, and despite its small size it is of great nodal importance with railway links to South Africa and the main port at Walvis Bay.

REGION 10—THE PLATEAU OF ANGOLA

RELIEF, STRUCTURE AND DRAINAGE

North of the Cunene and South-West Africa stretches the plateau of Angola, a continuation of the plateau of South-West Africa. Its physical character changes with the rapid increase of rainfall to the north, varying from 25 inches in the south to 50–60 inches in the Bihe Plateau and in the north. Altitude is generally over 4,000 feet but rises to 8,000 feet in the Bihe Plateau. The high grass savanna and bush of the plateau, from which rise smooth plugs of volcanic rock, is bordered northwards by a thick swathe of tropical forest marking the edge of the Congo basin. To the south are the seasonal swamps and sandy wastes of the Zambezi and Okovango rivers, while to the west is the sand drift of the coastal desert.

As in South-West Africa the plateau is close to the coast in the centre and the descent by a series of terraces is very steep. Drainage is radial with the Zambezi, Okovango and Cunene draining to the south, south-east and south-west and a number of short coastal rivers such as the Cuanza to the west. To the north and north-east drainage is to the Congo, the watershed between Congo and Zambezi drainage being the boundary between the regions of Southern Africa and the Congo basin.

POPULATION

With a total area of about 481,000 square miles Angola has a population of about 5 million, including nearly 200,000 Portuguese, the majority of whom have arrived as colonists since 1953 under the government-sponsored settlement schemes. The population density of only 10 per square mile is very low, a reflection of the ravages of the slave trade and Angola's economic subjection to Brazil in the past.

AGRICULTURE

Angola has suffered from centuries of neglect, and economic development only really began after the Second World War. Coffee, mainly high quality Robusta, is by far the most important crop, being responsible in 1962, when

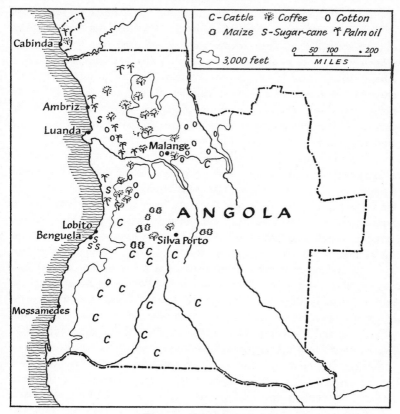

FIG. 79—Angola: crops

nearly 200,000 tons were exported, for almost half the total exports (Fig. 79), and this has provided much needed funds for development.

There are two main areas of production, on the Uige plateau some 200 miles north of Luanda, and near Gabela about 50 miles east of Porto Amboim. Sisal, the second largest export crop with a 1961 value of nearly £4 million is grown near the Benguela railway. Maize is the main African-grown crop on the plateau and much is exported. Other crops are beans, cassava, groundnuts and tobacco. Pastoralism is important in the south, much of which is suitable for ranching on a large scale.

Climatically the high plateau of Angola is suitable for *white settlement*. With a rapidly growing home population and thinly populated lands in Angola Portugal has endeavoured to settle some of her surplus farming population in Angola. The 1953 Development Plan fostered the emigration of Portuguese farmers and their families to the central and southern parts of the Angola plateau. The settlement at Cela was begun in 1952 and similar schemes have been started in the Cunene Valley and elsewhere. Lack of labour is one of Angola's major problems, and the settlement schemes have attempted to overcome this by relying on labour being provided by the Portuguese farmer's family and not by indigenous African labour. This is in contrast to the schemes for white settlement in East Africa, Zambia and Rhodesia which employ African labour in large numbers.

By 1960 less than 2,000 Portuguese had been settled at *Cela* in a dozen villages with an average holding of 15 acres irrigated land, 30 acres non-irrigated land and 300 acres of grazing per family. The scheme has proved far too expensive, and little seems to have been achieved as standards of farming are low. The hazards of unfamiliar tropical soils of poor quality, variable rainfall, and lack of skill and experience has proved too much for the average settler who has resorted to easily grown, low-value crops such as beans and maize instead of the high-value crops such as tobacco and coffee. Plans have therefore been revised by the Portuguese government and no more small holdings are being created. Instead, farms of up to 300 acres in size with 40 acres of irrigated land are now planned and the emphasis will be on dairy and beef cattle. This seems to indicate that there is little future over most of Africa for the peasant holdings typical of parts of Europe.

Settlement schemes for Africans have also been established, one of the largest being on the Uige plateau in the north. Here over a thousand Africans have been given small, terraced holdings on which cassava, maize and groundnuts are grown. The potential for the future economic growth of agriculture in Angola is great, as only a small part of the potential cultivable area is now being used.

Nova Lisboa, the second largest town in Angola, has large railway workshops and other industries.

MINERALS

Minerals are more valuable in Angola than in Mozambique and industrial diamonds are mined in the Lunda district near Dundo, which accounted for about a sixth of exports. Other minerals are iron ore, copper and manganese. The output of iron ore was nearly 1 million tons in 1964. In an attempt to solve the problem of lack of power, as Angola has no good quality coal, *hydro-electric* schemes have been developed at Mabubas, Biopio and Matala to supply power to Luanda, Lobito and Benguela, and Sa de Bandeira and Mossamedes in the south.

The pace of Angola's economic development has slackened somewhat recently with the fall in coffee prices and the disrupting effect of the rebellion in 1961–2.

REGION 11—THE WEST COAST

This is a very long, narrow strip of territory stretching for about 2,000 miles from the south-western Cape to the mouth of the Congo river. Most of the region is desert with less than 5 inches of rainfall and even Luanda has only 11 inches. The prevailing *aridity* is associated with the cool Benguela current which also causes cool temperatures for the latitude. The highest temperatures occur in winter when Berg winds warmed by descent from the upper air raise temperatures to over 90° F. (32° C.).

The *population* is very small except for local concentrations at ports such as Walvis Bay and Luanda, also the capital of Angola, in the diamond mining areas of the Cape or where irrigation enables crops such as sugar-cane and coffee to be grown. Population also increases on the better watered slopes which grow coffee, sugar, sisal and cotton and are marginal to the plateau in Angola.

AGRICULTURE AND FISHING

Agriculture is only of importance to the north of Lobito Bay and even here irrigation is needed during the long dry season. Sugar is the leading crop and is grown near the Catumbela river and Porto Amboin as a plantation crop. Sisal, cotton and palm oil are also produced in the northern coastal area. The Cabinda enclave north of the Congo river, which is politically part of Angola, produces timber, the most valuable export, coffee, palm oil and other crops.

Fishing is perhaps the most important economic activity along the coast. The Benguela current although denying an agricultural harvest on shore has provided a bountiful aqueous harvest of pilchards, tuna, mackerel, sardines and other fish which feed on the abundant plankton attracted by the phosphates and nitrates of the cold current. A string of small fishing ports with processing plants has developed along the coast from the Berg river mouth to Benguela.

South Africa is one of the most important fishing countries in the Southern Hemisphere, the fishing industry having grown rapidly since 1950. In South African waters off the west coast of the Cape Province in

1961 over ½ million tons of fish were caught to a value of about £20 million, approximately 80% by weight being pilchards. Deep-sea fishing, stock fish forming three-quarters of the catch, carried out by trawlers based on Cape Town caters largely for the local market. Inshore fishing is concerned mainly with pilchards and maasbankers and most of its products are exported. It is possible to fish for nearly all the year, a factor of considerable economic importance, as seasons for these fish differ. The bulk of the South African catch is canned, the rest being made into fish oil and meal and nearly 90% of these fish products are exported. In 1963 total earnings from fish products came to over £20 million.

In South-West Africa the fishing industry centred at Walvis Bay and Luderitz plays a valuable part in the economy of the country. Pilchards, and rock lobsters are the main varieties of fish caught. Fishing is now the second largest industry in South-West Africa and in 1964 650,000 tons of fish were caught.

In Angolan waters sardines, mackerel and tuna are the main varieties of fish caught. The fish are processed at Baia dos Tigres, Porto Alexandre, Mossamedes and Benguela, over 80% of the output being fish meal and fish oil. Much of the fish meal is exported to the United States of America, while sun-dried fish are exported to African markets in the interior of Angola and the Congo where proteins are scarce.

MINERALS

There are rich *diamond* fields in Namaqualand and in South-West Africa for a distance of 300 miles from the mouth of the Orange river to Conception Bay. Diamonds are found in marine terraces near the mouth of the Orange river. They are mined easily by the mechanical stripping of the overburden and removal of the diamondiferous gravel which is sorted and washed before diamonds are extracted. More than 20 tons of sand and gravel have to be removed to obtain 1 carat, so sparse is the distribution of diamonds. In the area north of the Orange river diamonds are obtained from detritus blown by the prevailing south-westerly winds into long hollows. The mining of diamonds has also recently started on the seabed off the coast of South-West Africa and in 1963 two dredges were used offshore. Diamonds from the marine gravels of the west coast, the world's most important single source, are good quality gem stones. The Namaqualand fields are operated by the State, production and sale being carefully controlled. South Africa's total annual production of diamonds is worth nearly £20 million, while South-West Africa's 1964 production was valued at £30 million.

The only other mineral of importance in the coastal region is *petroleum* at Benfica near Luanda. Production has been disappointing and amounts only to about ½ million tons per annum. In 1958 an oil refinery was constructed at Luanda which now supplies the Angolan market. *Salt* is obtained from coastal lagoons in various parts of the Western Cape, South-West Africa and Angola.

TOWNS AND COMMUNICATIONS

In comparison to the east coast, the west coast, apart from Lobito Bay has few *ports* of commercial importance. The western spur of the high plateau is much poorer in natural resources and far less economically developed than the eastern. Walvis Bay is the main port for South-West Africa, handling nearly 1 million tons in 1961, while Luderitz Bay in the south is less important but like Walvis Bay has fish canneries. Both these ports are connected by rail with the plateau and to South Africa by a line that crosses the Orange river at Upington and connects with the Cape railway at De Aar.

Lobito Bay is the terminus of the only *railway* to span the Angolan plateau and tap the mineralized region of Katanga and the Copper Belt of Zambia. The railway is handicapped by the very steep haul on to the plateau, climbing 3,000 feet in less than 40 miles, and by the lack of coal. In 1960 only three trains a week ran from Elizabethville to connect with the Angolan section, and the journey to the coast took three days. This railway took many years to build and was only completed in 1932, but has been of great value in opening up the agricultural potential of the Bihé Plateau and supplying a more direct route for mineral exports from Katanga and the Copper Belt of Zambia. The use of the railway for mineral exports from Central Africa was restricted when the Congo was under Belgian control, as they preferred to develop their own system of rail and river communications. Political factors also dictated the movement of copper from Zambia via Rhodesian Railways rather than through Lobito Bay, despite the fact that the port is some 3,000 miles nearer European and American markets than Beira and Lourenço Marques. However, by a recent agreement with the authorities a certain proportion of minerals from Zambia is being sent out via Lobito Bay.[12] Lobito Bay has a good natural harbour with deep water alongside protected by a sandspit, and can accommodate eight ships. In 1961, despite its locational advantages, it only handled about ¾ million tons of transit traffic.

[12] In 1960 the Rhodesian Railways and the copper companies concluded a new agreement whereby the amount of copper moving over the Benguela line was restricted to less than 36,000 tons.

The minor ports of Luanda, Mossamedes and Porto Amboim serve a local hinterland to which they are connected by short railway lines. *Luanda*, with a population of 200,000 (three-quarters African), is the largest town in Angola and the capital. It was the site of the first Portuguese settlement in 1575 and is situated north of the Cuanza river long used as a routeway to the interior. Its prosperity increased greatly with the coffee boom. New blocks of flats, and new, brightly painted villas cater for Luanda's élite, while concrete houses are being built to cater for the expanding African population, the majority of whom are still housed in mud and wattle huts. The port of Luanda has more local traffic than Lobito, exporting four-fifths of Angola's coffee and handling the bulk of the imports.

REGION 12—THE MIDDLE LIMPOPO AND ZAMBEZI VALLEYS AND THE LOW VELD OF SOUTH AFRICA

RELIEF, STRUCTURE AND DRAINAGE

Zambia, Rhodesia, South Africa, and Mozambique share political control of these subregions. The Limpopo and Zambezi valleys are both great down-faulted depressions creating hot lowland embayments cutting into the cooler heights of the plateau. These rivers have eroded back into the soft Karoo sediments let down in great faults below the plateau. The middle courses of both rivers are similar in their semicircular courses, but are very different in régime and size of catchment area. The drainage basin of the Zambezi has an area of 540,000 square miles and, including tributaries, a run-off of 80 million acre feet, while the Limpopo with its 'grey-green greasy banks' is merely a string of pools during the dry season. Its annual run-off is only 6 million acre feet.

The Limpopo, rising in the Witwatersrand at a height of nearly 6,000 feet, enters the Mozambique Plain at a height of less than 1,000 feet and for much of its course is below 2,000 feet. The valley is underlain by Old Granite and forms a level plain broken only by isolated inselbergs. The largest tributaries rising in the high veld of South Africa are the Crocodile, Magalakwin, Letaba and Olifants.

The Limpopo Valley curves round onto the low veld which also forms part of this region. The low veld is better watered than the dry Limpopo valley and crossed by a number of perennial rivers.

CLIMATE AND VEGETATION

The higher rainfall of up to 60 inches in places has caused the Old Granite of the low veld to be eroded into a rolling landscape of rounded hills and pronounced valleys. When viewed from the heights of the plateau in summer the eye of the onlooker is drawn out onto what seems a warm green sea from which rise a succession of islands. Massive baobabs are a feature of the Limpopo Valley. Mean temperatures vary from about 60° to 80° F. (15°–27° C.) and are considerably higher than on the plateau.

AGRICULTURE

Development of the low veld has for long been hindered by high temperatures and humidity in the summer, and by malaria and other diseases and pests. However, since 1945 development has been very rapid with a wide variety of crops grown under irrigation in the valleys of the Crocodile river and its tributaries and the Groot Letaba. Citrus is the most important product followed by other tropical and subtropical fruits such as mangoes, pawpaws, avocados and bananas, and vegetables such as tomatoes, green beans and peas. In the non-irrigated areas maize and groundnuts are grown. In the more isolated areas *cattle ranching* is the main activity. This is somewhat of a pioneer venture in the Limpopo Valley and low veld of Rhodesia. Farms are large, many being over 6,000 acres in size. A number of dams for watering stock and irrigating fodder crops such as Napier grass and paspalum are being built along the Limpopo, but in the main cattle have to rely on natural grazing. The Afrikander, which has developed as a distinctive breed over a fairly long period, is the main breed of cattle in the bush-veld and Limpopo Valley. With its thick skin and light colour it can withstand the heat and disease of these areas better than the exotic breeds such as Herefords, and Aberdeen Angus. A new breed, the Mara, has been evolved, which is five-eighths Afrikander and three-eighths Shorthorn, and which it is hoped will be similar to the Santa Gertrudis of Texas and reach maturity early in the hot bushveld region.

POPULATION

African reserves are significant in the north and along the slopes of the Zoutpansberg, which derives its name from a large saltpan at the western edge of the mountains. The Shangaan and Venda people live here and their domed and verandahed huts are not unlike those of the Tswana in Botswana. Agriculture is still traditional with the growing of maize and other crops and the rearing of livestock, and land use differs radically from the variegated pattern of agriculture in the European areas. The Venda were great iron workers in pre-European days and spun wild cotton and wove loose fabrics from it. They used copper rods for currency and were unique in South Africa for hollowing canoes out of logs and carving their hut doors.

The density of population in the eastern part of the low veld and over much of the Limpopo Valley is very slight because of heat, aridity, malaria and the setting aside of a large area in the north-east as a game reserve for wild animals, known as the Kruger National Park. This was established in 1898 and now covers an area of 8,000 square miles. It abounds with animals of all kinds including wildebeest, zebra, impala, kudu, eland, elephant, hippopotamus, giraffe, buffalo, lions, leopards, etc., and is a great tourist attraction from June to October.

MINERALS

The mining of *copper* began at Messina in 1904. There are now four producing mines. Although the ores average only 2% copper compared to the 5% of the Copper Belt they are exceptionally pure and mining has increased in importance since the war. In 1965 South Africa's total output was about 66,000 tons, valued at £16 million (nearly all exported).

THE ZAMBEZI AND THE VICTORIA FALLS

The three-pronged subregion of the middle and lower Zambezi and the Luangwa rivers cuts into the Rhodesian plateau. The great troughs of the Zambezi and Luangwa are formidable barriers to movement and until the Kariba Dam was built the Zambezi was a backward and isolated area, as is the Luangwa still. The *middle Zambezi* begins below the Victoria Falls and flows for some 800 miles to the Kebrabasa Rapids where the lower course begins. These rapids indicate where a proto-Zambezi cut back to capture the Luangwa and then worked back to the Victoria Falls. The river leaves the plateau here and flows in a broad valley, becoming braided and up to five miles wide in places. During the wet season the separate streams unite to form a broad flood of water. The major left-bank tributaries are the Kafue, Luangwa and Shire, the last being the southern outlet of Lake Malawi.

The middle course of the river, in contrast to its upper and lower courses, lies in a great depression let down 2,000 feet below the level of the surrounding plateau. Steep hills and ridges border the river which flows in a well defined course.

The *Victoria Falls* (Plate XXXI) were discovered by Livingstone in 1855. They plunge nearly 400 feet into a chasm nearly a mile wide. The

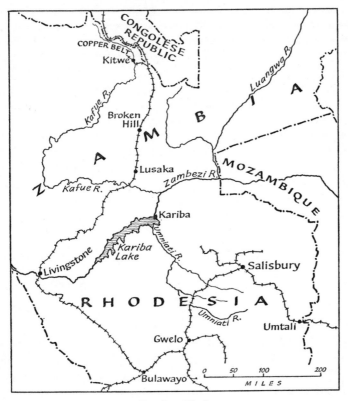

FIG. 80—Kariba

river then boils its way through a narrow, sharply angled gorge whose course is controlled by successive fault lines. The Falls themselves are caused by a barrier of Tertiary basalt and the gorge is cut in a band of lava.

KARIBA

The idea of harnessing the power of the Falls and the Zambezi has long been mooted. It was only considered seriously after the Second World War with the economic boom in what was then the Rhodesias, the lack of cheap power and the advent of Federation. The *Kariba Gorge* and Kafue schemes offered rival merits, but the decision was made in favour of the former. A dam has been built where the Zambezi narrows and cuts through a rocky barrier, the dam wall being 420 feet high. A lake 175 miles long has formed behind the dam (Fig. 80). Work was commenced in 1955 and completed in 1960, despite unexpected floods of great magnitude which halted work for a while. The carrying out of the whole scheme in a wild and isolated area was a massive undertaking. It involved the building of roads and a bridge over the river, an air strip, townships, power lines to the Copper Belt and high veld of Rhodesia and the resettlement of 50,000 backward Valley Tonga. Kariba (Plate XXXI) is located advantageously to meet the power needs of the economically developed areas of Zambia and Rhodesia and now supplies much of their power from six sets of generators.

Below the Kebrabasa Rapids the river is navigable to the delta for shallow draught vessels, but is now little used since the building of the Sena bridge and the railway to Malawi. The Karoo measures near Tete have comparatively large reserves of coal and production is now about 300,000 tons per annum. Sugar plantations are important in the lower Zambezi valley with cotton as a subsidiary crop.

REGION 13—THE RAILWAY ZONE AND COPPER BELT OF ZAMBIA

RELIEF AND STRUCTURE

Zambia's railway belt with its line of European farms, mining areas and towns forms a separate region stretching from Livingstone to the Copper Belt. This is a narrow zone some 40,000 square miles in area and with a population of nearly ¾ million. The total area of Zambia is about 285,000 square miles and the total population 3½ million. It occupies the higher ground of the plateau overlooking the Zambezi for a distance of some 300 miles before turning north near Lusaka and following the watershed between the Kafue and the Luangwa.

Most of the region is above 3,000 feet in altitude and consists of level surfaces composed of ancient granites and gneisses with Karoo sedimentaries to the east and Kalahari sands to the west.

CLIMATE AND VEGETATION

Rainfall increases towards the north, which has about 50 inches a year concentrated in a few summer months. Temperatures are modified considerably by altitude, although Livingstone at 3,000 feet is 6° F. (− 3·3° C.) warmer than Abercorn at 5,100 feet, which has an annual average temperature of about 67° F. (19° C.). The *vegetation* consists of a ubiquitous, deciduous woodland, *Brachystegia-Isoberlinia*, which clothes the plains with a dense covering of bush, becoming thicker in the north. *Soils* generally are poor and lacking in humus and mineral nutrients, particularly in the heavy rainfall area of the Copper Belt where soils are highly leached.

AGRICULTURE

From Kalomo to Broken Hill European farms occupy a zone along the railway where rainfall is lower and soils more fertile. The number of European farms has increased considerably with the growth of markets in the Copper Belt. Over 1,000 European farmers rear cattle and grow maize, tobacco, millet, groundnuts and potatoes. Maize is grown on about 100,000 to 120,000 acres with a fairly high yield of 10 bags to the acre. Adverse rainfall can, however, halve the yield per acre, as for example in the 1957–8 season when it was only 5·6 bags to the acre compared to 10·3 in the 1958–9 season. An increasing amount of maize is grown by Africans near the railway belt where ploughing by oxen replaces the traditional methods of mounding or chitimene. The old system of mixed cropping is being replaced by the monoculture of maize in large fields. There is an increasing use of kraal manure, green manure and fertilizers to maintain soil fertility; but sheet erosion has become serious in parts of the Mazabuka district.

Tobacco is the major export crop and is grown mainly on European estates. As in Rhodesia, Virginian flue-cured tobacco is the variety mostly grown. In 1959 there were nearly 14,000 acres of Virginian tobacco with a production of over 12 million pounds compared to about 22 million pounds in 1964. Because of the great distance from markets and high costs of transport tobacco is one of the few crops that can be grown profitably for export. A large new sugar estate was started in 1964 at Mazabuka. Some 5 million pounds of cotton were produced in 1965 in the Central and Southern provinces.

Because of the widespread distribution of tsetse fly, which affects over half the country, cattle rearing is limited to small areas such as Barotseland, the southern areas and the railway belt. There are less than 200,000 European-owned cattle compared to over 1 million African-owned cattle. There is a considerable production of beef, and dairying has expanded to serve the market of the Copper Belt.

MINERALS

The most valuable economic activity in Zambia is mining and the economy of the country depends very largely on exports of copper. Other minerals mined on a much smaller scale are cobalt which is produced in association with copper at certain mines; zinc and lead from Broken Hill and manganese from the Mkushi and Fort Roseberry districts. Figures of production for 1964 were as follows: cobalt £1·5 million, zinc £5·1 million, lead £1 million; that is, a total of about £7 million compared to a copper output worth £140 million. In 1965 mining contributed 47% of the national gross domestic product.

The exploitation of copper is comparatively recent, as it was only in 1925 that extensive deposits of sulphide ores with a fairly high copper content of 2% to 5% were found near the Belgian Congo border.[13] Reserves are considerable and are now estimated to last for 70 years and probably for much longer. Development was rapid after an initial setback in 1931, and production reached nearly $\frac{1}{4}$ million tons by 1936. By 1964 production was over 600,000 tons, valued at £140 million. The amount of electrolytic compared to blister copper has steadily increased and is now about three-quarters of total output. Mining is deep-level, except at Nchanga, compared to the opencast mines of Katanga. Although the sulphide ores from the Copper Belt are more difficult to extract they are more easily refined than the oxide ores of Katanga.

THE COPPER BELT

The Copper Belt lies on the Congo-Zambezi watershed just south of Katanga at an altitude of a little over 4,000 feet, and has an area of nearly 2,000 square miles (Fig. 81). There are seven producing mines and associated towns situated in two groups separated by the valley of the Kafue river. To the north are the towns of Ndola and Mufulira, and 20 miles to the south are the towns of Luanshya, Kalulushi, Kitwe, Chingola and Bancroft. These towns began as urban settlements associated with the various copper mines: for example, Ndola with the Bwana Mkubwa mine, now closed, Luanshya with Roan Antelope, Kitwe with Nkana, Kalulushi with Chibuluma and Chingola with Nchanga. Problems of administration have developed as the towns are administratively separated from, but economically dependent on, the mines.

The *population* of the Copper Belt has grown rapidly from a total of 87,400 including 7,600 Europeans in 1939 to nearly 370,000 including 42,000 Europeans in 1961. Over half the total white population of 77,000 in Zambia and about one-eighth of the African population live on the Copper Belt. This mushrooming, mining-urban community, which may soon number $\frac{1}{2}$ million people, is composed of several separate clusters with

[13] Katanga mines started producing in 1912. Copper deposits here are mostly oxides located on or near the surface, so that copper is exploited by opencast mining. Copper mining in Zambia started at Nkana in 1921.

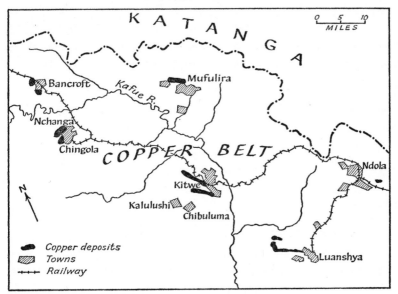

Fig. 81—Zambia: the Copper Belt

Kitwe-Nkana and Ndola as the largest with populations of about 130,000 and 100,000 respectively. *Kitwe* in the centre of the Copper Belt is likely to become the leading town, as *Ndola* is eccentric at the eastern edge. Each association of town and mine is well planned, with separate residential areas for Africans and Europeans and a flourishing commercial core with wide streets and a few large buildings. However, the very speed with which this large urban community has grown up in what was virtually empty bush lends a somewhat transient air to the rash of buildings and mining gear that has spread over the waste and been cut out of the bush.

There are now few problems of power and water, power coming from Katanga and Kariba. But there are social problems of labour supply and relations between European and African mine-workers complicated by the great differences in their way of life and background. The African has been subjected to the revolutionary change in one generation from a backward rural and agricultural environment to a complex, modern, mining and urban society in which attitude to work and customs are so different.

There is a long rail haul of 1,300 miles to Lobito Bay and nearly 1,500 miles to Beira, so that transport costs are high. Until recently most of the copper was exported via Beira, but an increasing proportion is now being sent via Lourenço Marques.

The other main towns of the railway zone are Broken Hill, Lusaka and Livingstone. The capital was moved in 1939 from Livingstone to *Lusaka* which is in a higher, healthier and more central position. Lusaka has grown rapidly and it now has the largest European population in a Zambian town with 14,000 out of a total population of about 140,000. It is in the middle of the European farming area and near the Kafue river. It is an important route centre and also has some light industries.

REGION 14—THE PLATEAU OF NORTH-WESTERN AND NORTH-EASTERN ZAMBIA

RELIEF AND STRUCTURE

The bulk of Zambia away from the railway belt is included in this region. It can be divided into two great lobes linked by the railway belt. The Upper Zambezi is included, but to the east the Luangwa Valley forms a barrier zone, a hot unhealthy area with a very sparse population, separating the Northern and Eastern Provinces of Zambia. The Eastern Province has closer links with Malawi and is therefore included in the next region. The northern boundary is the Congo-Zambezi watershed, an area of indeterminate drainage with many grassy flat-floored valleys, known as damboes, and swamps threading their way through the bush. Lake Bangweulu and the swamps at its southern end cover an area of nearly 4,000 square miles.

Most of the region consists of the usual expanse of ancient *crystalline plateau* between 3,000 and 4,000 feet in altitude, the surface falling from the Congo-Zambezi watershed in the north to the plains of Barotseland. Higher land occurs in the Muchinga mountains overlooking the Luangwa rift.

CLIMATE AND VEGETATION

Rainfall is less than 15 inches in parts of the Zambezi Valley, but rises to 60 inches in parts of the north. Temperatures vary according to altitude; Mongu with an elevation of 3,459 feet ranges from a mean of about 64° F. (18° C.) in July to 73° F. (23° C.) in January, while Kasama at 4,544 feet only ranges from 63° F. (17° C.) in July to 68° F. (20° C.) in January. The vegetation as in the railway belt consists of savanna with thick thorn and sparse, coarse grass in the drier areas of the south with denser bush in the higher rainfall areas further north. *Soils* vary from the sandy soils of Kalahari type to fertile loams, but in the main are shallow and poor.

POPULATION

Most of Zambia is very sparsely populated with less than 12 persons per square mile, and with a backward subsistence economy. Labour migration occurs on a comparatively large scale despite the slight rural density of population, as the environment is so unpromising. Local concentrations occur along the Copper Belt and in the Mazabuka district attracted by European enterprise. Of the total area of 288,000 square miles only about 2% is alienated to Europeans, nearly all of the rest consisting of land reserved for the Africans, with Barotseland occupying about 57,000 square miles.

AGRICULTURE

Only a small part of the plateau is under cultivation and even pastoralism is comparatively unimportant, as over half the country is infested by tsetse fly. Seen from the air, the impression is of a vast sea of trees and bush broken occasionally by patches which have been burnt and cleared for cultivation. Traditional methods of agriculture are still continued over large areas, modified in part by favourable physical conditions, access to markets and labour migration. Two areas are of particular interest—Barotseland and Bembaland.

BAROTSELAND

The Protectorate of Barotseland is centred on the Upper Zambezi Valley. The dominant tribe are the Lozi who inhabit the Central Barotse Plain, 120 miles long by 24 miles wide. Heavy summer rains in the head-waters of the Zambezi, impermeable clays below the Kalahari sands and a gentle gradient cause widespread flooding from January until June. The flood is of vital importance to the Lozi, forcing people, livestock and wild animals to move from the Plain to higher land from January onwards.

The *Lozi* had a remarkably varied environment with many resources. They were cultivators and collectors, cattle-keepers, fishermen and hunters and had plenty of meat and fish. They cultivated many kinds of cereals, root crops, pulses, cucurbits, etc., which ripened at different periods in a number of different gardens, so that the hunger season was very short. Their chief crop kaffircorn was grown on mounds. Cattle were used for manuring mound and ridge gardens. Nevertheless, the Lozi lack the great interest shown in cattle by tribes such as the Zulu, and their standard of cattle-keeping is generally low.

An elaborate system of land use has thus evolved based on varying conditions of water level, soil and grass, with a complicated movement of people and stock regulated by flood conditions. The system has undergone considerable changes with the growth of a money economy and migration of the men to work in the European areas. Migration is usually for a long period, so that work such as draining channels for seepage cultivation, or making fish traps, has decreased with a fall in production and variety of food. However, women are able to work mound and other gardens, so that the position is not as serious as in the areas of slash-and-burn cultivation in the bush where far fewer men are available to cut and burn the trees prior to cultivation of the land. Population is tending to concentrate more on the margin of the Plain where missions, government stations and stores are situated, and the social importance of mound ownership has decreased.

Bembaland

On the sparsely populated plateau of the northern part of the region with poor soils and a high rainfall various forms of the *chitimene* system or ash-and-hoe cultivation are practised. One of the most interesting and efficient variants is practised by the *Bemba* who live in the Northern Province in an area of very poor soils.

Millet, sweet potatoes and cassava are planted in cleared plots which are moved every few years because of soil exhaustion. Only a small density of population can be supported, as the material from 6–10 acres of pollarded trees is needed to provide ash for 1 acre of land. With the absence of men for long periods, a modification of the system has developed by which the trees are cut down completely, and this has caused the degeneration of large areas from forest to poor scrub and grassland.

Fishing is important in both these areas, especially on the lakes, the estimated production for 1965 being worth about £1 million. Over 60% of all fish sold is dried fish.

The effects of the rapid growth of the Copper Belt towns and widespread social and economic changes that have occurred in Central Africa have been felt in the remotest rural areas. Tribal customs and beliefs, for example rituals concerning the ancestor cult, are no longer appropriate in a mining and urban economy or are adapted to suit the changing conditions. New customs and institutions, such as the burial societies and new churches, have evolved to bridge the wide gap between tribal and urban life. Attempts are being made to broaden the economy of the rural areas and stimulate village life so as to decrease the flow of people to the urban areas, but so far these have only touched the fringe of the problem. Zambia is fortunate in having an adequate revenue from copper to help in this rural betterment programme, but is hampered by the isolation and poverty of natural resources over much of the plateau. It is estimated that nearly ½ million acres could be irrigated on the Kafue Flats. Experiments have shown that high yields of wheat, barley, cotton and groundnuts can be obtained. The latest development plan recommends the establishment of a polder of about 20,000 acres to support nearly 15,000 people on 6–18 acre holdings.

REGION 15—THE MALAWI RIFT AND BORDERING HIGHLANDS

RELIEF, STRUCTURE AND CLIMATE

This is a small region which includes the triangular wedge of North-Eastern Zambia between the Luangwa river and the Malawi border. The total area of Malawi is 46,000 square miles, which includes nearly 9,000 square miles of water. The outstanding physical feature is the Rift Valley with scarped highlands to east and west forming the uplifted edges of the plateau. The height of the plateau is 3,000 to 4,000 feet, rising to over 8,000 feet in the Nyika plateau. The Shire Highlands in the south are much lower and only 2,000 to 3,500 feet, but with heights of 7,000 feet on Zomba mountain and 10,000 on Mount Mlanje. The rifted depression containing Lake Malawi is at a height of about 1,500 feet, but drops to only a few hundred feet above sea level in the south where the Shire river leaves Malawi.

The bordering highlands rise steeply from the lake which is long and narrow and almost fjord-like in appearance. In some places there is a lake-shore plain several miles wide, but in other places the escarpment rises almost sheer from the lake. There are great variations in relief, climate and soil. Rainfall varies from 30 to 80 inches. The *climate* of the Lake differs from that of the plateau. Strong winds blow up and down the lake funnelled by the steep edges of the plateau and the lake waters can become rough. Temperatures are high and range from about 50° to 100° F. (10°–38° C.). The plateau has registered 26° F. (−3° C.). The south and south-east facing slopes of the mountain and hill masses on the plateau, such as Mount Mlanje, are cooler and moister than elsewhere with appreciable amounts of precipitation throughout the year.

POPULATION

The pronounced variations of rainfall and soil cause considerable differences over a short distance in population density and type of crop grown.

The total population of Malawi is about 4 million, including about 9,500 Europeans. There are densities of over 200 per square mile in the Shire Highlands and along parts of the lake shore. Population tends to concentrate along the lake shore, rivers draining into the lake and the upper Shire valley, while low densities occur in upland areas such as the Nyika plateau or on sandy soils developed on granites and gneisses.

TOWNS

The main urban centre in Malawi is Blantyre-Limbe with a population of about 100,000. It is the seat of the new university and has agricultural processing and other industries. Lilongwe is the second largest town. It is situated more centrally and the capital may be shifted here from Zomba in the south.

AGRICULTURE

Agriculture is the predominant economic activity, as there are few mineral resources and very slight industrial development. Exports of labour on a large scale are necessary to balance the economy. In 1957 emigrants' remittances brought in £1·6 million. It is estimated that nearly 260,000 Africans from Malawi went abroad to seek work in 1960, three-quarters of them to Rhodesia, compared with about 190,000 employed in Malawi, nearly half of them in agriculture, forestry and fishing.

The area of greatest density and economic development is in the *Shire or Blantyre Highlands*. This was the area traversed by Livingstone and where in 1874 the Church of Scotland established a mission. In 1891 a British Protectorate was proclaimed. A few Europeans settled in this area, but even by 1936 there were less than 2,000 white people in Malawi compared with 55,000 in Rhodesia. Nevertheless, the tiny European farming community has played an important part in the economic development of the country, and in 1959 produced cash crops worth £4 millon compared with only £3½ million for African sales.

Tea is the major European-produced crop and is grown on about 30,000 acres almost entirely in the high rainfall area of Cholo and Mlanje, which has 45–50 inches of rain. Experiments have been carried out to grow tea in the Northern Province as well. Exports of tea have risen from 15 million pounds weight in 1954 to 27 million in 1964 and employment is given to some 30,000 Africans. Coffee was important initially, but declined with the growth of the tea industry. Interest has revived and there are increased plantings of Arabica coffee in the Northern and Southern Provinces and on European estates in the Shire Highlands. Tung is grown on European estates, but acreage has declined from 21,500 acres in 1954 to 13,000 acres in 1964 because of falling prices and high production costs. Tung oil is used mainly in the varnish and paint industry.

80% of the Africans are engaged in subsistence agriculture, and production of subsistence crops was estimated at about £14 million in 1959. *Tobacco* is mainly an African-grown crop and total production in 1964 was nearly 28 million pounds, which represented about one-eleventh of Rhodesia's production. Europeans produce flue-cured tobacco, while Africans concentrate on fire-cured tobacco which is about 60% of Malawi's tobacco output. Tobacco is grown on small plots in the African areas and quality is quite good, although prices obtained for fire-cured were less than half those for flue-cured during the 1959/60 season. Cotton is another cash crop grown in the Shire Valley and Central Province. Yields fluctuate greatly because of pests, adverse weather and poor farming methods. In 1956 the yield per acre was only 172 pounds, compared with 595 pounds per acre in 1959. Most of the crop is exported to Rhodesia for use in the Gatooma cotton mills. Groundnuts are an important crop in the Eastern Province of Zambia.

In a concerted attempt to widen the basis of the economy and reduce the flow of labour from Malawi particular attention is being paid to other African cash crops, for example coffee and rice. Afforestation has considerable potential for the high uninhabited Nyika plateau and Mlanje mountain in the south. By the end of 1961 the Forestry Department had finished the planting of 28,000 acres of hardwoods and softwoods. The Shire Valley Scheme aims at harnessing the water power potential of the Shire rapids and at the reclamation and irrigation of the Elephant Marsh near Chiromo.[14] Crops such as cotton and tobacco could be grown under irrigation. Sugar is now being grown under irrigation in the lower Shire Valley. Fishing on Lake Malawi has increased, but there is danger of the southern part of the lake being overfished. Production in 1964 amounted to 14,600 tons. Fish are an important source of protein as meat is in short supply. Malawi has few livestock compared with Rhodesia which has 3½ million cattle. In 1964 there were only about 400,000 cattle, nearly ½ million goats and 70,000 sheep. Malawi is thus predominantly arable.

The break-up of Federation has caused numerous problems for the former partners. Trade between Zambia and Rhodesia has decreased considerably, and new arteries of trade are being developed through

[14] A barrage 700 feet long and 80 feet high is to be built, which will control the level of Lake Malawi which has varied greatly over the years; for example, in 1937 it was nearly 20 feet higher than in 1915.

Tanzania. However, Malawi's outlets are still to the south and large numbers of Africans from Malawi continue to work in Rhodesia.

REGION 16—NORTHERN MOZAMBIQUE

RELIEF, STRUCTURE AND CLIMATE

The last region, northern Mozambique, lies to the north of the Zambezi river and west of Lake Malawi. It has been considered as a separate region mainly for historical and political reasons, as under Portuguese control it has been developed in a very different manner from its neighbours, Malawi and Tanzania. *Physically* much of the region consists of the great African plateau with considerable areas above 3,000 feet and up to 5,000–6,000 feet in the Namuli highlands and near the lake. The eastern edge consists of plateau slopes and a narrow strip along the coast below 600 feet in height. The north, however, lacks the true coastal plain of southern Mozambique. Temperatures along the coast are high, but decrease over the misty plateau slopes east of Lake Malawi and rainfall increases to 45–60 inches.

AGRICULTURE

Until recently agriculture in the north was neglected, except for a few Portuguese sisal estates along the coast and plantations of coconut trees near Quelimome. Since 1940 development has been fairly rapid and it is now the major agricultural region producing a great variety of crops.

Most of the commercially grown cassava, potatoes, cashew nuts and groundnuts and, in particular, three-quarters of the cotton come from the north. Cotton grown by Africans is the most valuable commercial crop. In 1964 over 33,000 tons of cotton valued at £7 million were exported, mostly to Portugal. This was followed by exports of copra and coir worth over £6 million in 1964. Another African-grown crop is cashew nuts, particularly along the coast, most of which is exported to India. There are over 20 European sisal plantations along the coast with exports worth nearly £2 million in 1961. Tea is a European crop of increasing importance which is grown between Milange and Gurùe near the Malawi border. Production in 1964 was over 10,000 tons.

The north lacks the important transit trade and port traffic of the south, although the harbour of Nacala has been developed and the railway line from here extended over 400 miles and not far from its destination, Metangula on Lake Malawi.

Dislike of Portuguese rule and, until recently, lack of economic opportunity in the north has caused the permanent migration of over 100,000 Africans from northern Mozambique into southern Malawi. Despite the Portuguese administration's claim of non-racialism there is still a great difference between the political and social status of the Portuguese citizens or assimilados and the natives who comprise the great majority of the population and are governed strictly by traditional custom. In 1960 after more than four centuries of Portuguese rule there were less than 10,000 assimilados out of a total population of nearly 6 million. Contract labour is obtained for private employers, for work in the Angolan diamond industry and on the Rand mines. Dislike of labour regulations has caused widespread migration from Angola and Mozambique into neighbouring territories. In Nyasaland (Malawi) in 1933 the total number of immigrants from Mozambique was estimated at over 400,000.

Major reasons for the lack of economic development in Angola and Mozambique are the poverty of the mother country, Portugal's reluctance to allow foreign investment in her colonies and the meagre amount of Portuguese capital available for colonial development. Spurred on by the Angolan revolt and independence in nearby states, Portugal has belatedly implemented a measure of political and economic reform, such as the granting of citizenship to all inhabitants and the gradual relaxation of tariffs on trade and agriculture. These changes may have come too late and African governments in exile for Angola and Mozambique are planning for eventual independence.

CHAPTER THIRTEEN

Malagasy

MALAGASY is one of the world's largest islands and is nearly a 1,000 miles long with an area of about 230,000 square miles and population of about 5·8 million (Fig. 82). It is situated between latitudes 12° and 20° S. and at its narrowest point is about 240 miles from the African mainland. Despite its comparative accessibility to Africa its culture appears to be more Polynesian and Malayan than African. Economically too there are very few links with Africa and nearly three-quarters of the total trade is with France. There is a high degree of dependence on agriculture and trade is still fairly limited relative to the size of the island and its population. Handicapped by isolation and an expanding population, this old eroded island with two-thirds of its surface covered by laterite faces many problems in its new state of independence.

RELIEF AND STRUCTURE

Physically the island is like Africa in miniature, with a high central plateau falling steeply to a narrow and regular eastern coastal plain and with a broader western coastal plain. Malagasy was probably joined to Africa before the break-up of Gondwanaland. Land connections in early Quaternary times enabled lemurs to cross over to the island from Africa and there are now thirty-nine species of Lemuroidea, although there are no large mammalia because of the sundering of this land connection. The core of the island is Archaean and these crystalline rocks form the central plateau with many volcanic and granitic masses rising several thousand feet above the general level of the plateau at 3,000 to 4,000 feet. The highest part of the plateau near the northern tip of the island is the volcanic mass of Tsara-tanana which reaches a height of over 9,000 feet. The bulk of the plateau is covered by almost sterile laterites which greatly reduces the amount of cultivable land.

The eastern edge of the plateau is faulted, complex and polycyclic, and plunges by steep scarps to a narrow coastal plain and then to depths of over 13,000 feet below the sea. The coastline is extremely regular and, except for Diego Suarez in the far north, lacks natural harbours. The western coastal plain is much broader and more indented, and is composed largely of

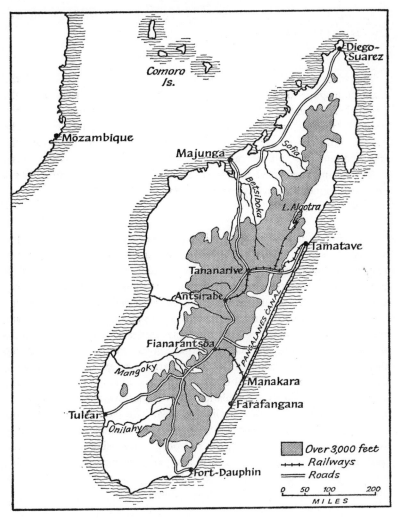

Fig. 82—Malagasy: relief and communications

sedimentary rocks with limestones, sandstones and clays of Jurassic to
Tertiary age. The plateau drops more gradually to the west than to the east
by a series of terraces. There is a close comparison between the erosion
cycles of Africa and Malagasy which has five erosion surfaces. The Jurassic
surface is the highest at altitudes of 6,200 and 7,300 feet, followed by late
Cretaceous residuals rising 700–1,700 feet above the Tertiary, the most

extensive of the surfaces at 3,300 feet. The youngest surfaces are those of the end Tertiary and Quaternary.

DRAINAGE

The main watershed is near the eastern edge of the plateau and the main drainage is to the west and north-west, the principal river being the Betsiboka. Because of rapids, as they debouch from the plateau on to the coastal plain, the west-flowing rivers are of little navigable value. The mouths of the shorter eastern rivers are, however, often sand-barred and the waters often splay out into a series of lagoons. These lagoons have been artificially interconnected to form the Canal des Pangalanes, providing a useful continuous waterway for 400 miles.

CLIMATE

The climate is essentially tropical and maritime, most of the island receiving 40 to 60 inches of rainfall per annum. There are a number of climatic regions varying according to altitude and aspect. Rainfall is heaviest along the east coast as it lies athwart the south-east trades. Some areas receive over a 100 inches of rainfall, which is well distributed, although the bulk of it comes in summer. Temperatures and humidity are high and the climate is not unlike parts of the south and east African coast. Conditions on the plateau are also somewhat similar to the African plateau, though with cooler temperatures; Tananarive has an annual average of 65° F. (18° C.) and a rainfall of over 40 inches which is restricted to the summer. The bulk of the population and the most advanced tribes live in the more invigorating plateau climatic area. The west coast has a more seasonal climate than the east coast; Majunga, for example, has 60 inches of rainfall from October to April, the rest of the year being dry. The south-western tip of the island is semi-arid and parts receive less than 15 inches per annum. Hurricanes are a menace along the east coast and high winds and heavy seas restrict Arab dhows to the west coast.

VEGETATION

Long interference by man has greatly affected the vegetation climax and forest now covers little more than 10% of the island (Fig. 83). Fire has caused the destruction of the plateau forest which is replaced by savanna with poor quality grasses and savoka or brushwood. Large areas of lateritic soils have been uncovered and hardpan exposed which is of little use for agriculture. The great expanse of red latosols has given Malagasy the name of the Red Isle. There are a few areas of more fertile soils, particularly the

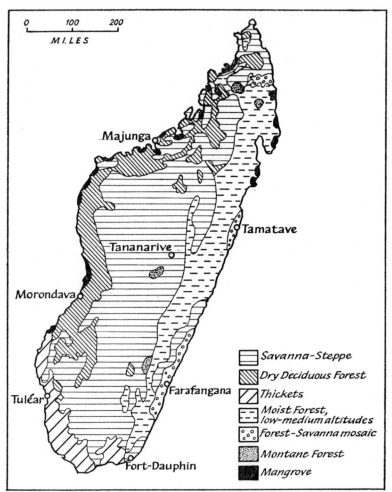

FIG. 83—Malagasy: vegetation (from UNESCO *Vegetation Map of Africa*)

volcanic soils near Tananarive and the alluvial soils of the western river valleys. To the south and south-west lack of rain has caused a highly contrasted form of vegetation, which is drought-resistant, which includes cactus and aloes and is somewhat reminiscent of the eastern Karoo.

HISTORY

European contact with Malagasy (formerly—until 1960—Madagascar)

427

dates from the sixteenth century with the growth of the Far Eastern trade. Neither the Portuguese nor the Dutch attempted to settle on the island as they were more interested in the African mainland. The French, however, established themselves at Fort Dauphin in 1642. British links were developed at the beginning of the nineteenth century and stations of the London Missionary Society were opened in the 1820s. The dominant tribe, the Merina,[1] gradually extended their hold over most of the island by the nineteenth century, and during the 1840s their queen tried to resist the spread of Christianity by persecuting the Christians and by closing the island to foreign traders. Relations with the French and English were re-established in the 1860s and the Rev. William Ellis, who visited the Merina (Hova) queen in 1856, describes her gold crown, scarlet umbrella and dress of white satin lamba, 'which, considering the lamba as the national Hova costume, was quite a queenly dress'. French interest in the island increased and, despite Merina opposition, the French occupied it in 1896. Galliéni became the first governor and, in contrast to the usual methods of French administration, established a form of indirect rule. There was a serious revolt led by the Merina in 1947–8, but the island is now an independent republic and member of the French Community.

POPULATION

There are conflicting theories as to the origin of the population. The most important groups are the Merina, who form nearly a quarter of the island's population, are dominant politically and live in the central part of the plateau, and the Betsileo, most like the Merina, who live to the south and are skilful rice cultivators. The Merina are of Malayan origin and may have come to the island over a long period. One school of thought suggests that the bulk of the people are migrants from Africa. To support the Polynesian-Melanesian background are factors such as language, concentration on rice cultivation and the use of square houses and vegetable fibre for clothes. Opposed to this is recent physiological evidence of the coastal tribes and the cattle culture which suggests an African background. The less developed people such as the Sakalava and Bara live on the lower and hotter areas to the south and west. In addition to the indigenous population there are nearly 100,000 Europeans and Comoriens, many of the Europeans being of mixed blood and having come from neighbouring islands. There are also small numbers of Indians and Chinese who are prominent in commerce. There has been a movement of population from the more densely peopled areas of the central plateau, where densities reach over 50 per square mile, to the

[1] Incorrectly termed the Hova.

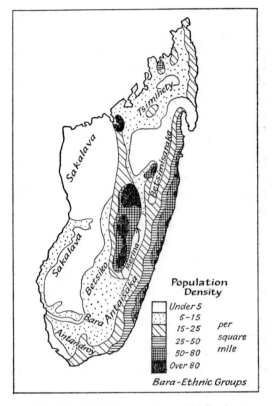

FIG. 84—Malagasy: population distribution

empty and more fertile western coastal plain (Fig. 84). The population, which was only 3·8 million in 1938, grew very slowly until the late 1940s and then suddenly increased, largely because of improved health measures; the rate of increase is $2\frac{1}{2}\%$ per annum now.

REGIONS

The effects of varying relief, drainage and climate have produced a number of distinct regions. There are, in fact, a large number of minor regions many of them isolated from each other by the inhibiting influence of the highly dissected landscape. The plateau has been cut into separate blocks by the westerly streams and there are a number of alluvial basins and depressions, some occupied by lakes such as Lake Alaotra.

The eastern coastal belt

The eastern coastal belt is the first region, nearly 1,000 miles long and only 10–50 miles wide. This region, with its high temperatures and abundant rainfall, has a tropical vegetation with raffia palms, mango trees, bamboos and patches of rain-forest. It is densely populated, the main tribe being the Betsimisaraka, numbering some ¾ million. It contains the principal port, Tamatave, and the fine natural harbour of Diego Suarez. *Tamatave*, connected by rail with Tananarive the capital, is the only port with modern facilities, with three deep-water quays, and handles much of Malagasy's trade. However, the total trade handled by Malagasy ports was less than ¾ million tons in the early 1960s. The island has 13 ports in all, many of minor importance, this large number reflecting the dispersed nature of crops grown for export, the lack of inter-regional contact and the backward nature of the economy. The regularity of the east coast, its lack of harbours and protection for shipping, the dangers of hurricanes and coral reefs has further handicapped the growth of trade. *Diego Suarez* with its sheltered harbour and wide encircling bay lies too far north to be of commercial value, and is only used as a naval station. The Canal des Pangalanes stretches for about 400 miles from Tamatave to Farafangana and provides a useful lateral waterway. Sections of it, however, are unimproved and shallow and this limits its overall use.[2] There is a short railway from the small port of *Manakara* to *Fianarantsoa*, where robusta coffee is of major importance. *Coffee* is grown on about 400,000 acres in the whole of the island and accounts for about a third of the exports. Coffee exports are only about 50,000 tons a year, and have not expanded as much as in Angola and other African countries. Other cash crops on the east coast are cloves, vanilla (Malagasy is a leading producer of these two crops) sugar and pepper.

The plateau

The main region is the plateau, covering two-thirds of the island's area having nearly half the population, and situated at an altitude of 3,000 to 4,000 feet. It is over 700 miles long and nearly 250 miles wide, and is deeply seamed by the west-flowing rivers. The surface of the plateau is far more undulating and dissected than the African plateau, and there are numerous rounded hills, interspersed with volcanic cones and granitic bosses. Much of the plateau slopes steeply and this, combined with the widespread exposure of lateritic soils, causes cultivation to be strictly limited in area to only about 3% of the island.

[2] The name of the canal is derived from the sills or pangalanes separating the lagoon.

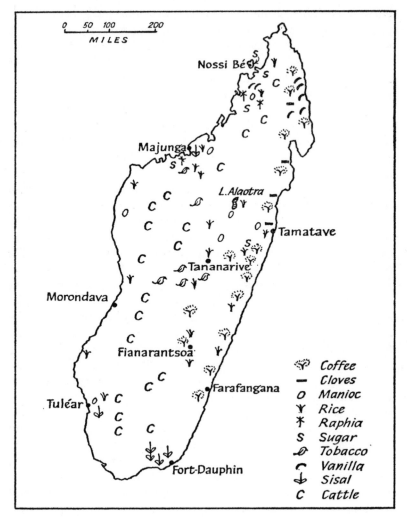

FIG. 85—Malagasy: crops and cattle

Agriculture is by far the most important activity, providing over 90% of the exports and supporting over 90% of the population. Agricultural methods are generally backward and inefficient, and most of the island is farmed in small peasant plots on a shifting basis. Plantations have been developed in a number of areas such as around Lake Alaotra, where rice and manioc are produced, and along the Mahavery river where sugar is

431

grown in estates. Sisal, tobacco, vanilla and essential oils are other Euro-pean-produced crops. The range of tropical crops produced and exported is considerable and includes coffee, vanilla, rice, sugar, cloves, raffia, tobacco, beans, sisal, groundnuts and pepper (Fig. 85). However, despite the wide range of exports they are on too small a scale to free the economy from the hazards of fluctuating prices.

Paddy rice is the main crop and is grown on about 40% of the cultivable area. It is widely distributed on the island, except in the dry south, in deltas and basins where it can be irrigated. The most favourable area is the Lake Alaotra basin where nearly 300,000 acres could be cultivated. It is the major crop of the Merina and Betsileo, who cultivate it on terraced hill-slopes and upland plains and basins, and who are skilled in the arts of terracing and water control. Nevertheless yields are poor and techniques backward. Although the Merina possess oxen there are no ploughs, and the main work of cultivation is done by means of a long-handled spade, the oxen being used to stamp the mud flat before transplanting the rice. The type of landscape created in the rice-growing areas of the plateau is similar to the terraced hill-slopes of the Far East. Rice was exported in the early years of the nineteenth century, and still provides a substantial surplus for export, despite the growth of population. Cassava or manioc occupies about 3 million acres, much of it near Tananarive. It is used as food and fodder for livestock and also provides starch and tapioca for export.

There are over 6 million *cattle* of poor quality and of the humpbacked zebu variety. As the island has no tsetse fly they can be kept almost any-where, although the bulk of the cattle are found on the plateau. Many of the Malagasy have the characteristic African attitude to cattle which they value for social and ritual reasons. However, there has long been an export of cattle to Mauritius and Réunion, and many of the larger towns have meat-packing stations. There are also exports of hides and skins. The value of pastoral products has dropped considerably since the war, as the Malagasy are generally little concerned with the quality and improvement of the animals. There are also about ¾ million sheep and goats, and 300,000 to 400,000 pigs.

The main *minerals* on the plateau are mica and graphite; the latter is exploited widely and provides the main mineral export with over 19,000 tons in 1963. Malagasy is handicapped by the lack of large exploitable deposits of minerals such as gold, copper, coal or iron ore, which have proved of great value in the economic growth of South Africa, Zambia and Rhodesia. There are small coal reserves in the south-west. These are of medium quality and not exploited as yet.

Tananarive is the main town on the plateau. It is also the largest town in the island and capital of Malagasy with a population of nearly $\frac{1}{4}$ million. It is centrally situated on a ridge in the highlands and on the edge of the Imerina plain. It is connected by rail with Tamatave, the main port on the east coast, and with Antsirabe which is about a hundred miles to the south. Antanarivo was the Merina capital in the nineteenth century, and the Rev. William Ellis, who visited it in 1856, describes the queen's palace on the highest part of the hill overlooking the rice fields of the plain below and wooden houses with high, narrow, thatched roofs built on artificially levelled terraces. Now large modern buildings and broad streets replace the old houses and narrow lanes.

The western coastal plain and low plateau

The third region consists of the wider western coastal plain and low plateau, drained by the island's largest rivers, which have divided up the region into a number of river basins. Compared with the east coast the west coast has a lower and less well distributed rainfall. The valleys of the north and north-west have some of the island's most fertile soils, alluvial and volcanic, and a high rainfall, and produce a variety of *crops*—sugar-cane, coffee, vanilla, pepper, cloves, tobacco, groundnuts, cassava and cotton. The island of Nossi-Bé produces sugar, pepper, vanilla and perfume essences. Sugar is grown by modern methods on large plantations in the Mahavy valley. Cassava is used both as food and fodder for livestock and provides starch and tapioca for export. In the south and south-west the drier climate encourages the growth of sisal and cotton. Cattle are also important and cattle-keeping forms the main activity, but as on the plateau quality of stock is low. Cattle rieving and grass firing are also serious problems.

The main *tribes* are the Sakalava, numbering some 330,000, who conquered much of the west coast during the sixteenth century, and the Tsimihety numbering about 400,000 who inhabit the province of Majunga.

Majunga, with a population of 34,000, is situated at the mouth of the Betsiboka river and is Malagasy's second largest port. It has no deep-water quays and ships have to anchor in the roadstead, while goods are handled by lighters. Links with other parts of the island are poor, and as yet Majunga serves a comparatively small hinterland. Because of the lack of adequate communications elsewhere along the west coast, which has no railways, a number of other small ports, such as Nossi-Bé and Tuléar, have developed slowly to serve restricted local regions. Air communications have grown rapidly, and there are now regular routes to various parts of the island, eighteen weekly services connecting Tananarive with ports and towns in the interior. There are also five airlines linking Tananarive by various routes with Paris.

Malagasy suffers from isolation and great distances from world markets. Although little more than 200 miles from the east coast of Africa economic links are very slight, as the east coast produces somewhat similar tropical crops also marketed in Europe. The variety of cash crops is too great and individual quantities produced too small. Many crops have stagnated or even declined in importance since the 1930s. Lacking valuable minerals large amounts of capital are essential for development and economic ties with France will continue to be strong. In future with the further growth of Malagasy and Merina nationalism increasing political links may well be sought with east and south Africa, and the connection with France may decline in importance. The familiar problems of an increasing population and soil poverty also have to be faced, despite the great area of the Red Isle and the small size of the population at present.

PART III: CONCLUSION

Conclusion

THE new African states south of the Sahara face crucial problems of backward farming, illiteracy, poverty, increasing population, low living standards, great ethnic and linguistic differences, the legacy of colonial policy and the artificiality of colonial boundaries. In the struggle for independence all the anti-colonial forces tended to merge, and any opposition to this by dissident African politicians was regarded as treachery. To some extent this has overlapped into the post-colonial phase, so that any attempt to oppose the ruling party, supposed to represent the will of the majority, is suppressed as inimical to the interests of the state. The new state, confronted by a mass of problems, also regards opposition as injurious to its welfare. Under these circumstances one-party rule or rule by the army has developed in many African states.

The development of the mass party is justified by some African politicians on the grounds that the artificially created African state includes large antipathetic groups who can only be linked together under the cloak of one and not a number of parties. They also argue that the problems of economic development can only be solved by a strong internal government on an individual territorial basis. An opposing school of thought emphasizes the need for closer regional economic cooperation. Partly as a result of this the Organization of African Unity was set up at Addis Ababa in May 1965. Its charter seeks to establish the idea of a special relationship between the African states, and bases their claim on membership of a geographical unity, the African continent.

Pan-Africanism has its roots partly in the philosophy of negritude which developed as a counterblast to white colonialism, but this aspect has to be played down as there are a number of non-negro states in the Pan-African movement. It possesses some common bond in the legacy of colonialism, but since 1963 most African countries have found internal problems more pressing than the wider concept of external and regional unity. Apart from the Pan-African movement and the dreams of a few politicians such as Dr Nkrumah for a United States of Africa there is little to link the states more closely. There is great physical and human diversity, and even a region such as the Guinea Coast of West Africa, with marked physical unity, has experienced considerable political fragmentation. Also there is not the special appeal of religion as in Pan-Arabism, or the bonds of history and sentiment as in the Commonwealth.

436

Various attempts to achieve unity have proved abortive, such as the Mali Federation, the Federation of the Rhodesias and Nyasaland and the possible federation of the three East African states of Uganda, Kenya and Tanzania, while plans for closer economic cooperation in the Maghreb and the Nile Valley have also been unsuccessful. The reasons are obvious: in most countries the politicians have concentrated on consolidating internal power first rather than pursuing the aim of a united Africa.

Therefore, despite plans for the promotion of interterritorial trade and economic development on a broad regional basis, little has been achieved. The late Sir Abubakar Tafawa Balewa, then Prime Minister of Nigeria, said at the 1964 Cairo Conference that customs barriers should be broken down, that development plans for external trade, transport and communications should be coordinated and that a regional payments union should be established. Some success has been achieved in East Africa with the decentralization of industries, e.g. a tyre factory has been allocated to Tanzania on an East African basis under the terms of the Kampala Agreement. But elsewhere the fragmented communications and the multiplicity of unconnected railway lines over the African continent make it extremely difficult for Africa's fifty states to develop new arteries of trade on a wide regional front. In addition the effects of former colonial economic links are still strong and cause a channelling of trade to the metropolitan country.

The strongest bond of unity among the African states is their aversion to colonialism, but most African countries lack the necessary force to back up their opposition to the white controlled south, and countries such as Botswana and Malawi have to tread cautiously. Only a limited success has been achieved by the Afro-Asian block in the U.N. in their endeavours to oppose the white dominated south. Despite repeated warnings about South West Africa nothing has been done to further the political growth of her African inhabitants. South Africa's trade has increased despite the boycott imposed by many African countries. Economic sanctions are of doubtful value against a country such as Rhodesia where the white minority is determined to stay in power. African leaders may fulminate vehemently against colonialism, but seem to be too heavily involved in their own affairs to do much about altering the status quo in the white south.

The new Africa, therefore faces numerous problems that can only be solved by establishing wider regional links particularly in social and economic fields, by a deep knowledge of and sympathy for the mass of the people and their forms of social organization and customs, and by generous help from the developed countries.

Appendixes

A TABLE OF GEOLOGICAL FORMATIONS

	IN EUROPE	IN SOUTH AFRICA
Quaternary Era	{ Recent Period { Pleistocene Period	Recent Period Pleistocene Period

$1\frac{1}{2}$ million years ago

	IN EUROPE	IN SOUTH AFRICA
Tertiary Era	{ Pliocene Period { Miocene Period { Oligocene Period { Eocene Period	Alexandria Formation

70 million years ago

	IN EUROPE	IN SOUTH AFRICA	
Secondary Era	{ Cretaceous Period { Jurassic Period { Triassic Period	Cretaceous Period Stormberg series Beaufort series	} Karoo } System

225 million years ago

	IN EUROPE	IN SOUTH AFRICA	
Primary Era	{ Permian Period { Carboniferous Period { Devonian Period { Silurian Period { Ordorician Period { Cambrian Period	Ecca series Dwyka series Witteberg series Bokkeveld series Table Mountain series Waterberg system	} } } Cape } System

600 million years ago

Pre-Cambrian Era	Bushveld Igneous Complex Transvaal system Ventersdorp system Witwatersrand system

(?) 3,000 million years ago

Archaean Era	(?) The Old Granite Swaziland system

(?) 4,500 million years ago

The correlations between European and South African formations are approximate.

Appendix II

BIBLIOGRAPHY

The following general geographies of Africa as a whole, Tropical Africa or Africa south of the Sahara are recommended:

FITZGERALD, W. *Africa*, London, Methuen, 9th ed., 1961.

GROVE, A. T. *Africa South of the Sahara*, Oxford, Clarendon Press, 1966.

HANCE, W. *The Geography of Modern Africa*, London, Columbia University Press, 1964.

HARRISON, CHURCH, R. J. and others. *Africa and the Islands*, London, Longmans, 1964.

KIMBLE, G. H. T. *Tropical Africa*, New York, The Twentieth Century Fund, 2 vols., 1960.

STAMP, D. *Africa*, London, Chapman and Hall, 2nd ed., 1964.

SUGGATE, L. S. *Africa*, London, Harrap, 6th ed., 1961.

Statistics can be obtained from sources such as *The United Nations Statistical Yearbook, The United Nations Yearbook of International Trade Statistics, The Statesman's Yearbook* and monthly and annual publications by Barclays Bank D.C.O. Also useful are *The Times* supplements such as the supplement on Kenya, 20 October 1966.

CHAPTER ONE (Discovery, Exploration and Colonization)

BOVILL, E. W. *The Golden Trade of the Moors*, London, Oxford University Press, 1958.

DAVIDSON, B. *Old Africa Rediscovered*, London, Gollancz, 1959.

FAGE, J. D. *An Atlas of African History*, London, Arnold, 1958.

OLIVER, R. H. AND FAGE, J. D. *A Short History of Africa*, London, Penguin Africa Library, 2 vols, 1962.

POLLOCK, N. C. and AGNEW, S. *An Historical Geography of South Africa*, London, Longmans, 1963.

PRESTAGE, E. *The Portuguese Pioneers*, London, Black, 1933.

SIMMONS, J. *Livingstone and Africa*, London, English Universities Press, 1955.

WOOD, H. J. *Exploration and Discovery*, London, Hutchinson, 1951.

CHAPTER TWO (Physical Environment)

AUBREVILLE, A. and others. *Vegetation Map of Africa South of the Tropic of Cancer*, UNESCO, Oxford University Press, 1958.

COLE, M. *South Africa*, London, Methuen, 1961.

FURON, R. *Geology of Africa* (trans. by A. Hallam and L. A. Stevens), Edinburgh and London, Oliver and Boyd, 1963.

HOLMES, A. *Principles of Physical Geology*, London, Nelson, 2nd ed., 1965.

KENDREW, W. G. *The Climates of the Continents*, Oxford, Clarendon Press, 5th ed., 1961.

KING, L. C. *The Morphology of the Earth*, London, Oliver & Boyd, 1962.

MILLER, A. A. *Climatology*, London, Methuen, 9th ed., 1961.

MOHR, E. C. J. and VAN BAREN, F. A. *Tropical Soils*, London, Interscience, 1954.

WELLINGTON, J. H. *South Africa*, Cambridge University Press, Vol. I, 1955.

WORTHINGTON, E. B. *Science in the Development of Africa*, London Commission for Technical Cooperation in Africa South of the Sahara, 1958.

(For further details about the theories advanced by Gregory, du Toit, Dixey and other writers see Holmes, *Principles of Physical Geology*, and J. W. Gregory, *The Rift Valleys and Geology of East Africa*.)

CHAPTER THREE (Economic and Social Geography)

ALLAN, W. *The African Husbandman*, London, Oliver & Boyd, 1965.

BARBOUR, K. M. and PROTHERO, R. M. (Eds.) *Essays on African Population*, London, Routledge, 1961.

BASCOM, W. R. and HERSKOVITS, M. J. (Eds.) *Continuity and Change in African Culture*, Chicago, University of Chicago Press, 1959.

BEAUJEU-GARNIER, J. *Geographie de la Population*, Paris, Librarie de Médicis, tôme 2, 1956.

BENNETT, M. K. *An Agroclimatic Mapping of Africa*, Food Research Institute Studies, Nov. 1962, p. 195.

BOHANNON, P. J. and DALTON, G. (Eds.) *Markets in Africa*, Evanston, North-western University Press, 1962.

DESHLER, W. 'Cattle in Africa: Distribution, Types and Problems,' *The Geographical Review*, **53** (1963), 52–8.

GANN, L. H. and DUIGNAN, P. *White Settlers in Tropical Africa*, London, Penguin Africa Series, 1962.

GEEEN, L. P., and FAIR, T. J. D. *Development in Africa*, Johannesburg, Witwatersrand University Press, 1962.

GREENBERG, J. H. *The Languages of Africa*, Bloomington, Indiana University Research Centre in Anthropology, Folklore and Linguistics, 1963.

HAILEY, LORD. *An African Survey*, London, Oxford University Press, 1957.

HANCE, W. *African Economic Development*, New York, Harper, 1958.

HANCE, W. *The Geography of Modern Africa*, London, Columbia University Press, 1964.

HERSKOVITS, M. J. *The Human Factors in Changing Africa*, New York, Knopf, 1962.

HUNTER, G. *The New Societies of Tropical Africa*, London, Oxford University Press, 1962.

JABAVU, N. *Drawn in Colour*, London, Murray, 1960.

KIMBLE, G. H. T. *Tropical Africa*, New York, The Twentieth Century Fund, 1960.

LEE, D. H. K. *Climate and Economic Development in the Tropics*, New York, Council of Foreign Relations, 1957.

LORIMER, F. *Demographic Information on Tropical Africa*, Boston, Boston University Press, 1961.

MAIR, L. P. *Studies in Applied Anthropology*, London, Athlone Press, 1957.

MURDOCK, G. P. *Africa: Its Peoples and their Culture History*, New York, McGraw-Hill, 1959.

Oxford Regional Economic Atlas of Africa. Oxford, Clarendon Press, 1965.

PHILLIPS, J. *Agriculture and Ecology in Africa*, London, Faber, 1959.

SOUTHALL, A. (Ed.) *Social Change in Modern Africa*, London, Oxford University Press, 1961.

STEEL, R. W. and PROTHERO, R. M. *Geographers and the Tropics*, London, Longmans, 1964.

UNITED AFRICA COMPANY. *Statistical and Economic Review*.

UNITED NATIONS, Department of Economic and Social Affairs. *Economic Survey of Africa since 1950*, New York, 1959.

UNITED NATIONS, Food and Agriculture Organization. F.A.O. Africa Survey, *Report on the Possibilities of African Rural Development in Relation to Economic and Social Growth*, Rome, 1961.

There is a great deal of information about economic and social change in Africa and only some of the more recent sources have been suggested above. The majority of African countries have published a variety of government publications relating to economic development and there are a number of economic development reports prepared by the International Bank for Reconstruction and Development. Also useful are the monthly, annual and other reports published by Barclays Bank, D.C.O., and articles in magazines such as *African Affairs*.

CHAPTER FOUR (Political Geography)

BARBOUR, K. M. and PROTHERO, R. M. (Eds.) *Essays on African Population*, London, Routledge, 1961.

BAULIN, J. *The Arab Role in Africa*, Harmondsworth, Penguin, 1962.

CARTER, G. M. *Independence for Africa*, London, Thames & Hudson, 1961.

CARTER, G. M. (Ed.) *African One-Party States*, Ithaca, Cornell University Press, 1962.

DUFFY, J. *Portugal in Africa*, Harmondsworth, Penguin, 1962.

FORDHAM, P. *The Geography of African Affairs*, Harmondsworth, Penguin, 1965.

FORTESCUE, M. M. and EVANS PRITCHARD, E. E. *African Political Systems*, London, Oxford University Press, 1961.

HAILEY, LORD. *An African Survey*, London, Oxford University Press, 1957.

HAMDAN, C. 'The Political Map of the New Africa,' *The Geographical Review*, **53** (1963), 418—39.

HODGKIN, T. *African Political Parties*, Harmondsworth, Penguin, 1961.

HOSKYNS, C. 'The African States and the United Nations,' *International Affairs*, **40** (1964), 466—80.

LEGUM, C. *Pan-Africanism*, London, Pall Mall Press, 1962.

LUCAS, C. P. *The Partition and Colonization of Africa*, London, Oxford University Press, 1922.

McKAY, V. *Africa in World Politics*, New York, Evanston, Harper & Row, 1963.

MARTELLI, G. 'Portugal and the United Nations', *International Affairs*, **40** (1964), 453–65.

MBOYA, T. 'The Party System and Democracy in Africa', *Foreign Affairs*, **53** (1963), 650—58.

MPHAHLELE, 'The Fabric of African Cultures', *Foreign Affairs*, **42** (1964), 614–27.

RICHMOND, A. *The Colour Problem*, Harmondsworth, Penguin, 1954.

RIVKIN, A. *Africa and the West*, London, Thames & Hudson, 1962.

SITHOLE, N. *African Nationalism*, Cape Town, Oxford University Press, 1960.

CHAPTER FIVE (The Maghreb)

AWAD, H. 'Morocco's Expanding Towns', *The Geographical Journal*, **130** (1964), 49–64.

Atlas of the Arab World and Middle East, London, Macmillan, 1960.

BARBOUR, N. (Ed.) *A Survey of North-West Africa*, London, Oxford University Press, 2nd ed., 1962.

BEHR, E. *The Algerian Problem*, Harmondsworth, Penguin, 1961.

BIROT, P. and DRESCH, J. *La Mediterrannée Occidentale*, Paris, Presses Universitaires de France, 1953.

CAPOT-REY, R. *Le Sahara Français*, Paris, Presses Universitaires de France, 1953.

CÉLÉRIER, J. *Le Maroc*, Paris, Berger Levrault, 1948.

DRESCH, J. *L'Agriculture en Afrique du Nord*, Paris, Centre de Documentation Universitaire, 1956.

DESPOIS, J. *La Tunisie*, Paris, Armand Colin, 1961.

DRESCH, J. (et al), *Industrialisation au Maghreb*, Paris, Maspero, 1963.

GALLISOT, R. *L'Economie de L'Afrique du Nord*, Paris, Presses Universitaires de France, 1961.

GENDARME, R. *L'Economie de L'Algérie*, Paris, Armand Colin, 1959.

HOUSTON, J. *Western Mediterranean World*, London, Longmans, 1964.

MISKELL, M. W. *Northern Morocco: A cultural Geography*, Berkeley, University of California Press, 1961.

NAVAL INTELLIGENCE DIVISION. Geographical Handbook Series, *Algeria, Morocco and Tunisia*, 1941–5.

OXFORD REGIONAL ECONOMIC ATLAS. *The Middle East and North Africa*, Oxford, Clarendon Press, 1960.

There are also atlases of Algeria and Tunisia published in 1934 and 1936 respectively, and an atlas of Morocco in preparation. The excursion guides of the International Geological Congress, Algiers, 1952, are useful.

CHAPTER SIX (The Sahara)

AMBASSADE DE FRANCE, SERVICE DE PRESSE ET D'INFORMATION. *The Islamic Republic of Mauretania*, New York, November 1960.

BENJAMIN, T. *Trade Routes of Algeria and the Sahara*, Berkeley, University of California Press, 1957.

BOAHEN, A. A. *Britain, the Sahara and the Western Sudan, 1788–1861*, London, Oxford University Press, 1964.

BOVILL, E. W. *The Golden Trade of the Moors*, London, Oxford University Press, 1958.

BRIGGS, L. C. *Tribes of the Sahara*, Cambridge, Harvard University Press, 1960.

CLARKE, J. 'Oil in Libya', *Economic Geography*, **39** (1963), 40–59.

FURON, L. *Le Sahara*, Paris, Payot, 1957.

GERSTER, G. *Sahara*, London, Barrie & Rockliffe, 1960.

GROVE, A. J. 'Geomorphology of the Tibesti Region, with special reference to Western Tibesti', *Geographical Journal*, **126** (1960) 18–31.

INTERNATIONAL BANK FOR RECONSTRUCTION AND DEVELOPMENT. *The Economic Development of Libya*, Baltimore, Johns Hopkins University Press, 1960.

MEIGIS, P. 'Outlook for Arid North Africa: The Sahara', *Focus*, Vol. V, No. 4, 1954.

MONOD, T. and TOUPET, C. 'Land Use in the Sahara-Sahel Region', in *A History of Land Use in Arid Regions*, UNESCO, 1961, 239–53.

OWEN, R. *Libya, a Brief Political and Economic Survey*, London, Oxford University Press, 1961.

PETROLEUM INFORMATION BUREAU. *Newsletter*, London, September 1965.

WILLIMOT, S. G. and CLARKE, J. I. (Eds.) *Field Studies in Libya*, Durham, Durham University Press, 1960.

CHAPTER SEVEN (The Nile Valley)

AYROUT, H. *The Fellaheen* (trans. by J. A. Williams), Boston, Beacon Press, 1963.

BARBOUR, K. M. *The Republic of the Sudan*, London, University of London Press, 1961.

BUTT, A. *The Nilotes of the Anglo-Egyptian Sudan and Uganda*, London, International African Institute, 1952.

ELISOFON, E. *The Nile*, London, Thames & Hudson, 1964 [mainly photographs].

GAITSKELL, A. *Gezira: a story of Development in the Sudan*, London, Faber, 1959.

HURST, H. E. *The Nile*, London, Constable, 1952.

ISSAWI, C. *Egypt at Mid-Century*, London, Royal Institute of International Affairs, 1954.

ISSAWI, C. *Egypt in Revolution, An Economic Analysis*, New York, Oxford University Press, 1963.

LACOUTURE, J. and S. *Egypt in Transition* (trans. by F. Scarfe), New York, Criterion, 1959.

LEBON, J. H. G., and ROBERTSON, V. C. 'The Jebel Marra, Darfur and its Region, *The Geographical Journal*,' **127** (1961), 30–49.

MOOREHEAD, A. *The White Nile*, London, Hamish Hamilton, 1960.

OXFORD REGIONAL ECONOMIC ATLAS. *The Middle East and North Africa*. Oxford, Clarendon Press, 1960.

PLATT, R. R. and HEFFNY, M. B. *Egypt: A Compendium*, New York, The American Geographical Society, 1958.

REPUBLIC OF THE SUDAN. Department of Agriculture, Agricultural Economics Division. *Agricultural Statistics, 1960–61*, Khartoum, 1962.

TOTHILL, J. D. (Ed.) *Agriculture in the Sudan*, London, Oxford University Press, 1948.

CHAPTER EIGHT (West Africa)

AMBASSADE DE FRANCE, SERVICE DE PRESSE ET D'INFORMATION, *The Republic of Dahomey*. New York, November, 1960; *The Republic of the Ivory Coast*. New York, November, 1960; *The Republic of the Niger*. New York, October 1960; *The Republic of Senegal*. New York, December 1960; *The Republic of the Upper Volta*. New York, October 1960.

ACQUAH, L., *Accra Survey*, London, University of London Press, 1958.

ATLAS OF SIERRA LEONE. Freetown, Survey and Lands Department, 1953.

BANTON, M. *West African City*, London, Oxford University Press, 1957.

BOATENG, E. A. *A Geography of Ghana*, Cambridge, Cambridge University Press, 2nd ed., 1966.

BUCHANAN, K. M. and PUGH, J. C. *Land and People in Nigeria*, London, University of London Press, 1959.

BURNS, A. *History of Nigeria*, London, Allen & Unwin, (6th ed.) 1963.

COLEMAN, J. S. *Nigeria: Background to Nationalism*, Berkeley, University of California Press, 1958.

DIKE, K. O. *Trade and Politics in the Niger Delta*, Oxford, Clarendon Press, 1956.

HARRISON CHURCH, F. J. *West Africa*, London, Longmans, 4th ed., 1963.

HILL, P. *The Gold Coast Cocoa Farmer: A Preliminary Survey*, London, Oxford University Press, 1957.

HILTON, T. E. *Ghana Population Atlas*, published on behalf of the University College of Ghana by Nelson, 1960.

HODDER, B. W. 'Traditional Markets in Africa South of the Sahara', *Institute of British Geographers, Transactions*, **36** (1965), 97–105.

HODGKIN, T. *Nigerian Perspectives*, London, Oxford University Press, 1960.

HUNTER, J. M. 'Cocoa Migration and Patterns of Land Ownership in the Densu Valley near Suhum, Ghana,' *Institute of British Geographers, Transactions*, **33** (1963), 61–87.

JACKSON, Sir R. The Volta River Project, *Progress*, **450** (1964), 146–61.

JOHNSTON, B. F. *The Staple Food Economies of Western Tropical Africa*, Stanford, Stanford University Press, 1958.

KIRK-GREENE, A. H. M. *Barth's Travels in Nigeria*, London, Oxford University Press, 1962.

LEWIS, R. *Sierra Leone*, London, Her Majesty's Stationery Office, 1954.

MORGAN, W. D. 'Food Imports of West Africa,' *Economic Geography*, **39** (1963), 351–62.

NIGERIA. Federal Ministry of Commerce and Industry, *Handbook of Commerce and Industry in Nigeria*, 1961.

PEDLER, F. J. *Economic Geography of West Africa*, London, Longmans, 1955

PORTER, P. W. 'Liberia,' *Focus*, American Geographical Society, 1961.

POST, K. *The New States of West Africa*, London, Penguin Africa Library, 1964.

RICHARD-MOLARD, J. *Afrique Occidentale Française*, Paris, Berger-Levrault, 1956.

The Sociological Review, New Series, July 1959, Special Number on Urbanism in West Africa.

THOMPSON, V. and ADLOFF, R. *French West Africa*, London, Allen & Unwin, 1956.

TRIMMINGHAM, J. S. *Islam in West Africa*, Oxford, Clarendon Press, 1959.

WILLS, J. B. *Agriculture and Land Use in Ghana*, London, Oxford University Press, 1962.

CHAPTER NINE (The Congo Basin)

BELGIAN CONGO AND RUANDA-URUNDI. Information and Public Relations Office, *Belgian Congo*, 2 vols, Brussels, Inforcongo, 1960.

FABRI, M. and MAYER J. *La Population Futur du Congo*, Brussels, Centre d'Etudes des Problèmes Sociaux, 1960.

GOUROU, P. *La densité de la population rurale au Congo belge*, Brussels Académie Royale des Sciences coloniales, 1955.

HANCE, W. A. and VAN DONGEN, I. S. 'Matadi, Focus of Belgian African Transport,' *Annals of the Association of American Geographers*, **48** (1958), 41–72.

HILLING, D. 'The Changing Economy of Gabon', *Geography*, **48** (1963), 155–65.

HOSKYNS, C. *The Congo since Independence*, London, Oxford University Press, 1965.

LEFEVER, E. W. *Crisis in the Congo*, Washington, The Brooking Institution, 1965.

PATTEN, G. F. 'Republic of the Congo (Brazzaville)', *Focus*, American Geographical Society, 1962.

Pouquet, F. *Afrique Equatoriale Française*, Paris, Presses Universitaires de France, 1954.

Robert, M. *Géologie et Géographie du Katanga*, Brussels, Union Minière du Haut Katanga, 1956.

Thompson, V. and Adloff, R. *The Emerging States of French Equatorial Africa*, Stanford, Stanford University Press, 1960.

Wigny, P., 'Belgium and the Congo', *International Affairs*, 37 (1961), 273–84.

Chapter Ten (The Eastern Horn)

Busk, D. *The Fountain of the Sun*, London, Parrish, 1957.

Cheeseman, R. E. *Lake Tana and the Blue Nile*, London, Macmillan, 1936.

Drysdale, J. *The Somali Dispute*, London and Dunmow, Pall Mall Press, 1964.

Greenfield, R. *Ethiopia*, London, Pall Mall Press, 1965.

Huffnagel, H. P. *Agriculture in Ethiopia*, Rome, F.A.O., 1961.

Hunt, J. A. *A General Survey of the Somaliland Protectorate 1944–50*, Hargeisa, 1951.

International Bank for Reconstruction and Development. *The Economy of the Trust Territory of Somalia*, Washington, 1957.

Johnston, M. 'Addis Ababa from the Air', *Ethiopian Observer*, VI (1962), 17–32.

Karp, M. *The Economics of Trusteeship in Somalia*, Boston, Boston University Press, 1960.

Last, G. C. 'Introductory Notes on the Geography of Ethiopia', *Ethiopian Observer*, VI (1962), 82–134.

Lewis, I. M. *A Pastoral Democracy*, London, Oxford University Press, 1961.

Lewis, I. M. *The Modern History of Somaliland from Nation to State*, London, Weidenfeld, 1965.

Luther, E. W. *Ethiopia Today*, Stanford, Stanford University Press, 1958.

Simoons, F. J. *North-West Ethiopia, Peoples and Economy*, Madison, University of Wisconsin Press, 1960.

Ullendorf, E. *The Ethiopians*, London, Oxford University Press, 1959.

Chapter Eleven (East Africa)

African Land Development in Kenya, 1946–62. Nairobi, Ministry of Agriculture, 1962.

Atlas of Kenya. Nairobi, The Survey of Kenya, 1959.

Atlas of Tanganyika. Dar es Salaam, Survey Division, Department of Lands and Mines, 1956.

Atlas of Uganda. Department of Lands and Surveys, 1962.

BENNETT, G. *Kenya, A Political History*, London, Oxford University Press, 1963.

DE BLIJ, H. J. *Dar es Salaam, A Study in Urban Geography*, Chicago, Northwestern University Press, 1963.

COLE, S. *The Prehistory of East Africa*, London, Weidenfeld, 1964.

ECONOMIST INTELLIGENCE UNIT. *The Economy of East Africa: A Study of Trends*, London, 1955.

FORD, V. C. R. *The Trade of Lake Victoria*, East African Institute of Social Research, 1955.

HICKMAN, G. M. and DICKINS, W. H. G. *The Lands and Peoples of East Africa*, London, Longmans, 1960.

HOLLINGSWORTH, L. W. *The Asians in East Africa*, London, Macmillan, 1960.

HOYLE, B. S. 'The Economic Development of Jinja, Uganda,' *The Geographical Review*, **53** (1963), 377–88.

GORDON BROWN, A. (Ed.), *Year Book and Guide to East Africa*, published for the Union Castle Mail Steamship Co., London (annual).

GREGORY, J. W. *The Rift Valleys and Geology of East Africa*, London, Seeley Service, 1921.

INGRAMS, H. *Uganda*, London, Her Majesty's Stationery Office, 1960.

INTERNATIONAL BANK FOR RECONSTRUCTION AND DEVELOPMENT:
The Economic Development of Kenya, Baltimore, Johns Hopkins University Press, 1963.
The Economic Development of Tanganyika, Baltimore, Johns Hopkins University Press, 1961.
The Economic Development of Uganda, Baltimore, Johns Hopkins University Press, 1962.

MATHESON, J. K. and BOVILL, E. W. *East African Agriculture*, London, Oxford University Press, 1950.

McMASTER, D. N. *A Subsistence Crop Geography of Uganda*, Bude Geographical Publications Ltd, 1962.

MEINERTZHAGEN, R. *Kenya Diary, 1902–6*, London, Oliver & Boyd, 1957.

MOFFETT, J. P. *Handbook of Tanganyika*, Dar es Salaam, Government Printer, 1958.

MORGAN, W. T. W. 'The White Highlands of Kenya,' *The Geographical Journal*, **129** (1963), 140–55.

O'CONNOR, A. M. 'New Railway Construction and the Pattern of Economic

Development in East Africa,' Institute of British Geographers, *Transactions* **36** (1965), 21–30.

OLIVER R. and MATHEW, G. *History of East Africa*, Vol. I, London, Oxford University Press, 1963.

OMMANEY, F. D. *The Island of Cloves*, London, Longmans, 1955.

POLLOCK, N. C. 'Industrial Development of East Africa', *Economic Geography*, **36** (1960), 344–54.

PRITCHARD, J. M. *A Geography of East Africa*, London, Dent, 1962.

Reports of the East African Royal Commission, 1953–55, Cmd. 9475, Her Majesty's Stationery Office, 1955.

RUSSEL, E. W. (Ed.) *The Natural Resources of East Africa*, Nairobi, D. A. Hawkins, 1962.

SELWYN, P. and WATSON, T. Y. *Report on the Economic Development of the Zanzibar Protectorate*, Zanzibar, Government Printer, 1962.

SOFER, C. and R. *Jinja Transformed*, East African Studies, No. 4, East African Institute of Social Research, 1959.

WALMSLEY: R. *Nairobi: The Geography of a New City*, East African Local Studies, No. 1, 1957.

CHAPTER TWELVE (Southern Africa)

ACOCKS, J. P. H. 'The Veld Types of South Africa', Botanical Survey of South Africa, Memorandum No. 28, 1953.

ASHTON, H. *The Basuto*, New York, Oxford University Press, 1952.

BARBER, W. J. *The Economy of British Central Africa*, London, Oxford University Press, 1961.

BARBOUR, K. M. and PROTHERO, R. M. (Eds.) *Essays on African Population*, London, Routledge, 1961.

Basutoland, Bechuanaland and Swaziland. Economic Survey Mission Report, Her Majesty's Stationery Office, 1960.

BRELSFORD, W. V. (Ed.) *Handbook to the Federation of Rhodesia and Nyasaland*, London, Cassell, 1960.

COLE, M. *South Africa*, London, Methuen, 1961.

DE BLIJ, H. J. *Africa South*, Evanston, Northwestern University Press, 1962.

DAVIES, R. J. 'Social Distance and the Distribution of Occupational Categories in Johannesburg and Pretoria', *The South African Geographical Journal*, **46** (1964), 24–39.

DE KIEWIET, C. W. *A History of South Africa*, London, Oxford University Press, 1946.

DEBENHAM, F. *Nyasaland*, London, Her Majesty's Stationery Office, 1955.

DOVETON, D. M. 'The Human Geography of Swaziland', *Institute of British Geographers, Publication No. 8*, London, 1937.

DUFFY, J. *Portugal in Africa*, Harmondsworth, Penguin, 1962.

DU TOIT, A. L. *The Geology of South Africa*, London, Oliver & Boyd, (3rd ed.) 1954.

EGERTON, F. C. C. *Angola in Perspective*, London, Routledge, 1957.

FAIR, T. J. D. and GREEN, L. P. 'Development of the Bantu Homelands', *Optima*, 12 (1962), 7–19.

GANN, L. H. *A History of Northern Rhodesia*, London, Chatto, 1964.

GREEN, L. P. and FAIR, T. J. D. *Development in Africa*, Johannesburg, Witwatersrand University Press, 1962.

GREEN, L. P. and FAIR T. J. D. 'Preparing for Swaziland's Future Economic Growth', *Optima*, 10 (1960), 194–206.

GORDON BROWN, A. (Ed.) *The Southern African Yearbook*, published for the Union Castle Mail Steamship Co., London (annual).

HALL, R. *Zambia*, London, Pall Mall Press, 1965.

HANCE, W. A. and VAN DONGEN, I. S. 'Beira, Mozambique, Gateway to Central Africa', *Annals of the Association of American Geographers*, 47 (1957), 307–35.

HAZLEWOOD, A. and HENDERSON, P. D. *Nyasaland: The Economics of Federation*, Oxford, Blackwell, 1960.

HELLMANN, E. (Ed.) *Handbook of Race Relations in South Africa*, London, Oxford University Press, 1949.

HOBART HOUGHTON, P. *The South African Economy*, Cape Town, Oxford University Press, 1964.

HORRELL, M. *A Survey of Race Relations in South Africa, 1964*, South African Institute of Race Relations, Johannesburg, 1965.

INSTITUTE OF RACE RELATIONS. *Angola: A Symposium, Views of a Revolt*, London, 1962.

KEATLEY, P. *The Politics of Partnership*, Harmondsworth, Penguin, 1963.

KING, L. C. *South African Scenery*, London, Oliver & Boyd (3rd ed.), 1963.

MARQUARD, L. *The Peoples and Policies of South Africa*, London, Oxford University Press, 3rd ed., 1962.

MAYER, P. *Townsmen or Tribesmen*, Cape Town, Oxford University Press, 1961.

POLLOCK, N. C. and AGNEW S. *An Historical Geography of South Africa*, London, Longmans, 1963.

PIKE, J. G. and RIMMINGTON, G. T. *Malawi*, London, Oxford University Press, 1965.

PRESCOTT, J. R. V. 'Population Distribution in Southern Rhodesia', *The Geographical Review*, **52** (1962), 559–65.

REEVE, W. H. 'Progress and Geographical Significance of the Kariba Dam', *Geographical Journal*, **126** (1960), 140–6.

SILLERY, A. *The Bechuanaland Protectorate*, London, Oxford University Press, 1952.

SPENCE, C. F. *Mozambique*, London, Bailey & Swinfen, 1963.

TALBOT, A. M. and W. J. *Atlas of South Africa*, Pretoria, Government Printer, 1960.

TRAPNELL, G. G. *The Soils, Vegetation and Agriculture of North-Eastern Rhodesia*, Lusaka, Government Printer, 1953.

Summary of the report of the Commission for the socio-economic development of the Bantu areas within the Union of South Africa (*Tomlinson Report*). Pretoria, Government Printer, 1955.

WALKER, E. A. *History of Southern Africa*, London, Longmans, 1957.

WELLINGTON, J. H. *Southern Africa*, Cambridge, Cambridge University Press, 2 vols, 1955.

TALBOT, W. J. *Swartland and Sandveld*, Cape Town, Oxford University Press, 1947.

Only a few of the many articles in the *Geographical Journal, Geography*, the *South African Geographical Journal, Optima* and other geographical journals and magazines have been mentioned.

CHAPTER THIRTEEN (Malagasy)

AMBASSADE DE FRANCE. SERVICE DE PRESSE ET D'INFORMATION, *The Malagasy Republic*, New York, September 1960.

CHEVALIER, L. *Madagascar: Populations et Ressources*, Paris, Presses Universitaires de France, 1952.

DESCHAMPS, H. *Histoire de Madagascar*, Paris, Berger Levrault, 1961.

HANCE, W. A. 'Transportation in Madagascar', *The Geographical Review*, **48** (1958), 45–68.

HATZFELD, O. *Madagascar*, Paris, Presses Universitaires de France, 1960.

ROBEQUAIN, C. *Madagascar et les bases dispersées de l'Union Française*, Paris, Presses Universitaires de France, 1958.

THIERRY, S. *Madagascar*, Paris, Editions du Seuil, 1963.

Index

Figures in *italics* refer to the map on the specified page.
References in Roman numerals are to the photographic plates.

INDEX

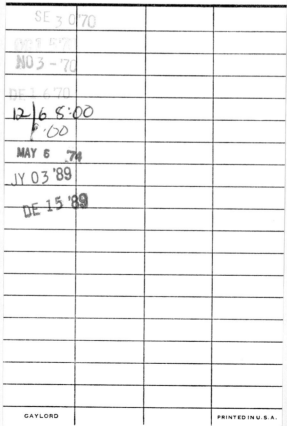

DATE DUE

SE 3 0 '70		
OC 9 5 7		
NO 3 - '70		
DE 1 6 '70		
12/6 8:00		
8:00		
MAY 6 74		
JY 03 '89		
DE 15 '89		

GAYLORD   PRINTED IN U.S.A.

916
P.777a

24768

Pollock, Norman C.
Africa

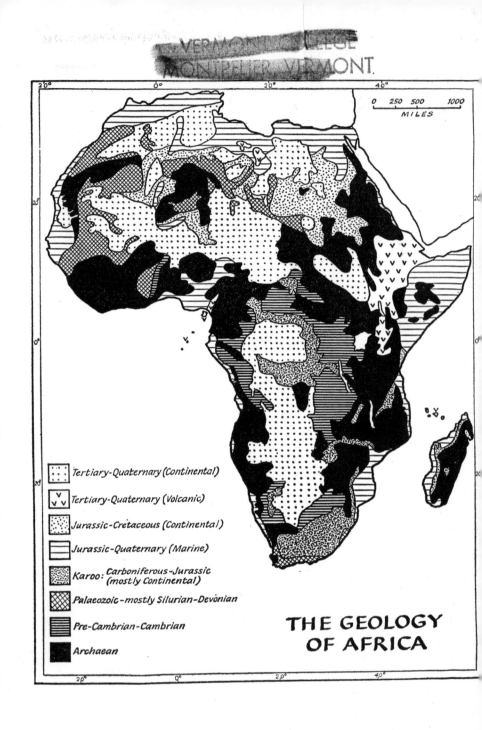

0   250   500   1000
MILES

Tertiary-Quaternary (Continental)

Tertiary-Quaternary (Volcanic)

Jurassic-Cretaceous (Continental)

Jurassic-Quaternary (Marine)

Karoo: Carboniferous-Jurassic (mostly Continental)

Palaeozoic-mostly Silurian-Devonian

Pre-Cambrian-Cambrian

Archaean

THE GEOLOGY
OF AFRICA